Springer Series in Wood Science

Editor: T. E. Timell

Springer

Berlin
Heidelberg
New York
Hong Kong
London
Milan
Paris
Tokyo

Springer Series in Wood Science

Editor: T. E. Timell

L. W. Roberts/p. B. Gahan/R. Aloni
Vascular Differentiation and Plant Growth Regulators (1988)
C. Skaar
Wood-Water Relations (1988)
J. M. Harris
Spiral Grain and Wave Phenomena in Wood Formation (1989)
B. J. Zobel/J. P. van Buijtenen
Wood Variation (1989)
P. Hakkila
Utilization of Residual Forest Biomass (1989)
J. W. Rowe (Ed.)
Natural Products of Woody Plants (1989)
K.-E. L. Eriksson/R. A. Blanchette/P. Ander
Microbial and Enzymatic Degradation of Wood and Wood Components (1990)
R. A. Blanchette/A. R. Biggs (Eds.)
Defense Mechanisms of Woody Plants Againts Fungi (1992)
S. Y. Lin/C. W. Dence (Eds.)
Methods in Lignin Chemistry (1992)
G. Torgovnikov
Dielectric Porperties of Wood and Wood-Based Materials (1993)
F. H. Schweingruber
Trees and Wood in Dendrochronology (1993)
P. R. Larson
The Vascular Cambium: Development and Structure (1994)
M.-S. Ilvessalo-Pfäffli
Fiber Atlas: Identification of Papermaking Fibers (1995)
B. J. Zobel/J. B. Jett
Genetics of Wood Production (1995)
C. Matteck/H. Kubler
Wood – The Internal Optimization of Trees (1995)
T. Higuchi
Biochemistry and Molecular Biology of Wood (1997)
B. J. Zobel/J. R. Sprague
Juvenile Wood in Forest Trees (1998)
E. Sjöström/R. Alén (Eds.)
Analytical Methods in Wood Chemistry, Pulping, and Papermaking (1999)
R. B. Keey/T. A. G. Langrish/J. C. F. Walker
Kiln-Drying of Lumber (2000)
S. Carlquist
Comparative Wood Anatomy, 2nd ed. (2001)
M. T. Tyree/M. H. Zimmermann
Xylem Structure and the Ascent of Sap, 2nd ed. (2002)
T. Koshijima/T. Watanabe
Association Between Lignin and Carbohydrates in Wood and Other Plant Tissues (2003)
V. Bucur
Nondestructive Characterization and Imaging of Wood (2003)

Voichita Bucur

Nondestructive Characterization and Imaging of Wood

With 201 Figures, 24 of them also in color and 49 Tables

Springer

Prof. VOICHITA BUCUR
Institut National de la Recherche Agronomique
Centre de Recherches Forestières de Nancy
Laboratoire d'Etudes et Recherches sur le Matériau Bois
54280 Champenoux
France

Series Editor:
T. E. TIMELL
State University of New York
College of Environment Science and Forestry
Syracuse, NY 13210, USA

Cover: Transverse section of *Pinus lambertiana* wood. Courtesy of Dr. Carl de Zeeuw, SUNY college of Environmental Science and Forestry, Syracuse, New York

ISSN 1431-8563
ISBN 3-540-43840-8 Springer-Verlag Berlin Heidelberg New York

Library of Congress Cataloging-in-Publication Data

Bucur, Voichita.
Nondestructive characterization and imaging of wood / V. Bucur.
p. cm. – (Springer series in wood science; 760)
Includes bibliographical references and index.
ISBN 3540438408 (alk. paper)
1. Wood–Testing. 2. Non-destructive testing. I. Title. II. Series.
TA420.B92 2003 620.1′27–dc21

Typesetting: SNP Best-set Typesetter Ltd., Hong Kong
Cover Design: Design & Production, Heidelberg
31/3150 – 5 4 3 2 1 0 – Printed on acid-free paper

Hommage à la mémoire
de mes parents qui m'ont appris
à penser et à réfléchir avec
enthousiasme.

Foreword

This book on the *Nondestructive Characterization and Imaging of Wood* by Professor Voichita Bucur is truly the most outstanding reference on the subject ever written. Since the origins of mankind, wood has played a key role in the history of humans and other living creatures, ranging from provision of life from trees giving air, heat, light, and food to nourish their bodies to structures to protect them from the elements. Wood has also played a key role in one of the world's primary religions.

Nondestructive diagnostics methods have long found application in medical practice for examination of the human body in order to detect life-threatening abnormalities and permit diagnosis to extend life. Nondestructive testing has been used for many years to insure the safety of machinery, aircraft, railroads, tunnels, buildings and many other structures.

Therefore, it is timely for a treatise, like the present one, to be written describing how wood can be characterized without employing destructive test methods. Since wood is so valuable to mankind, it is important to know the latest methods to nondestructively characterize wood for all practical applications.

Starting with a historical review of nondestructive evaluation of wood, the author proceeds with a systematic description of the nondestructive techniques used to image the structural elements of wood. The author has organized the book according to the wavelength of the interrogating nondestructive characterization technique used to create images of the internal structure. The book proceeds from high resolution computed tomography using ionizing X-ray and gamma radiation, through thermography, thermal wave imaging, microwaves, ultrasonic waves and nuclear magnetic resonance. Although theoretical approaches are presented, emphasis is placed on practical experimental applications which are extremely useful in many fields of human endeavor.

Professor Bucur is an internationally known expert on the mechanical properties of wood and nondestructive techniques for the characterization of wood. Not only are the techniques she describes scientifically important, but often the images she provides are beautiful works of art.

Baltimore, December 2002

Professor Robert E. Green Jr.
Johns Hopkins University, USA, Center for Nondestructive Evaluation

Preface

In writing this book, I have attempted to provide a comprehensive account of the principles, practice and achievements of imaging of the internal structure of wood by various techniques developed and used during the last decades of the twentieth century. This book is written from the point of view of one who sees the development of these techniques principally as procedures for non-destructive evaluation. I hope that it will be of particular interest to those who are concerned with the study of wood as an engineering material, either for fundamental research or practical application. The technical level and scope are such that it should be of interest to scientists and students. Wherever possible, the use of complicated mathematics was avoided in favor of physical reasoning. The book can serve as a guide to the literature on the subject and cites more than 500 references.

At the present time, two fields claim predominant attention: the imaging of the internal structure of wood with methods based on the propagation of electromagnetic waves, elastic waves, and heat; and imaging with nuclear particle radiation-based methods. For this reason, the book is divided into two main parts. The first part is presented in five chapters and deals with the imaging of wood structure via wave propagation. These chapters are presented in order of increasing wavelength, and hence resolution of the wave phenomena used for imaging. X-ray computed tomography (10^{-12} to 10^{-10} m wavelength, 1 nm resolution) is followed by thermal wave imaging (10^{-6} to 10^{-4} m wavelength, 1 mm resolution), microwave radiation (10^{-4} to 10^{-1} m wavelength), ultrasonic, and nuclear magnetic resonance methods (10^{-3} to 10^{-1} m wavelength, 1 mm resolution).

The second part of the book is devoted to neutron imaging of wood structure. The task of writing this has been difficult in attempting to convey some sense of the relative importance of the various contributions to this rapidly developing subject.

First of all, I am very much indebted to Professor Timell for the opportunity to write this book for the Springer Series in Wood Science.

I am very much indebted to the following colleagues and friends for reading the manuscript and making comments for the improvement of the comprehension of the text: Dr Günter S. Bauer, Germany, Professor Frank Beall, Berkeley, USA, Dr. Ion Paul Beldie, Berlin, Germany, Dr. Harald Berndt, Oakland, USA, Professor Gerd Busse, Stuttgart, Germany, Dr. Giorgio Catena, Rome, Italy, Dr. Laurent Chrusciel, Nancy, France, Professor Colin Gough, Birmingham, UK, Professor Adolf Habermehl, Marburg, Germany, Professor Phong M. Luong, Paris-Palaiseau, France, Dr. Fréderic Mothe, Nancy, France, Professor Peter

Niemz, Zürich, Switzerland, Dr. Patrick Rasolofosaon, Paris-Malmaison, France, Dr. Robert Ross, Madison, USA, Professor Christen Skaar, Blacksburg, USA.

The bibliography was prepared with the kind cooperation of various librarians in France: Marie Annick Bruthiaux, University Henri Poincaré, Nancy, Marie Jeanne Lionnet, curator of the library of "Ecole Nationale des Eaux et Forêt de Nancy", David Gasparotto, librarian at the same institution, and Angelica Onteniente from the library of the "Institut National de la Recherche Agronomique" in Versailles. The difficult task of carefully checking all bibliographic references was accomplished by David Gasparotto, whom I wish to acknowledge. Thanks are also due to my colleagues Simone Garros, André Perrin, Pierre Gelhaye, Christian Herbé, Claude Houssement and Etienne Farré for their continuing help and technical assistance for more than 20 years, as well as to Corrine Courtheoux and Yvonne Sapirstein for every day help and assistance during the writing of this book.

I wish to acknowledge my Ph.D. students present in the laboratory during the writing of this book: Nadia Mouchot, Adrian Hapca, Saeed Kazemi, Roberto Martinis for their generous cooperation and assistance with handling modern computer techniques for writing the text and for the illustrations. Their comments and criticism were very stimulating. The final electronic version of the manuscript was accomplished with the very enthusiastic help of Adrian Hapca.

I am also very grateful to my brother-in-law, Constantin Spandonide, for his significant contribution in scanning and printing the numerous figures of this book. Thanks are due to my sister, Despina Spandonide, and to my nephew, Bruno Spandonide, for continuous and particularly enthusiastic encouragement during the writing of this book.

I am indebted to many individuals and organizations for permission to reproduce figures and tables. In each case, the sources are indicated in the text and in the figure legend. I am especially conscious of my indebtedness to my colleagues and friends from all over the world who followed the writing of this book with interest.

I wish to thank Professor R. Green, Jr., Baltimore, USA, for his enthusiastic support of my project – in writing this book and for agreeing to write the Foreword of this book – from the beginning of this project, in autumn 1999, when we were at a conference in Brazil.

Last, but not least, I wish to thank Springer-Verlag, the publisher of this book, for producing the printed volume, and the following French institutions for sponsoring the color figures: CUGN (Communauté Urbaine du Grand Nancy); INRA (Institut National de la Recherche Agronomique); Forests and Natural Environment Department, Communication Department and Michelle Cussenot; Forest Research Center in Nancy, Director Dr. Gilbert Aussenac; and the Wood Research Laboratory (UMR 1093) associated with University Henri Poincaré, Nancy I, Director Professor Xavier Deglise.

Nancy, France, December 2002

VOICHITA BUCUR

Contents

1 Introduction

1.1 Brief Historical Review of Nondestructive Evaluation of Wood and Aim of the Book

Nondestructive evaluation of the physical properties of wood has its origin in the need to solve practical problems without destruction of the integrity of the object under inspection. It is generally accepted that the terms "nondestructive evaluation of wood properties" and "nondestructive testing of wood properties" can be used interchangeably. Beall (1996), Ross and Pellerin (1991) and Bodig (1994) selected the term "nondestructive evaluation" as more attractive since "it represents the process by which selected physical properties of a material is being assessed without damage or alteration to its end-use capabilities."

The earliest nondestructive evaluation of wood was visual inspection, largely used for the selection of timber used as load-bearing members for specific applications. Even today, this method is extensively used for the grading of wood products for lumber, poles, plywood, etc. and for the detection of biological degradation of these products.

The development of scientific nondestructive methods became possible in the early 20th century with the development of the theory of elasticity and of the instrumentation for the measurement of wood properties. The interest of scientists was initially focused on the determination of the modulus of elasticity by static methods (Hörig 1935; Kollmann 1951). Later, the use of acoustic vibrations for dynamic methods to determine elastic constants of wood were published in Europe (Barducci and Pasqualini 1948; Hearmon 1948, Kollmann and Krech 1960) in North America (Jayne 1955; James 1959), and in Japan (Fukada et al. 1956).

The enthusiasm for the development of X-ray techniques for evaluation of the internal structure of wood in the 1960s had a beneficial influence on the development of microdensitometry in different laboratories all over the world (Polge 1978). The X-ray diffraction technique was developed in Japan for the study of the crystallinity of cellulose in wood (Fukada 1965; Fukada et al. 1956).

Pioneering results of dynamic tests using vibrational methods were reported in the early 1960s by Hoyle (1961), Senft et al. (1962), and Pellerin (1965) for the nondestructive testing of structural lumber. The development of ultrasonic techniques for elastic characterization of wood was promoted in the

United Kingdom by Hearmon (1965). Since the late 1970s, the activity of laboratories all over the world on nondestructive testing of wood has been stimulated by symposia organized in the USA at Washington State University by Pellerin and coworkers in collaboration with the Forest Products Laboratory, Madison (Ross and Pellerin 1991, 1994) and since 1996 alternatively in Europe and in the USA.

Since 1970, reference books have been published in the field of mechanical characterization of wood (Jayne 1972; Bodig and Jayne 1982), dielectric properties of wood (Torgovnikov 1993), and acoustical properties of wood (Bucur 1995). Schniewind (1981) was the editor of the first encyclopedia of wood science. The publication of an especially important series of books in the *Springer Series in Wood Science* was initiated in 1983. The last 20 years of the 20th century have been characterized by an extraordinary development of different methodologies for nondestructive evaluation of wood products, which has succeeded in the imaging of the internal structure of wood at both the macroscopic and microscopic levels.

Due to the extensive literature on this subject, this book will provide an overview of wood structure imaging techniques and the corresponding basic concepts related to the nondestructive characterization of this material that permitted the development of the modern imaging procedures. These outstanding modern wood structure imaging techniques are a logical extension of existing nondestructive methods developed previously and has benefited from their breadth.

1.2 General Concepts of Nondestructive Testing of Wood

To promote the efficient use of wood materials in the future, three major areas need to be addressed:

- development of nondestructive techniques for the evaluation of different properties such as: physical, mechanical, chemical, aesthetic, etc.,
- improvement of natural qualities of wood through the modification of properties with different treatments, and
- creating new products using wood as a major raw material, corresponding to the requirements of a modern society.

The development of nondestructive techniques has as it principal purpose to reduce the uncertainty of wood products characteristics as influenced by wood's biological nature. Despite the great attention given to quality control in the development of manufacturing processes for glue laminated timber, laminated veneer lumber or plywood, for particleboard and other wood-based composites, interfaces are still the weakest link in the performance of these products. To ensure interfacial integrity it is important to develop methods for nondestructive inspection of wood-based composites and their components (solid wood and adhesives).

The interfacial discontinuities, delaminations, cracks, porosity or density variations may be detected by nondestructive techniques such as acoustic, microwave, thermal, radiographic, or classical static methods. These methods can help in the understanding of material behavior under different environmental conditions, but difficulties still remain for detection and for quantitative description of structural discontinuities and defects. It is important to relate the nondestructive measurements to the mechanical properties of wood and wood-based composites. The basis of such relations is the dependence of the interfacial strength, on one or more mechanical characteristics related to the nature and morphology of defects. There is also a need for the development of nondestructive techniques in industry. The development of these techniques will lead to intelligent manufacturing processes for wood products, resulting in processes that will identify defects without characterizing "good material" as defective.

The second approach is the quality assessment of improved wood products from different chemical or mechanical treatments, which requires the development of nondestructive techniques appropriate to each material. Analytical assessment techniques will be used to gain an understanding of failure mechanisms in wood composites. The combination of physical testing and modeling should then yield an improved understanding of the factors that determine the quality of products.

The third approach is to create new products using wood as a major raw material, corresponding to the requirements of modern society. The market for wood-based composites is a growing one worldwide, with new production plants coming on line. Probably "biomimetics" will be a source of inspiration for designing new composites. Progress can be expected from a "better microarchitecture and from the mastering of smaller and smaller scales for the basic constituents, and their spatial arrangement" (Roux 1998). In many applications, the nonlinear behavior of solid wood and of its composites must be considered. The sensitivity of new nondestructive evaluation techniques to defects provides an opportunity to control the statistical properties of structures.

Given the hierarchical structure of wood it is obvious that one should seek multiscale characterization tools. The use of multiscale approaches and the prediction of the behavior of very complex systems through probabilistic approaches and small-scale measurements must be emphasized. The problem of selecting the most relevant scale for the study of the properties of the equivalent medium has to be addressed for all applications.

The importance of the microstructure must be underlined because of its basic role in nondestructive evaluation of the properties of materials. New materials are currently "designed with a view towards developing microstructural conditions that more effectively allow the material to meet application specific structural requirements" (Frantziskonis and Blodgett 1998). The mechanics of heterogeneous media require the definition of the representative elementary volume (Bourbié et al. 1987; Chelidze et al. 1998). This volume must be larger than the size of the elementary heterogeneity and, in the case of wood, larger than the width of the annual ring, fiber length, etc (Fig. 1.1). Under this assumption at different scales, the sample can be considered quasi-homogeneous. This approach can be applied to any physical property and the principle of physical analogies can be used. However, experimental studies as well as theories have confirmed the dependence of the properties of heterogeneous media on the scale of observation and size of the system. It is reasonable to expect that in the near future the new approach to the mechanics of heterogeneous materials will be that of fractal mechanics (Sahimi and Arbabi 1993). It is also generally accepted that the control of mechanical properties of new wood-based composites such as glued laminated timber and oriented strandboard, can be performed using probabilistic design (Castera 1998). At times, the deterministic approach may be better than the probabilistic one, because it is difficult to find a general correlation between micro- and macro-scale behavior of wood-based composites. A possible explanation of this can be found in the multiplicity of the factors influencing the phenomena studied.

Today, the technologies for wood-based composites are based on quality control with nondestructive evaluation techniques, which play an increasing, important role in adapting the market to the change of timber resources. In the future, it will be important to develop new nondestructive techniques and devices for quality control of new wood composites produced with a more diverse raw material supply. As noted by Youngquist and Hamilton (1999), this is the challenge for the 21st century.

1.3 Classification of Nondestructive Techniques for Wood Quality Assessment

The characterization of wood properties is critical for the understanding of material behavior and performance under operating conditions. Tailoring the properties of new wood-based composites is essential for advanced product design. The composites of the future will have to be made from such new resources as underutilized species, recycled wood, and municipal waste, which will be a mixture of wood, paper, plastic, and agricultural residues (Maloney 1992; Peterson 1993; Greaves 1998; Bowyer 2000).

The need to characterize wood-based composites for a myriad of applications has spurred the development of many new methods and instruments. An ideal characterization tool would provide data about the properties that are related to micro and macro structure without destructive sectioning. Such data can only be obtained using nondestructive methodologies. Nondestructive evaluation of wood and wood-based composites enables the determination of

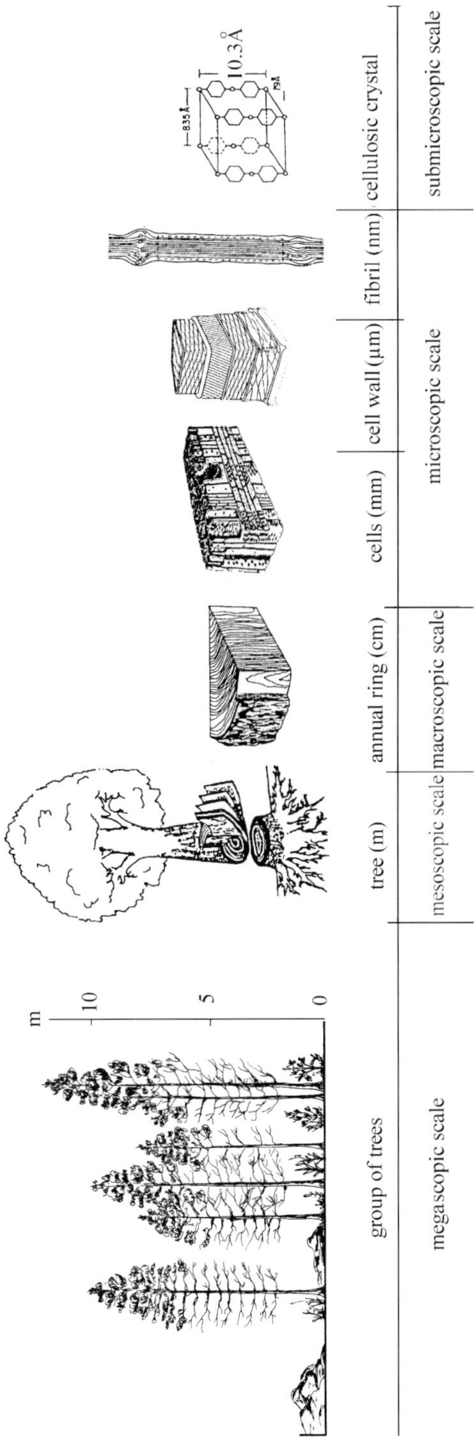

Fig. 1.1. Hierarchical structure of wood. (Data from Bucur 1995, with permission)

material parameters at nearly any point, line, surface or volume element of interest, and at nearly any state during the life of the material. The large number of potential methods for nondestructive evaluation of wood requires a synergism of many scientific and engineering disciplines. Beall (1996) summarized the nondestructive evaluation opportunities, and the needs for wood products, as can be seen from Table 1.1. As noted, the development and the implementation of nondestructive techniques depend on comprehensive monitoring, and on integrated and joint-venture solutions. "Comprehensive monitoring implies multiple sensing methods to determine a fuller spectrum of properties that could be part of an in-process technology or one in service."Having in mind the multiple sensing methods for the determination of one parameter", Sobue (1993) proposed the classification of nondestructive methods (Table 1.2) as a function of the physical property of wood and the necessity for application (moisture content, density, stress rating, detection of knots, grain orientation, etc.).

In the case of industrial wood-based composite products, the design and production have always been considered the preeminent engineering challenges, with inspection and defect detection relegated to a subsidiary role. However, today there is an increasing awareness of the potential for in situ monitoring of processes by nondestructive methods with the objective of significant improvement of predictability of the quality of the products while optimizing utilization of raw material.

Table 1.1. Nondestructive evaluation opportunities and the needs in the wood products. (Beall 1996, with permission)

Material properties	Preprocess	Process	Product	Field
Defects				
Surface defects	X	X	X	X
Internal defects	X			X
Basic properties				
Grain orientation	X	X		
Density	X	X	X	
Moisture content	X	X	X	X
Strength/stiffness			X	X
Permeability/diffusion		X	X	
Surface roughness		X	X	
Geometry				
Profile, shape	X	X	X	
Thickness		X	X	
Bonding quality				
Composites		X	X	
Laminate materials		X	X	X
Fasteners			X	X

X indicates the need for nondestructive characterization

Table 1.2. Classification of nondestructive methods for solid wood as a function of its physical properties. (Sobue 1993; with perrmission)

Wood parameter	Properties measured	Techniques
Moisture content	Dielectric properties	Impedance DC and high frequency, microwave, infrared, NMR
	Thermal properties	Infrared
	Ionizing rays absorption	X-rays; gamma rays
	Elastic properties	Frequency resonance method, ultrasonic method
Density	Weight	Gravimetric method
	Ionizing rays	X-ray microdensitometry; attenuation of X, beta and gamma rays
	Dielectric properties	Microwave technique
	Mechanical properties	Pilodyne
Fiber direction	Elastic properties	Ultrasonic velocity
	Dielectric properties	Microwaves
	Light scattering	Laser, visible light
	Thermal properties	Infrared
Knots detection	Optical properties	CCD camera, colorimetry,
	Elastic properties	Ultrasonic CT
	Ionizing radiation	X-ray CT, gamma ray CT
	Thermal properties	Infrared thermography, laser thermography
	Electrical properties	Microwave, NMR
Decay detection	Elastic properties	Vibration, ultrasonic
	Mechanical properties	Penetration of pilodyn
	Dielectric properties	Microwave
	Ionizing radiation	X-ray CT
Mechanical capacity	Static properties	Bending stress grading, proof-loader
	Dynamic properties	Vibration and ultrasonic methods
	Acoustic properties	Acoustic emission

This book is an attempt to comprehensively review numerous aspects of imaging and nondestructive characterization of wood products, ranging from the tree scale to the most complex wood composite, and hence to provide a convenient point of reference. For this purpose, the various nondestructive methods can be classified according to the characteristic scale (i.e., frequency, energy or wavelength) of the radiation that interacts with wood specimens (Fig. 1.2).

The synthesis of a wide range of phenomena involving electric and magnetic fields is provided by the electromagnetic theory developed by Maxwell. The position of the spectral lines can be characterized by their frequency, η, wavelength, λ, and wavenumber, k, which are defined by Eq. (1.1):

$$c = \eta\lambda = \eta \, 2\Pi/k \tag{1.1}$$

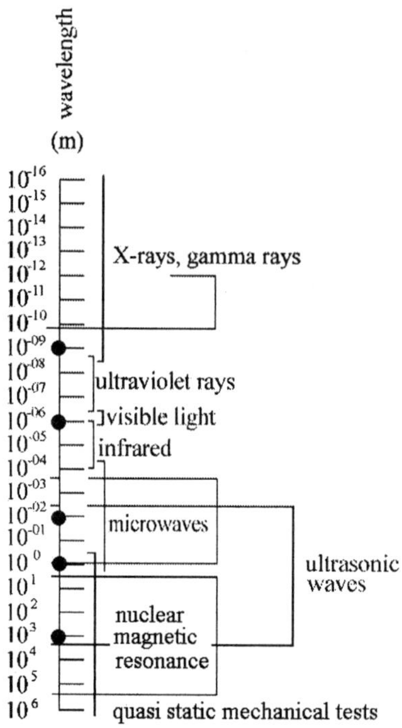

Fig. 1.2. Characteristics of the electromagnetic waves. (Data from Swanson and Hailey 1987; courtesy of Forest Products Society)

where c is the speed of propagation of the radiation.

The position of spectral lines of radiation can be characterized by their wavelength, as is shown in Fig. 1.2. The various radiation and therefore types of nondestructive methods can be classified according to the frequency of the radiation involved. This is the criterion chosen for the description of different wood imaging techniques in this book.

As can be seen, the highest transmitted energy, corresponding to the smallest wavelength (10^{-12} m) is obtained with X-ray methods. In the opposite position we have acoustic and radio waves, for which the wavelengths are in the range of 1 m. The interaction of X-rays with wood is accompanied by changes in the energy of the electrons. The visible and ultraviolet regions are related to transitions of the valence electrons, and the microwave and infrared regions are related to molecular rotational and vibrational energy changes. The frequency characteristics of the NMR region are low and in the range of 1 to 500 MHz and associated with transitions between energy levels corresponding to the magnetic states of atomic nuclei.

Wood is a natural composite that has an hierarchic structure, which is heterogeneous and anisotropic. Anisotropy and heterogeneity are not absolute characteristics, but are relative to a given physical property and to the scale length of the corresponding physical phenomenon, characterized by a wavelength (Bucur and Rasolofosaon 1998). Having in mind the hierarchical archi-

Table 1.3. Scale of wood structural characteristic and the required resolution of imaging techniques expressed by the wavelength

Imaging method with	Wavelength (m)	Scale of wood structure
X-ray	$10^{-12} \ldots 10^{-9}$	Submicroscopic, microscopic and macroscopic structure
Visible light	10^{-6}	Micro and macroscopic structure
Infrared	$10^{-5} \ldots 10^{-6}$	Macroscopic structure
Microwave	$10^{-3} \ldots 10^{-1}$	Macroscopic structure
Ultrasound	$10^{-2} \ldots 10^{3}$	Mesoscopic (tree), macroscopic and microscopic structure
Radio wave	$10^{3} \ldots 10^{5}$	Gigascopic (forests) and megascopic (group of trees) scale

tecture of wood, the selection of the most relevant technique for structure imaging is directly related to the resolution required by the selected method and consequently to the wavelength, as can be seen from Table 1.3.

The general understanding of the interaction between the electromagnetic waves and the wood material under inspection must be based on an accurate description of the phenomena, with an increasing degree of sophistication of the theoretical models. Consequently, the nondestructive evaluation of the properties of the media becomes an inverse problem.

The development of a corresponding theory that describes the interaction between the electromagnetic waves and the material requires three principal steps:

- observation of the phenomena to provide guidance for the theoretical approach;
- integration of the theory to the specific problem of observed phenomena;
- conduct of experiments that verify the first and the second steps.

The interaction between the second and the third step generates a model that can be validated by the experimental data of the first step. The second half of the 20th century has shown the development of a new branch of wood physics related to the nondestructive evaluation of its properties.

1.4 Imaging of the Internal Structure of Wood

The nondestructive imaging of wood structure is based on different methods utilizing typical wavelengths of elastic waves, penetrating radiation, light, electric and magnetic fields, etc. Computed tomography was developed on the mathematical basis established by Radon (1917). He demonstrated that from a complete set of projections of relevant physical variables it is possible to reconstruct the image of an object. The image can be reconstructed by mapping of different measured parameters or can be obtained by a tomographic technique, using algorithms and advanced computational procedures for data collection, image reconstruction, and display. Analytical methods, based on Fourier transform or algebraic iterative methods, using substantial projection data, are used for tomographic reconstruction. In the future, tomographic techniques are expected to become of general use.

The nondestructive techniques presented in this book for wood imaging are ionizing radiation (X-rays and gamma rays) having a resolution of 10^{-9} m, thermal waves with a resolution of 10^{-3} m, microwaves with resolutions from 10^{-4} to 10^{-1} m, and ultrasonic waves with a resolution of 10^{-3} to 10^{-2} m, or more, depending on frequency, for imaging wood structure at the macroscopic scale. The nuclear magnetic resonance window is superimposed on the ultrasonic window.

X-ray tomographic imaging can be produced with fixed or mobile equipment with one source and one detector acting in parallel by translation or with an array of transducers located around the sample.

High resolution ultrasonic tomographic techniques were developed as a complement to X-ray tomographic imaging. The ultrasonic waves propagating in wood are related to its elastic properties. The waves interact with the microstructure, and their characteristics are modified as they travel through the material due to reflection, scattering and attenuation. The detected signal is displayed, processed, and interpreted in terms of the internal structure of wood based on its relations to the input wave. Recent technological advances allow new capabilities for measurement of elastic properties and, consequently, the production of high-resolution ultrasonic images.

The microwave frequency region is bounded by the radio wave region (upper bound) and by the infrared region (lowest bound). The nature of the interaction of microwaves with a wood structure suggests a certain similarity with ultrasonic waves. An advantage of the high velocity microwave propagation coupled with the noncontacting feature of the transducers is that it permits a rapid inspection limited only by mechanical considerations. Therefore, amplitude and phase measurements of waves transmitted or reflected by the specimens will contain information about the structure of internal flaws, inhomogeneities, moisture content of logs, lumber and wood-based composites.

Microwave imaging of vegetative material, including leaves, stems and trunks at various moisture contents and temperature conditions is possible over a wide range of frequencies. The probes can be either scatterometers, radar, or reflectometers and can operate in the proximity of the medium as open-ended coaxial lines, cavity resonators or antennas.

In wood science, thermal imaging is a relatively new field for ascertaining the integrity of subsurface structure. The scanning infrared imaging technique appears to be a very promising remote sensing method for thermographic inspection of trees, solid wood, and composites in situ. The NMR imaging method is totally nondestructive and noninvasive for wood. The availability in the future of low-cost devices will contribute to the widespread use of this method. Neutron imaging is one of the most recent nondestructive techniques developed for wood. High-resolution imaging (15 µm) of water distribution in wood can be obtained with small, clear specimens and with structural elements.

1.5 Summary and Outline of the Book

The purpose of this introductory chapter, which is concerned with the nondestructive characterization and imaging of wood, is to provide a brief overview of the material that will be discussed in the subsequent chapters. With computed tomography two- and three-dimensional images of the internal structure of objects can be obtained without physical sectioning. Tomograms obtained with different techniques, ranging from ionizing radiation (X-rays and gamma rays) to ultrasonics and nuclear magnetic resonance and nuclear imaging, provide excellent means of obtaining information about the internal structure of wood.

The criterion selected for the sequence of the chapters is the wavelength of the physical radiation, which is closely related to the scale of observation used for wood characterization and imaging. This means that the book will begin with X-ray and gamma ray imaging techniques, corresponding to the smallest wavelength and consequently to the highest resolution, followed by thermal imaging, microwave imaging, and ultrasonic and nuclear magnetic resonance imaging methods. The last chapter is devoted to neutron imaging.

The book offers an in-depth review of the state of the art of the use of nondestructive evaluation techniques as applied to the imaging of wood. The theoretical approach will be outlined, and emphasis will be placed on practical applications. For the readers who would like to have more detailed information about any particular technique, an extensive bibliography with a subject index concludes the book.

2 Ionizing Radiation Computed Tomography

2.1 Introduction

Computed tomography scanning with ionizing radiation provides three dimensional information about the internal inhomogeneous structure of the specimen under test in a nondestructive, noninvasive and rapid manner. This technique is a "cornerstone of materials characterization" (Kinney et al. 1988). The aim of computed tomography is to create a precise map of the internal inhomogeneities of the sample. This map is obtained through the determination of the attenuation coefficient of an ionizing radiation (X-ray, γ-ray) in a single thin layer of the sample. The tomogram shows this layer in a perpendicular position to the main axis of the object, while conventional X-ray technique, called Röntgen densitometry, produces an image parallel to the object under inspection. A conventional radiographic image is produced by sample translation and offers a rapid scan of the specimen under test, and, at the same time, locates the defect zones in one plane.

Tomographic images produced with ionizing radiation (X-rays and γ-rays) are called tomograms of slices or CT images and are obtained from the translation and rotation of the source and detectors around the specimen. The slices produce spatial information (in 3D) able to discern zones of low attenuation contrast. First-generation tomograms have been obtained with one source and detector acting in parallel, by translation. The sample was rotated by a 1°-step angle and the whole image was obtained for 180° of collected data. The second-generation tomograms were also obtained by translation, using an array of detectors that made simultaneous measurements through different angles during a single traverse inspection. The sample was rotated by the array beam angle. The third-generation tomograms are produced by a fanning movement, with a scanner provided with many detectors located on an arc focused at the X-ray source. The fourth generation of scanners also has a fan system of detection, and the detection array is located on a circle that surrounds the source and the sample.

In inhomogeneous media like wood, the attenuation coefficient depends on both the quantum energy of the ionizing radiation and the chemical composition of the sample. The electronic signals corresponding to images taken on wood specimens with X-rays in the Compton energy range enable the precise recording of mass density variations, clearly distinguished from the effects of high atomic number constituents. A characteristic signature of the specimen inspected with X-rays can be obtained (Polge 1966; Mothe et al. 1998).

Because the high quantum energy of radiation and the low atomic numbers of chemical constituents of wood, the attenuation phenomenon is caused mainly by the Compton effect and consequently is proportional to the mass density of the wood. It is generally accepted that the density of the cell wall in oven dry wood is constant (1500 kg/m^3). Therefore, the density variations observed on tomographic images are due to the distribution of various anatomic structural elements of the specimen under inspection and to the water content in the cell walls and in lumina.

The tomograms are obtained by calculation, using a sophisticated computer program (Kak and Slaney 1988; Habermehl and Ridder 1994; Grundberg et al. 1996; Schmolt et al.1999) that involves a complex technology. Each zone of the specimen having the same density is evaluated statistically and is represented by one color, or by a gray scale. More commonly, the red and the violet indicated low wood density and yellow high density.

This chapter discusses the ability of X-ray or gamma rays to provide a non-destructive method to identify and evaluate the internal structure of wood and wood-based composites. X-ray or gamma ray imaging is a rapid tool able to provide visual and analyzable internal structure of solid wood and of wood-based composites. The ability of this method to quantify spatially the density variations in solid wood, the presence of internal defects and inhomogeneities in lumber or in wood-based composites, the variation of moisture content in wood specimens, the distribution of water inside a trunk, etc. explains its important increasing implementation in wood science and technology.

During the last 20 years, Scandinavian countries and Germany in Europe, Canada, Japan, Australia, New Zealand and the United States have put stress on the understanding of the basic aspects of X-ray computed tomography in view of the development of specific wood, industrial scanners for routine quality control of wood products (Benson-Cooper et al. 1982; Habermehl 1982a,b; Mc Millin 1982; Cown and Clement 1983, Asplund and Johanson 1984; Onoe et al. 1984; Funt and Bryant 1985; Lindgren 1985; Harley and Morris 1988; Davis et al. 1989; Wagner et al. 1989a,b; Rinn 1991; Habermehl and Ridder 1992a,b, 1993,1995, 1998; Grundberg et al. 1995; Schmoldt 1996; Sugimori and Lam 1999; Oja et al. 2000).

The advantages of ionizing radiation computed tomography are numerous when compared with conventional radiography. Firstly, this technique eliminates the intermediate steps involving photographic film and optical densitometry and is able to make data available in real time. In process control and in manufacturing situations, the density feedback is very important for technological competitiveness. Secondly, an important advantage of direct scanning technique is the improvement of the calibration procedure, because the mass attenuation coefficient can be determined directly, using a scintillation detector with standard radiation pulse shaping and counting equipment, which allows the user to select the energy range counted (Fig. 2.1). Thirdly, a large

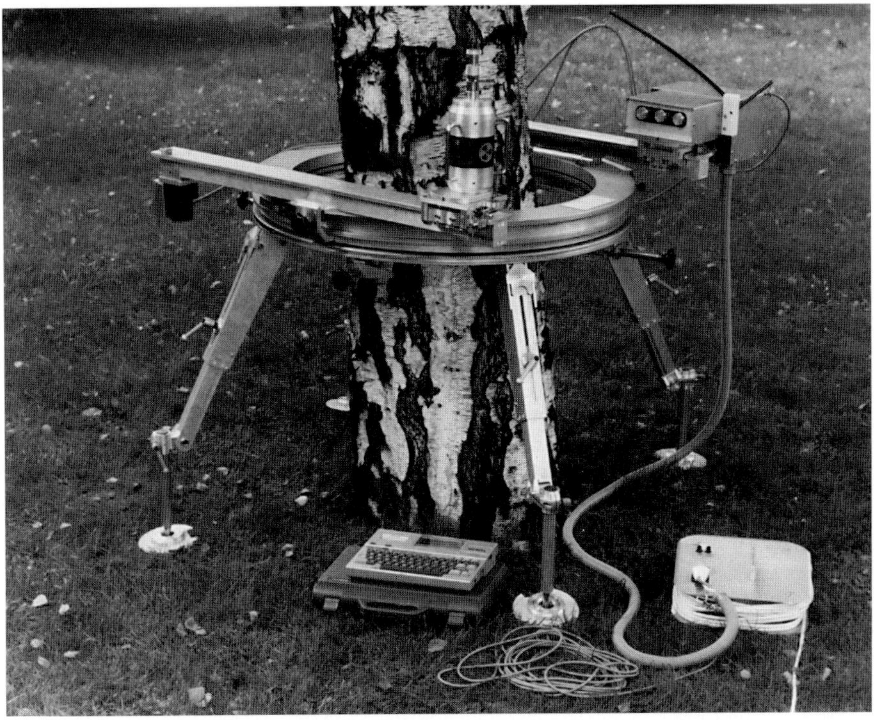

Fig. 2.1. X-ray device for computed tomography on a tree. (Habermehl, pers. comm., with permission)

volume of material can be inspected quickly and implementation of scanning technology in sawmills and other factories using wood materials will have an important payback for wood processors.

2.2 Basic Phenomena

The basic phenomena involved in X-ray reconstitutive tomography is illustrated in Fig. 2.2. The translation and rotation of a radiation beam, defined by collimators, are scanned across the sample. After each linear scan, the source-detector system rotates through an angle around an axis perpendicular to the chosen plane. The sequence continues for a complete rotation of 180°. For each beam path, the detector quantifies the attenuation coefficient of the radiation beam through the sample, also called the "object".

As noted by Liu et al. (1988), the term "attenuation coefficient of X-rays" used in this chapter refers to "total decreases in X-ray intensity, regardless of whether absorption or scattering have been the interaction mechanisms in the

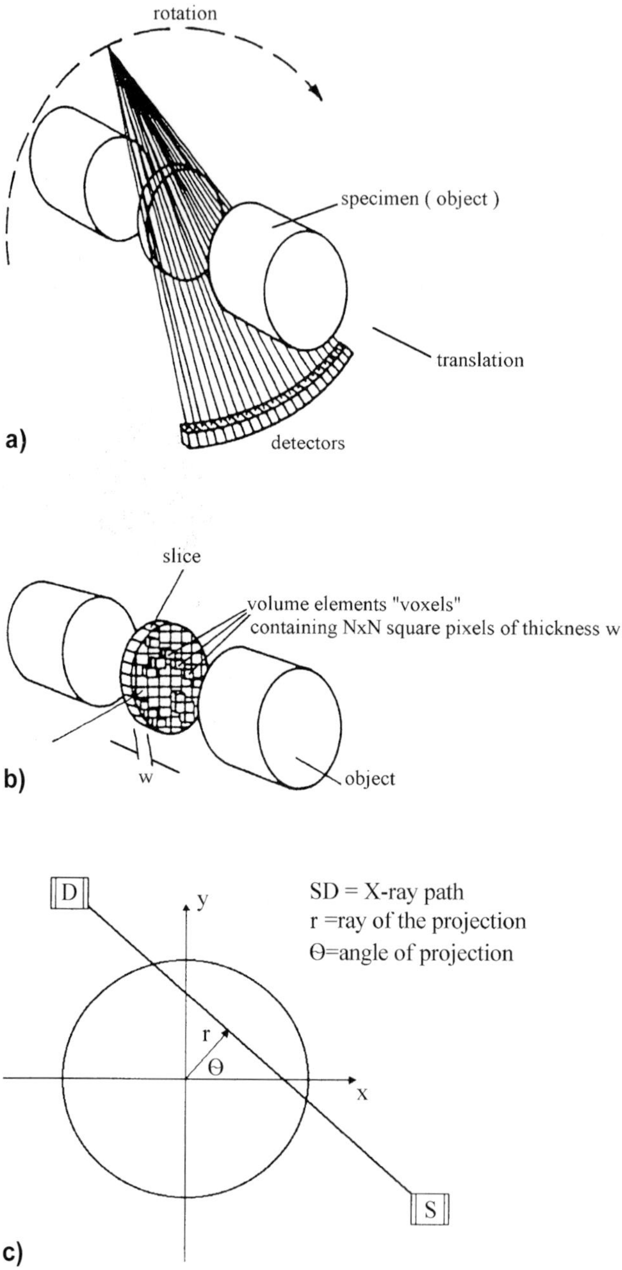

Fig. 2.2. Basic aspects involved in X-ray reconstitutive tomography. **a** Rotation of source-detector system around the specimen (Lindgren 1991b, with permission). **b** Volumetric element (voxel) in a slice (Lindgren 1991b, with permission). **c** Beam path geometry between the source and the detector. (Gilboy and Forster 1982, courtesy of Academic Press)

transmission equipment". The term "absorption coefficient", often used in the literature must be avoided. From a physical point of view, X-ray or gamma ray attenuation corresponds to the removal of quanta from the incident beam by whatever process reduces its intensity. Ionizing radiation computed tomography is a nondestructive radiation method that allows the conversion of attenuation coefficients into density data and then into images.

2.2.1 X-Ray Propagation in Solids

As described in reference books (Herman 1980; Kak and Slaney 1988) and articles (Polge 1966; Hägglund and Lindgreen 1985; Lindgren 1985, 1991a,b; Fioravanti and Ricci 1991; Habermehl and Ridder 1992 a,b; Lindgren et al. 1992), in the case of monoenergetic photons such as γ-rays or highly filtered X-rays, with rectilinear propagation in solids, the relationship between incident and transmitted ray intensities, the attenuation coefficient and the thickness of the sample, is:

$$I = I_O\, e^{-\mu t} \qquad (2.1)$$

where

I = intensity of the transmitted ray beam, passing through the sample
Io = intensity of the incident ray beam, passing through air only
μ = attenuation coefficient
t = thickness of sample

As ionizing radiation traverses the matter it is absorbed exponentially in accordance with the equation:

$$I/I_O = e^{-\mu t} \qquad (2.2)$$

The attenuation coefficient μ depends on the mass density of the sample ρ (kg/m^3) and on the mass attenuation coefficient μ′ (m^2/kg or cm^2/g as noted more often in the literature).

$$\mu = \mu'\rho \qquad (2.3)$$

The density is then

$$\rho = \mu/\mu' \qquad (2.3')$$

The density (Eq. 2.3′) of the inspected sample is the ratio between the linear attenuation coefficient μ and, μ′, the mass attenuation coefficient, and can be determined experimentally by measuring the difference between the intensity of the incident beam and of the absorbed beam.

The Eq. (2.1) can be rewritten as:

$$I/I_O = e^{-\mu\rho t} \qquad (2.4)$$

and the corresponding logarithm is:

$$(\ln I - \ln I_O)/t = \mu'\rho = \mu \qquad (2.5)$$

From Eq. (2.5) the attenuation coefficient μ may be calculated, as demonstrated by Olson and Arganbright (1981).

The mass attenuation coefficient of wood μ′, has been determined by several methods, using X-rays or gamma rays, and reported in the literature by many authors, for specimens of different thicknesses and with different energy levels of the radiation source, as can be see from Table 2.1. The mass attenuation coefficient is a basic parameter of wood. Moschler and Dougal (1988) noted the value $\mu' = 0.185\,cm^2/g$, Malan and Marais (1992) using gamma rays, reported values for μ′ ranging between $0.1858\,cm^2/g$ for air dry wood and $0.1797\,cm^2/g$ for oven dry wood, measured on 32 wood species having a gravimetric density ranging between 156 (balsa, *Ochroma pyramidalis*) and 1194 kg/m3 (tropical boer-bean, *Schotia* sp.). The experimental value of μ′ depends on the energy level of the source, on the moisture content of the wood and on the wood species. Theoretical values reported by Tiitta et al. (1996) is $\mu' = 0.192\,cm^2/g$ at 0% moisture content and 59.5 keV and, is $\mu' = 0.198\,cm^2/g$ at 100% moisture content.

Table 2.1. Wood mass attenuation coefficient μ′ (cm^2/g) as found in the literature

Coefficient μ′ (cm^2/g)	Energy source –	Moisture content (%)	Reference –	Differences (%)
0.191	Am^{241}	8	Moschler and Dougal (1988)	–
0.185	Am^{241}	0	Moschler and Dougal (1988)	–
0.189	Am^{241}	0, Particleboards	Ranta and May (1978)	2
0.191	Am^{241}	0, *Pinus elliotti*	Feraz (1976)	3
0.21–0.28	Am^{241}	0, 17 Species	Ferraz (1976)	12–34
0.192	0.060 Mev	0	Olson and Aganbright (1981)	3.6
18.22	Fe^{55}	8	Moschler and Dougal (1988)	–
17.72	Fe^{55}	0	Moschler and Dougal 1988	–
17.70	0.006 Mev	0%	Moschler and Dougal (1988)	0.1
0.185	Am^{241}	8, 31 Species	Malan and Marais (1992)	0
0.176	Am^{241}	8, Oak	Malan and Marais (1992)	–
0.0726	Ce^{137}	8, Oak	Malan and Marais (1992)	–
0.190	Am^{241}	8, Rose gum	Malan and Marais (1992)	–
0.082	Ce^{137}	8, Rose gum	Malan and Marais (1992)	–
0.198	Am^{241}	8, Stinkwood	Malan and Marais (1992)	–
0.081	Ce^{137}	8, Stinkwood	Malan and Marais (1992)	–
0.192	59.5 keV	0, Theoretic	Tiitta et al. (1996)	
0.198	59.5 keV	100, Theoretic	Tiitta et al. (1996)	
22.61	5.9 keV	0, Theoretic	Tiitta et al. (1996)	
23.42	5.9 keV	100, Theoretic	Tiitta et al. (1996)	

The differences were calculated considering the data reported by Moschler and Marais as reference

The attenuation coefficient in inhomogeneous solids is not constant and depends on the local coordinates x, y of the pixel. Consequently, the intensity of the ray beam is:

$$I = I_O\, e^{-\int_{SD} \mu(x,y)ds} \tag{2.6}$$

where SD are the ray paths and x, y the coordinates of the chosen slice (Fig. 2.2c). The integral is taken along the beam path SD, that is the distance between source and detector.

The ratio

$$I/I_O = e^{-\int_{SD} f(x,y)ds} \tag{2.7}$$

and the corresponding logarithm, noted p, is called the ray sum of the function f(x, y):

$$p = \ln I/I_O = -\int_{SD} f(x,y)ds \tag{2.8}$$

and is proportional to the average linear attenuation coefficient along the total beam path.

The tomograms are reconstructed with the measuring projection data f(x, y). A set of ray sums p(r, θ) at a given angle θ defines a projection that is used for the reconstructive radiographic image in polar coordinates. For a point (x, y) the thickness, w, of the pixel is:

$$w = y \cos\Theta + x \sin\Theta \tag{2.9}$$

A projection at the angle θ is called the set of ray sums with θ constant, for a range of r. The function f(x, y) in Cartesian coordinates or f(r, θ) in polar coordinates are the basic data for the reconstruction of the image, which theoretically requires an infinite number of ray sums. Practically only a finite number of ray sums are recorded. The image reconstruction is based on a square array containing $N \times N$ square pixels of thickness w. Each projection consists of S measurements at spacing w.

The spatial resolution of the image is determined by the beam width, which is related to the spacing between two neighboring rays in each projection. For medical scanners used also for wood scanning, the calculated X-ray linear attenuation coefficient in each voxel is normalized within the computer by the corresponding linear attenuation coefficient for water according to Eq. (2.10). This normalized value is referred to as the CT number, or Hounsfield number. The unit is the Hounsfield, noted [H]. The CT number is calculated as:

$$CTnumber = \frac{(\mu s - \mu w)}{\mu w} \times 1000 \tag{2.10}$$

where μw is the linear attenuation coefficient in water and μs is the attenuation coefficient for the x voxel in the cross-sectional slice of the tested material at an average photon energy of 73 keV, a typical value used for medical scanners.

2.2.2 Attenuation and Profile of the Inspected Solids

The attenuation values calculated for each pixel are translated into a gray scale or a color scale and transformed into a picture of the cross section of the sample on a TV screen. For medical scanners, the CT number ranges between −1000 for air to 0 for water and to +1000 for human bone. When working in the optimum range, the medical scanners exhibit a typical photon limited noise level of 0.5% in pixels of 1 mm^2 with 1-cm slice thickness. The image is often reconstructed with 512 × 512 pixels and saved on floppy discs. In this way, it is possible to use different digital image techniques such as geometrical transformations followed by subtraction of the image to measure the dynamic processes such as wood drying as a function of time or wood impregnation with preservative substances as a function of time or of thickness of the sample.

The relationship between wood density and the attenuation coefficient was established by Lindgren (1985) for pine (*Pinus sylvestris*) on a Philips 210 CT medical scanner and is expressed by Eq. (2.11) as:

$$\mu = 0.93\,\rho_o - 1001 \tag{2.11}$$

where ρ_o is the oven dry density (kg/m^3) and the attenuation μ is expressed in [H]. The corresponding correlation coefficient is r = 0.99 (Fig. 2.3). The relationship between the moisture content and the attenuation coefficient for $\rho_o = 430\,kg/m^3$ and for $\rho_o = 540\,kg/m^3$ was also established as can be seen from Fig. 2.4, using rectangular samples of 14 × 59 × 116 mm and 30 × 97 × 250 mm. The accuracy was 1.5% for density measurements and 2% for moisture content. It is well known that the attenuation measurements depend on different factors such as: the energy level of the source, the moisture content, the geometry of the specimen and the general noise observed on the tomograms.

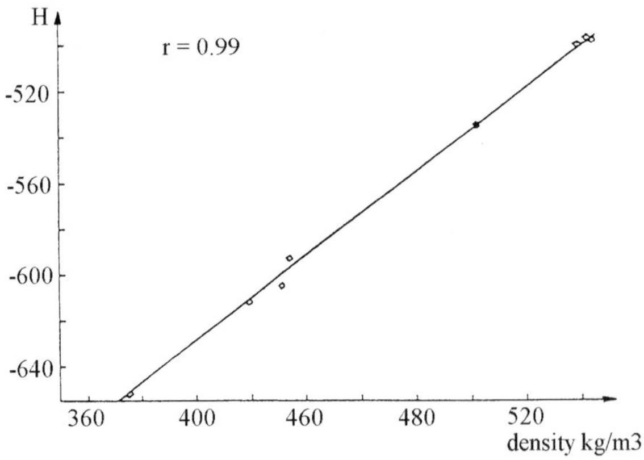

Fig. 2.3. Relationship between oven-dry density and attenuation coefficients in *Pinus sylvestris* sapwood (Lindgren 1985, with permission)

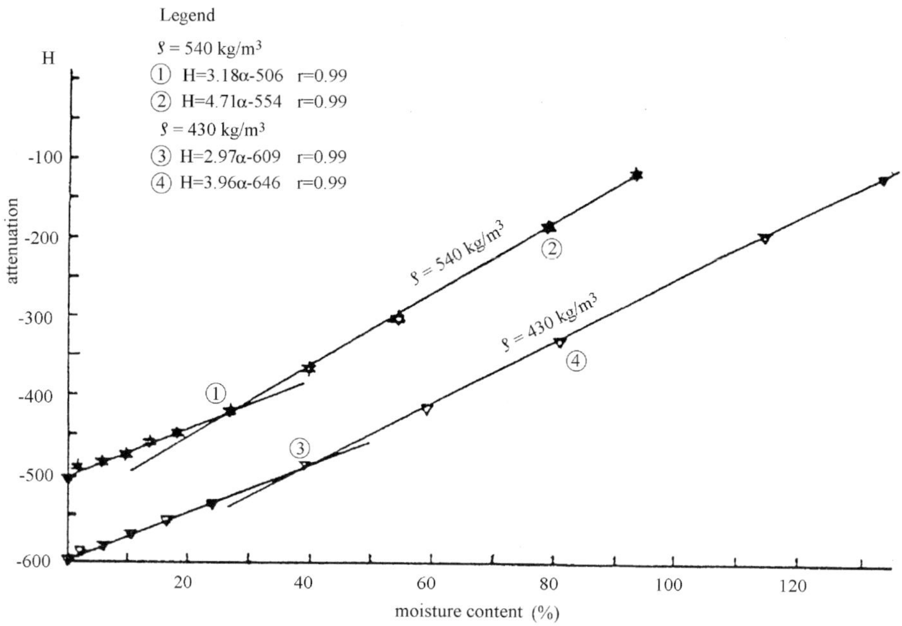

Fig. 2.4. Relationship between moisture content and attenuation for oven-dry wood at two constant densities of 430 kg/m^3 in *Pinus sylvestis*. (Lindgren 1985, with permission)

Based on the data existing in the literature, Gierlik and Dzbenski (1996) studied the influence of moisture content on the mass attenuation coefficient for four levels of moisture content of wood, namely 0, 9.5, 18.5 and 26.8%. A nonsignificant decreasing relationship was obtained.

The theoretical problems of the noise on the image were discussed by Gilboy (1984) for the conventional translate rotate scanning geometry, using the following characteristics: D, diameter of the sample; N, pixel size; w, the width of the pixel; S, ray sum; t, spacing between rays, usually the same as w; M, total number of projections; n, finite photon count per ray sum; k, a constant depending of the type of the filter used; T, total run time; μ, linear attenuation coefficient; A′, the effective activity.

The noise propagated to the image is estimated as a variation of the attenuation coefficient μ, and can be expressed by Eq. (2.12):

$$\Delta\mu = k(2\Pi/n\,Dt)^{1/2} \tag{2.12}$$

The relative accuracy of the measured attenuation coefficient is given in Eq. (2.13) as:

$$\Delta\mu/\mu = 4\Pi\,kN^2\cdot(D/t\ TA')^{1/2}\cdot\left[\sinh(\mu D)/(\mu D)^3\right]^{1/2} \tag{2.13}$$

The first part of this equation, $4\Pi\ kN^2.\ (D/tTA')^{1/2}$, depends on the geometrical factors, the total run time and the source intensity. The second term of Eq. (2.13), $[\sinh(\mu D)/(\mu D)^3]^{1/2}$, is called "sensitivity factor," and is accuracy dependent. The plot of the sensitivity factor versus the diameter of the sample to be inspected goes through a minimum and can indicate the choice of optimum photon energy required for the best image.

Lindgren (1991a) identified the noise on five specimens of *Pinus sylvestris* using a medical scanner. The noise in each pixel was ±4 CT numbers and a variation with ±1 CT number corresponded to a $\pm 2\,kg/m^3$ variation in dry wood density and $\pm 6\,kg/m^3$ variation in wet wood density. No difference in noise depending on annual ring width was observed as long as the same direction of rotation was used to obtain the images. The accuracy of the noise needs to be established each time measurements are performed on the same sample, for example, in studies of moisture content variation over time.

The image quality is dependent on spatial and density resolution and an improvement in one can only be achieved by a reduction in the other. For a given time of measurement, the use of a high intensity source (such as a gamma source) reduces the compromise between these two parameters. Gamma sources possess this important practical advantage and also have a constant photon energy over space and time. X-ray tubes produce an energy spectrum that is variable with time, due to supply fluctuations or to tube degradation. They also induce variations of the take-off angle for fan beam measurements. Beam hardening is affected by all these effects. Indeed, the effective energy of the radiation is a function of many parameters related to a specific scanner, and the corresponding resulting image is scanner dependent. Gilboy (1984) noted "the lack of precision in the X-beam energy becomes a greater problem when working in the photoelectric region where the strong energy dependence magnifies the effect of small energy changes."

Lindgren (1991a) noted that the total attenuation coefficient of a material μ is the sum of the attenuation coefficient du to the photo-electric effect, μ_p, and of the attenuation coefficient due to the Compton scattering effect of photons, μ_c, as can be seen from Eq. (2.14).

$$\mu = \mu_P + \mu_C \tag{2.14}$$

where μ_p and μ_c depend on the density of the material being tested, electron density, effective atomic number and photon energy. Consequently, it is possible to calculate the attenuation coefficient for any material if the chemical composition and density are known.

The attenuation coefficients of principal wood constituents are: 0.2634 for cellulose, 0.2655 for hemicellulose and 0.2608 for lignin. Lindgren (given citation) calculated the attenuation coefficient of wood and the corresponding CT numbers for different volumes of cellulose, hemicelluloses and lignin and for the oven-dry density $\rho_o = 500\,kg/m^3$, as shown in Table 2.2.

The decrease in the attenuation coefficient is related to the increasing lignin volume and the decreasing cellulose volume. A more detailed analysis consid-

Table 2.2. The influence of chemical constituents of wood on the attenuation coefficient and CT numbers, calculated for an oven-dry density of 0.5 g/cm^3. (Lindgren 1991a, with permission)

Cellulose (%)	Hemicellulose (%)	Lignin (%)	Attenuation coefficient	CT number
50	25	25	0.08775	−527
25	25	50	0.08753	−528
50	40	10	0.08799	−526
10	10	80	0.08715	−530

Table 2.3. Theoretical mass attenuation coefficients of wood calculated for two levels of photon energy and for the following mass fraction of elements: 50% C, 44% O, 5.7% H, 0.3% N (Tiitta et al. 1996) (10th International Symposium on Nondestructive Testing of Wood, 2002, Press Polytechniques et Universitaires Romandes, p. 189. Reproduced with permission of the editor. All rights reserved)

Material	Moisture content (%)	Mass attenuation 5.9 keV	Coefficient (cm^2/g) 59.5 keV
Wood	0	22.61	0.192
	5	22.69	0.193
	10	22.76	0.193
	20	22.88	0.194
	30	22.98	0.195
	50	23.15	0.196
	100	23.42	0.198
Water	–	24.23	0.205

ering the mass fraction of C, O, H and N was proposed by Tiitta et al. (1996). They calculated the mass attenuation coefficient of wood for two levels of photon energy and for moisture content ranging from 0 to 100%., as can be seen from Table 2.3. The mass fraction of the elements was: 50% C, 44% O, 5.7% H, 0.3% N. From this theoretical approach, it was confirmed that the photon energy level is a very important factor for the definition of mass attenuation in wood. For example, at 0% moisture content, the mass attenuation coefficient is 22.61 cm^2/g at 5.9 keV and 0.192 cm^2/g at 59.5 keV. It was also noted that at the same level of photon energy the mass attenuation coefficient increases slightly with moisture content. For example, at 59.5 keV and 10% moisture content, the mass attenuation coefficient is 0.193 cm^2/g and at 100% moisture content, it is 0.198 cm^2/g. A possible explanation of this statement is the small difference between the mass attenuation coefficient in water (0.205 cm^2/g) and in oven-dry wood (0.192 cm^2/g).

The linear regression equations established between the attenuation coefficients expressed in CT numbers and the measured density of oven dry wood and of green wood are given in Table 2.4. The proposed equations are calibra-

Table 2.4. Regression relationships between wood density and CT numbers for different moisture contents. (Lindgren 1991a).(with permission)

Variables	Regressions	Density Standard deviation	CT numbers Standard deviation
ρo and CT	ρo = 1053 + 1.052 CT; R = 0.99	±2.7 kg/m³	±4
ρu and CT	ρu = 0.993 CT + 1015; R = 0.99	±6.7 kg/m³	±13.4

tion equations. They must be established for each scanner used for laboratory or industrial measurements.

2.2.3 Reconstitutive Algorithms

As noted in reference books and articles (Herman 1980; Kak and Slaney 1988; Gilboy and Foster 1982; Gilboy 1984; Fioravanti and Ricci 1991; Schmoldt 1996; Bahyl 1998; Coles 1999), image reconstruction methods are:

- analytical methods, based on Fourier transforms and related techniques including filtered back-projection (convolution). Data obtained from convolution can be processed immediately as collected and the virtual image built by projections is available instantly after scanning. Finite Fourier transform (FFT) is periodical in the time domain, requires an infinite number of projections and gives a picture in the frequency domain in the Cartesian coordinate system. In practical situations, the number of projections is not infinite as it should be. The Radon transformation in the image reconstruction process overcomes this aspect. Convolution of the density function using a filter function realizes the filtering process of the transformed data, for a selected angle and position. A huge amount of computation is required with parallel projections or with fan beam projections. The effect of an increasing number of projections on image quality has been reported and seems to be limited to 45 projections, as noted by Gilboy (1984) for a specimen in light alloy casting. When data rates are low, this method can give a useful saving in measurement time.
- algebraic (iterative) methods need a very large amount of data for image reconstruction, and better results are obtained from relatively few projections.

Each attenuation value used for the image reconstruction is associated with a specific cross-sectional pixel size and finite slice thickness (creating the voxel). Pixel squared size is between 0.9 and 3 mm. Pixel resolution depends on the source-detector distance and on the number of detectors. Slice thickness can vary from 1.5 to 20 mm (Schmoldt 1996) and limits the sensitivity and resolution of scanning of heterogeneous samples. Indeed, a better coverage of the specimen is achieved with higher scanning frequencies, but the process requires a long exposure time to ionizing radiation and can reduce the life of the X-ray tube. The run time can be reduced by the use of a wider scanning beam, but this choice induces a loss in spatial resolution.

2.2.4 Treatment of Images

The images reconstructed by back projection of a filtered projection require attenuation values at any point in the image. These data can be numerically presented and statistically interpreted as density variations along any line segments across the image. Histograms of the attenuation values within any region can be plotted, and line integrals can be obtained (Funt and Bryant 1985). Groups of pixels can be then interpolated to give a degree of image smoothing. Faust et al. (1996) show a histogram of the image (Fig. 2.5) and threshold selection for several gray levels for knots and cracks, in a red oak log as: g_1 is the lowest gray level corresponding to pixels from air, cracks, large voids, parts of earlywood; g_2 is the gray level corresponding to pixels from decay, earlywood, and a few portions of latewood; g_3 is the gray-level corresponding to pixels from latewood and a few portions of knots; g_4 is the gray level corresponding to pixels from knots and some portions of latewood.

The scanning parameters are influenced by different factors, such as: wood species, size of the specimen (diameter of log or thickness of sample), level of contrast in density change necessary to discriminate various defects, defect size, end use of scan information, with large-scale or fine-scale details, and the speed of scanning required for the production context.

The huge amount of data must then be condensed, and only those internal features important for the subsequent processing must be identified automatically (Schmoldt 1996; Schmolt et al. 1997) with a machine vision using artificial neural networks. Javadpour et al. (1996) developed a specific artificial neural network for the identification and classification of knottiness of Irish Sitka spruce. The study was concerned with the optimal orientation of the log, from its modeled images as a function of internal defects, to yield a high quality lumber. Major difficulties are related to the high moisture content of green logs that render obscure the recognition of the knot pattern from surrounding clear wood. The knot is simulated as a branch of a right elliptic cone and is defined by several geometric parameters related to its shape and position in the stem. The model provided geometric parameters and density variations as a function of the age of the annual ring, within the trunk and internal multiple branches and, effectively simulates the profiles obtained by tomographic techniques. Using this model, it was possible, for a given stem, to obtain any number of internal transverse and longitudinal profiles of densities. The annual ring widths and the percentage of earlywood and latewood in each ring

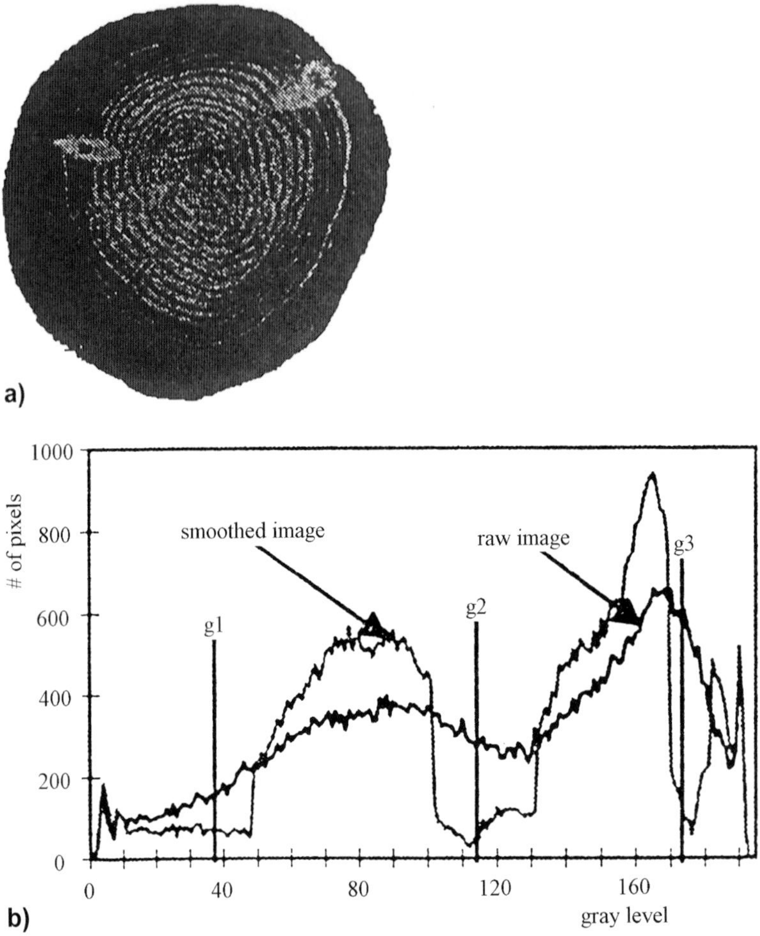

Fig. 2.5. Histogram of the image of knots and cracks in a red oak log (Faust et al. 1996). **a** CT image of red oak section; **b** histogram of the image and threshold selection. (10th International Symposium on Nondestructive Testing of Wood 2002 Press, Polytechniques et Universitaires Romandes, p. 203. Reproduced with permission of the editor. All rights reserved)

were incorporated into the model. This was a very original application of this model.

Faust et al. (1996) presented a flowchart for a system (Fig. 2.6) to detect knots and cracks in red oak, black walnut and hard maple, with 3D analysis of the position, orientation and shape of these features. The basis of this analysis is the density. The system first processes each single image and then relates processed results of each CT image across neighboring CT images to pinpoint defects. The most current algorithms link annual ring structure in an edge-detected image. Locating the pith is an important step in identification of

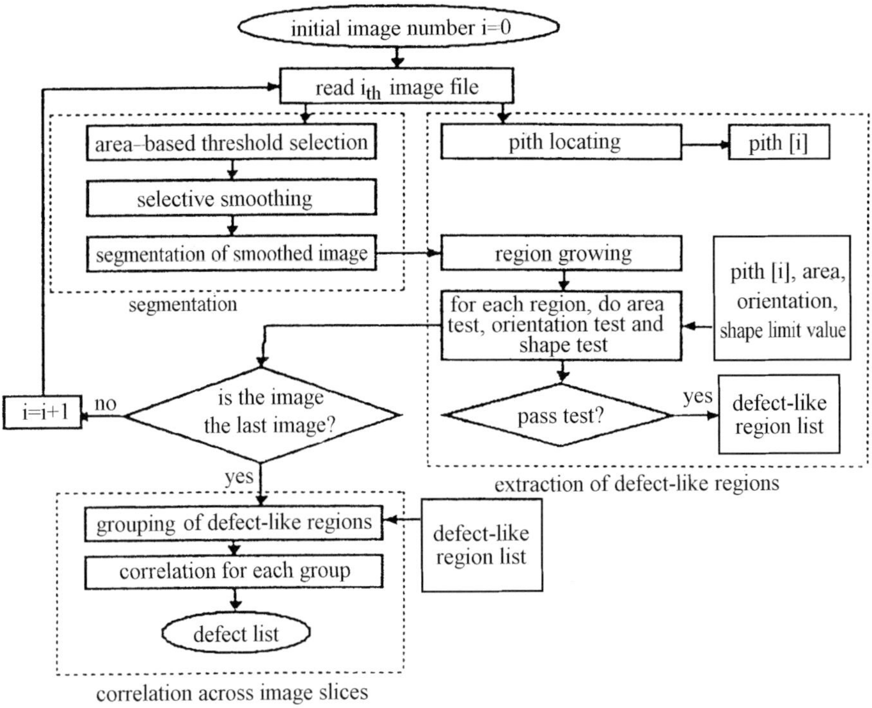

Fig. 2.6. Flow chart for the detection of cracks, knots, earlywood, and latewood in a hardwood (Faust et al. 1996) (10th International Symposium on Nondestructive Testing of Wood 2002, Press Polytechniques et Universitaires Romandes, p. 202.Reproduced with permission of the editor. All rights reserved)

Table 2.5. Precision of pith location in three hardwood logs using the algorithm proposed by Faust et al. (1996) with the precision within 5, 10 and 20 pixels. (10th International Symposium on Nondestructive Testing of Wood, 2002. Press Polytechniques et Universitaires Romandes, p. 205. Reproduced with permission of the editor. All rights reserved)

Species	Images with 5 pixels (%)	Pith location 10 pixels	At precision 20 pixels
Maple	54.5	77.3	99.0
Oak	60.5	83.0	98.9
Walnut	83.2	97.8	100.0

growth rings and the longitudinal axes of knots and cracks that normally pass through the pith. Table 2.5 shows the percentage of images with pith location precision with different numbers of pixels ranging from 5 to 20. The best location was obtained with 20 pixels for walnut, a semi-ring porous species.

2.3 Equipment for Imaging Techniques

In 1979, the works of Cormak in South Africa and Hounsfield in England were recognized with the award of the Nobel Prize for Medicine for the development of ionizing radiation computed tomography. Cormak (1963) manually scanned collimated beams of ^{60}Co gamma rays across aluminum and wood phantoms and detected the transmitted photons with a GM counter. He reconstructed the image of the scanned sample from one-dimensional projections viewed at different angles in terms of a map of linear attenuation coefficients, using a mathematical model he developed. Hounsfield (1972), using a much more intense X-ray source, produced images of a human head in a run time of several minutes, using a one-line minicomputer. This was the starting point for the development of X-ray computed tomography for medical use. It is also notable that today the unit used for X-ray imaging with medical scanners is the Hounsfield [H]. The Hounsfield scale for medical scanners is between −1000, corresponding to attenuation in air, and +1000, corresponding to attenuation in human bone, and is 0 for attenuation in water. This scale is linear between 1000 and 0, and its accuracy is ±2 H units.

Medical scanners were used in the 1980s for nondestructive inspection of different nonmedical objects, and new devices were developed for specific purposes. Medical scanners were first used for imaging wood specimens because the physical properties of wood do not differ greatly from that of the human body. The energy normally available on a medical scanner is about 70 keV, with soft X-rays. Presently, in wood science and technology, four generations of scanners are being used. A detailed description of each type of scanner will be given in the next chapter.

The development of the nonmedical scanners, first experimental and later commercial, started in the mid-1970s (Reimers et al. 1984). A wide range of applications were emphasized, ranging from an investigation of the foaming behavior of various coals at high temperature, concrete specimens, cracks and delaminations in metallic structures, defects in ceramics, fluid distributions inside porous simulated rock cores, wood structures and trees, thick plastics to a maximum 400 mm, crack detection in welded steel bridges, internal features in jet and automobile engines, fluid flow measurements in nuclear engineering, to a complete inspection of the Trident C-4 rocket motor which is nearly 2 m in diameter. However, the cost of the systems is a real barrier, and the choice of testing methods must be made chiefly on economic grounds. The industrial systems in the wood industry require high energy X-ray sources, accommodation for a wide range of specimen sizes and densities and operation in industrial environments with a speed imposed by effective wood processing.

The major system components of an X-ray scanner are: X-ray generation system, fan beam or parallel beam X-ray collimator, system for data collection and analysis, image display software, mechanical gantry with possibility of rotation or translation or of simultaneous translation and rotation of the sample under inspection.

2.3.1 Description of the Equipment

Fixed or portable equipment has been used for different purposes in wood science and technology. If the test object can be brought to the scanner, which is the case of logs and boards, fixed equipment is recommended. In this case, an X-ray source has commonly been used. The portable equipment was designed to be used for in situ inspection of trees, poles and building elements. Gamma rays, which are mono energetic, were used in this case because they have many advantages in the wide range of energies available, portability and cost. In spite of these superior characteristics, gamma ray sources are of low intensity, which is 10^4 times less than X-ray tube sources (Gilboy 1984). Malan and Maris (1992) have summarized the use of gamma rays for wood densitometry.

2.3.1.1 Fixed Equipment

Four types of scanner systems, referred to as "generations" have been developed, as can be seen in Fig. 2.7. Each type of the architecture of the scanner has advantages in terms of speed and artifact generation, which make it beneficial for specific applications.

The first-generation scanner (Fig. 2.7a) is a single-detector translate-rotate system with a single X-ray source detector. The reconstructed 2D tomograms are based on the attenuation coefficient measurements obtained when the specimen was rotated through 180°. The source-detector pair move in unison past the sample. The system is relatively inexpensive, simple in data collection and requires simple algorithms for tomogram reconstruction. Any size of specimen can be inspected. Despite its advantages, this scanner is speed limited and is used only for laboratory measurements.

The second-generation scanner (Fig. 2.7b) is a translate rotate system with an array of detectors, which makes simultaneous measurements through different angles in a single traverse. The geometry of the fan system is simple, data collection and reconstruction algorithms are all simple. Unlimited size samples can be inspected. This system is very slow because of the time needed for the mechanical operations with the sample and for the collection of unneeded data at the beginning and at the end of each traverse.

The third-generation (Fig. 2.7c) scanner rotate-only system has a detector array in an arc focused at the X-ray source. The data are collected in a fanning movement and for 180° data the sample is rotated 180°. The advantages of this system are: simultaneous collection of data through the entire sample for each

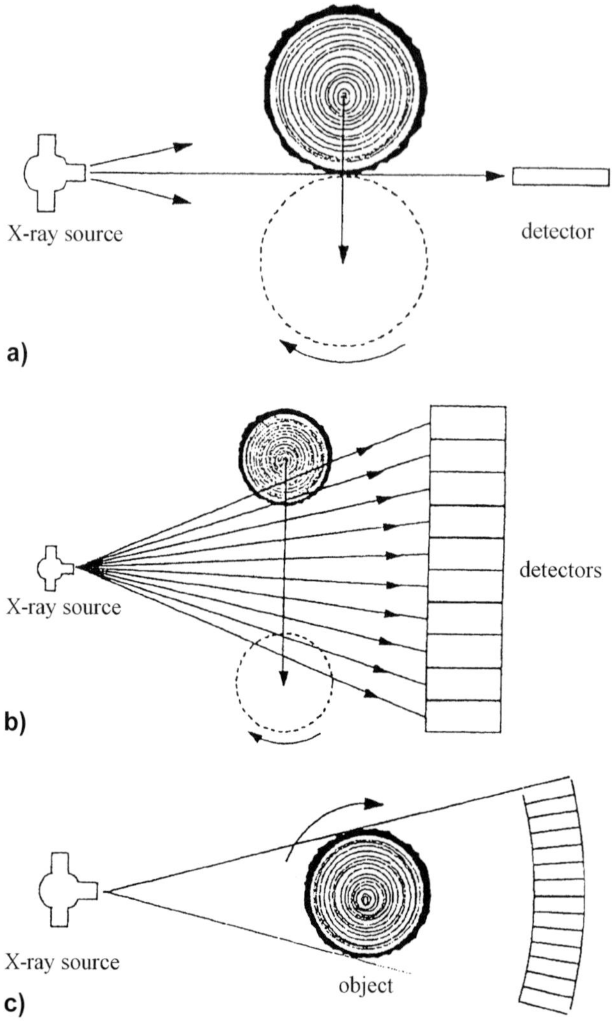

Fig. 2.7. Scanner system with detectors perpendicular to the axis of rotation (Schmoldt 1996). **a** First generation, a single detector, translate-rotate system (fan system). **b** Second generation, multidetectors, translate-rotate system. **c** Third generation, rotate only.

view, simple rotation of the sample, with a continuous mechanical motion, short scan times because of simultaneous collection of data. The limitations of the system are related to the single slice data collection, the diameter of the sample and number of the detectors, and high cost because of a large number of detectors required for large objects.

The fourth-generation scanner is the most rapid, where only the source moves (Fig. 2.7d). The data collection is based on a fan system, with detectors

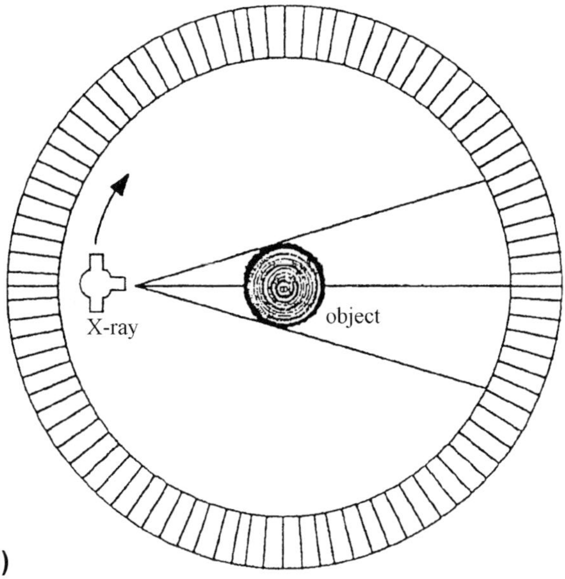

d)

Fig. 2.7. **d** Fourth generation, stationary detector, rotate only (with permission)

located on a circle that surrounds the X-ray source and the sample. The main advantage is the simultaneous data collection through the entire sample for each view, resulting in fast scans. Due to the very high cost, the system is rarely used.

To avoid the limitations of all four generations of scanners, an alternative was proposed with a scanner operating tangentially as can be seen from Fig. 2.8. The main advantage of this system is its ability to collect data for the entire volume of the object simultaneously from many cross-sectional slices. Most images are generated using data from a single detector. The speed is faster than that of the third- and fourth-generation systems. The mechanical motion system is very simple. The limitation of this system for the moment is the lack of a fast reconstruction algorithm.

The capability of the tangential system to generate images of any planar or curved surface within the object was demonstrated in laboratory work by Lindegaard-Andersen et al. (1990) for a block of pine wood of 60 × 60 × 51 mm. The distribution and shape of the annual rings were clearly identified. The detector array is placed parallel to the axis of rotation of the sample and perpendicular to the cross section. A fan-shaped X-ray beam is formed by the source and the detector array. The number and spacing of detectors determine how many tomograms can be collected simultaneously.

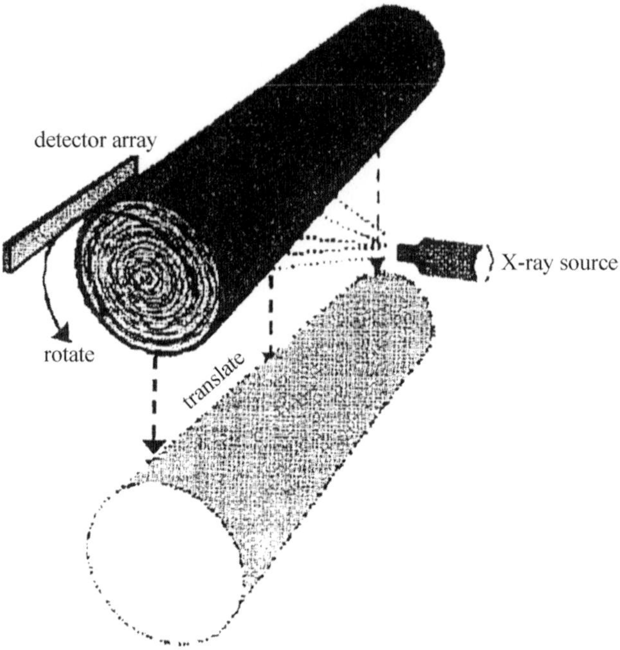

Fig. 2.8. Tangential scanning system. (Schmoldt et al. 1999, with permission)

2.3.1.2 Portable Equipment

Portable equipment for computed tomography requires an array of detectors. The X-ray tubes rotate in a plane perpendicular to the longitudinal axis of the sample. Usually a thin fan beam is intercepted by a detector array. The X-ray beam has a finite thickness that defines the "slice" plane or cross section that is cut through the sample. After a complete rotation of the X-ray tube and detectors, the computer receives the transmitted X-ray intensity for each narrow beam passed through the sample. From data obtained for all paths through the sample, at various angles of rotation, the computer calculates the average of the X-ray linear attenuation coefficient, μ, for each small volume element (voxel) within the slice.

2.3.2 Factors Affecting the Quality of the Image

Qualitative and quantitative information can be obtained from the X-ray computed tomography images. Qualitative applications require only visual assessment of the images and include evaluations of internal flaws and orientation of the samples. Quantitative applications of X-ray imaging are obtained in

Table 2.6. Optimal X-ray energy, combined effect and transmission probability for softwood species (1 mm thick) (Olson et al. 1988, courtesy of Wood Fiber Science)

Species	Energy (keV)	Low density (kg/m^3)	High density (kg/m^3)	Density range	Combined effect (m^3/kg)	Transmission probability (max ΔP)
Balsam fir	5.621	240	760	0.52	2220	0.40
Grand fir	5.531	250	680	0.43	2330	0.35
Sitka spruce	5.379	220	650	0.43	2520	0.38
Jack pine	5.266	240	580	0.34	2600	0.31
Shore pine	5.690	310	720	0.41	2060	0.30
Ponderosa pine	5.423	270	620	0.35	2380	0.30
Red pine	5.472	270	650	0.38	2310	0.31
Scots pine	5.127	230	520	0.29	2810	0.29

addition to qualitative ones and are related mainly to the measurement of the density, moisture distribution and porosity of the wood. To obtain accurate and reliable data from X-ray imaging, the scanning must be carried out in a very consistent manner as was demonstrated by Lindgren (1985, 1991a,b), Liu et al. (1988), Moschler and Dougal (1988), Olson et al. (1988), Fioravanti and Ricci (1991) and others.

2.3.2.1 Beam Path

X-rays travel in a straight line through materials. The monoenergetic beam will radiate perpendicularly on a piece of thin wood with uniform thickness and heterogeneous density structure, consisting of four major elements (H, C, N, O) and various ash ions (Ca, K, Mg, P, Si) present in minor quantities (0.15–0.37% based on oven dry weight). Two main parameters (Table 2.6) are related to the beam path as defined by Olson et al. (1988): the range of transmission probability and the combined effect. The range of transmission probability (ΔP) is defined as "the radiation resolution of a given wood X-ray experiment, in reference to the separability of transmission probabilities resulting from density variations in the radial direction of wood." The combined effect C is a global factor taking into consideration the combination of X-ray energy, wood specimen thickness and heterogeneities induced by the anatomic structure. Consequently, the maximum radiation resolution is determined by density variations in wood specimens, X-ray energy (keV), X-ray intensity (mA), spatial and contrast resolution, display contrast, and beam hardening. The quality of the tomographic image is directly related to the X-ray energy. Because X-ray attenuation is highly dependent on the energy of the X-ray used, it is always necessary to provide this information for laboratory or field measurements. Theoretical studies published by Olson et al. (1988) have

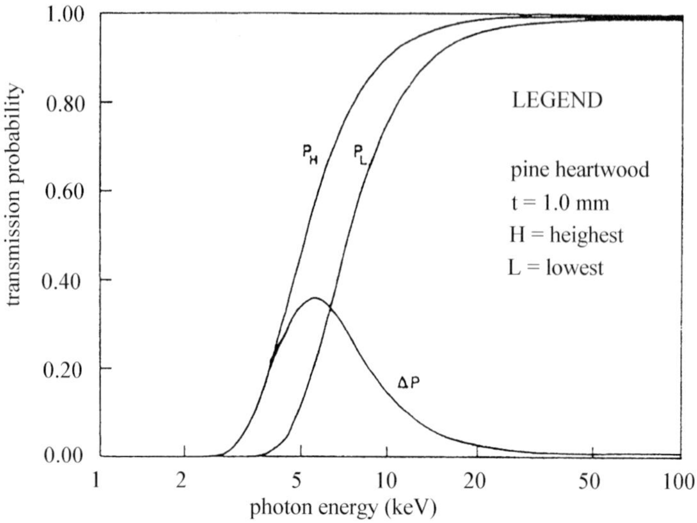

Fig. 2.9. Transmission probability and X-ray energy in pine heartwood of 1 mm thickness (Olson et al. 1988). *H* Highest value; *L* lowest value. (Courtesy of Wood Fiber Science)

demonstrated that accurate wood density measurements can be achieved only if the radiation used is in the optimal range. Values of optimal X-ray energy for eight species are given in Table 2.6 and are in the range of 5.127 and 5.69 keV for specimens of 1-mm thickness. The transmission probabilities under various combinations of radiation parameters are between 0.29 and 0.40. The transmission probability increases with increasing X-ray energy and decreases with increasing density of the sample. The relationship between maximum transmission probability and X-ray energy is shown in Fig. 2.9. The maximum transmission probability corresponds to the maximum value of ΔP, identified at the peak of about 6 keV. The use of a monoenergetic beam is recommended because the radiation resolution reaches its maximum when the polyenergetic X-ray is converted to photons of the same energy.

X-ray attenuation in wood densitometric analysis is influenced by the quantity of oxygen and carbon present in the specimen. The influence of photon energy on mass attenuation coefficients of C, O, N and H and on spruce heartwood is shown in Fig. 2.10. It is recommended to choose the photon energy in the linear part of the diagram and to avoid the abrupt zones corresponding to the presence of different elements in wood.

Lindgren (1991a) theoretically obtained an attenuation coefficient of 500 kg/m^3 for various contents of cellulose, hemicelluloses, lignin and $CaCO_3$ as can be seen from Table 2.7. Because of the low content of Ca (<1%), this ion has no significant effect on the attenuation coefficient and CT numbers. A similar situation was observed for other ions (Mg, Mn, K, etc.) present in wood in very small quantities.

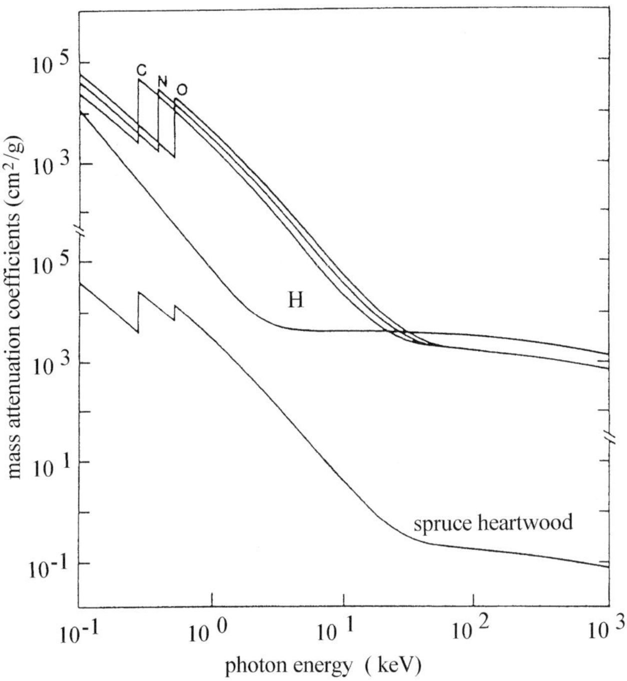

Fig. 2.10. The influence of photon energy on monoatomic attenuation coefficients for C, O, H, N and for spruce heartwood. (Liu et al. 1988, courtesy of Wood Fiber Science, with permission)

Table 2.7. The influence of $CaCO_3$ on the attenuation coefficient of wood and on CT numbers. (Lindgren 1991a, with permission)

Cellulose (%)	Hemicellulose (%)	Lignin (%)	$CaCO_3$ (%)	μ_{wood}	CT number
45	30	25	0	0.0878	–527
45	29.5	25	0.5	0.0882	–525
45	29	25	1	0.0886	–523

2.3.2.2 Spatial and Contrast Resolution

Spatial resolution is the ability of the system to distinguish two or more closely spaced high contrast details of the internal structure of the specimen under test. The spatial resolution of any imaging system is controlled by the accuracy of ray paths through the sample and is related to the following factors: operative size of the source and detector, distance between the source/detector and the sample, number of independent projections produced by the system, ability to separate overlapping features information, signal-to-noise ratio at small source and detector apertures (usually $<0.25\,\mu m$), and the cost of computation.

The contrast resolution is the ability of the scanning system to distinguish small differences in contrast in relation to the homogeneous contrast of the background and is directly related to the linear attenuation coefficient. The smallest difference in density of the sample that can be identified with a given system depends on the contrast resolution of the scanning system. A low-contrast detail of the structure can be identified if it is large, whereas small details can be identified if they have a large contrast relative to the homogeneous background. This is due to the partial volume effect. The smaller details of the structure must be of significant attenuation contrast to affect the voxel average. Lindgren (1991a) studied the relationship between wood density and the X-ray attenuation coefficients and CT numbers with a medical scanner and specified that the minimum volume required must exceed $0.005 \times 1 \times 1.5$ mm. For each scanning system, this relationship between the density and CT number must be determined precisely for the correct interpretation of data.

The range of CT numbers displayed in a given image is referred to as the image "window". This range is mapped and is represented by a discrete number of gray levels to create the image. A pixel (picture element) is the bidimensional representation of a voxel (volume element) on which two dimensions correspond to the pixel dimension, and the third dimension corresponds to the thickness of a slice (1–10 mm). The field of view (FOV) is equal to the maximum dimension of the scanned specimen. As noted previously, on the image each pixel represents the average value of the attenuation coefficient of the volumetric unit. For correct reproducibility of the image, the technical parameters to be defined are: dimension of the FOV, thickness of the slice, time of scanning and the algorithm of reconstruction. For calibration, FOV is usually taken between 3 and 50 cm. For image reconstruction FOV is related to the pixel size as can be seen from Table 2.8. The minimum FOV size of 30 mm is related to the representative size of the anatomic structure of wood. The selection of the thickness of the slice is also a function of the size of the pixel and the voxel as well as the intensity of the beam. The thickness of the slice gives the volumetric resolution of the image as shown in Table 2.9. The value attributed to the voxel corresponds to the average value of all anatomic elements from the voxel. Important errors can occur when the densitometric analysis is performed at the limit of an annual ring, that contains latewood and earlywood. For this reason it is recommended to select the minimum dimension of the pixel with the thickness of 1 mm, if possible. The selection of the intensity of the beam is a function of the slice thickness and of the volumet-

Table 2.8. The size of the pixel for the array of 512×512 as a function of the size of the FOV. (Fioravanti and Ricci 1991, with permission Holzforschung)

FOV (mm)	500	400	300	200	100	50	30
PIX (mm)	0.976	0.781	0.586	0.390	0.195	0.097	0.058

(pix = FOV/512)

ric resolution. For wood, the intensity can be higher than (130 mA) for a scanning time of 3 s. Fioravanti and Ricci (1991) published an example of the matrix of density (Fig. 2.11) with all the measured values for all voxels, given in a matrix of 11 lines × 5 columns. A specific voxel can be identified with the reconstruction matrix of 512 × 512 elements. The visualization with points is a concise representation of all experimental data, making possible the observation of the isodensity curves inside a selected annual ring (Fig. 2.12). The precision of the measurement of the attenuation coefficient, using X-ray computed tomography, is ±0.5%, compared with conventional radiographic techniques for which the precision is ±3%.

Table 2.9. Relationship between the voxel volume and the slice thickness (Fioravanti and Ricci 1991) (with permission)

Pixel (mm)	Voxel Slice thickness 1 mm	Volume Slice thickness 5 mm	(mm^3) Slice thickness 10 mm
0.586	0.343	1.717	3.434
0.390	0.152	0.760	1.524
0.195	0.038	0.190	0.380
0.097	0.009	0.047	0.094
0.058	0.003	0.017	0.033

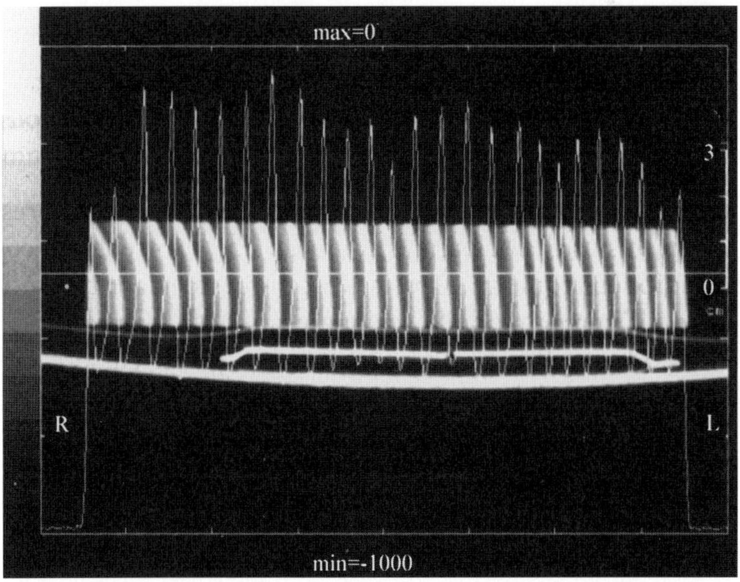

Fig. 2.11. The attenuation values superimposed on the tomographic image of annual rings pattern. (Fioravanti and Ricci 1991, with permission)

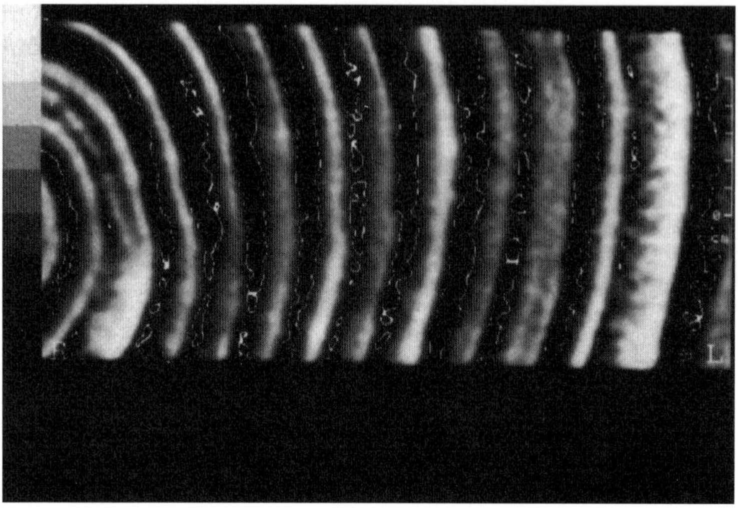

Fig. 2.12. Isodensity zones in an annual ring. (Fioravanti and Ricci 1991, with permission)

2.3.2.3 Anisotropic Direction

Theoretically, density is a scalar and not a vector and a priori is not influenced by the anisotropic direction of wood. Nevertheless, because wood is a natural composite, laboratory studies have reported the influence of the anisotropic direction on the measurement of attenuation coefficient (Moschler and Winistorfer 1990; Fioravanti and Ricci 1991; Malan and Marais 1992; Karsulovic et al. 1999). For precise densitometric measurements, it is recommended to orient the irradiation direction of the specimen with the longitudinal anisotropic direction of the wood. In this case, the number of photons absorbed by the cell wall is minimal. If scanning along radial or tangential directions is required, it is obvious that the scanning inspection must be performed parallel to the selected anisotropic direction. The attenuation coefficient depends on the thickness of the cell wall. Errors in attenuation measurements are created if the specimen is presented in an inclined position versus the incident radiation, because the experimental wall thickness is greater than the theoretical.

Table 2.10 gives the effect of anisotropic direction on the linear attenuation coefficient measured on 141 wood specimens (Malan and Marais 1992). The longitudinal values compared with those in radial and tangential directions are the lowest values. No significant differences were observed between the radial and the tangential coefficients. The linear relationships between the density and the linear attenuation coefficients in radial and tangential directions for *Pinus radiata* are given in Fig. 2.13. The mean value of the attenuation coefficient in the radial direction was 4% greater than in the tangential direction but the difference is not significant at the 95% confidence level. Moschler and

Table 2.10. The effect of anisotropic direction on the linear attenuation coefficient. (Malan and Marais 1992, with permission Holzforschung)

Species	Source	Mean linear Radial	Attenuation Tangential	Coefficient (cm^{-1}) Longitudinal
Coniferous	^{137}Cesium	0.0406	0.0400	0.0392
Hardwoods	^{137}Cesium	0.0567	0.0571	0.0571
Coniferous	^{241}Americium	0.0986	0.0997	0.0943
Hardwoods	^{241}Americium	0.1390	0.1386	0.1389

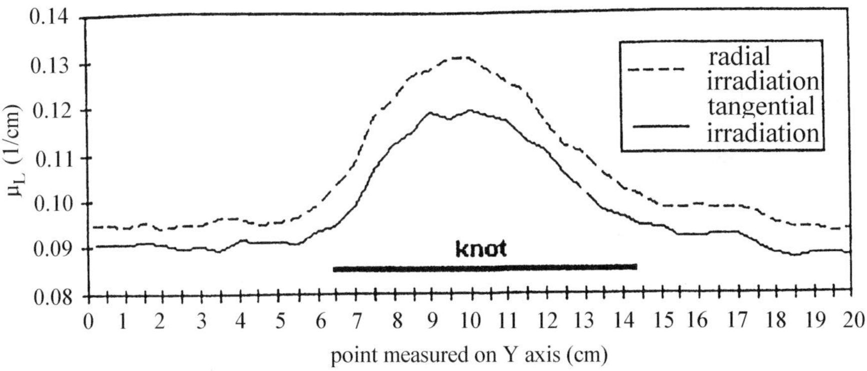

Fig. 2.13. Relationship between the density of *Pinus radiata* wood and linear attenuation coefficients in radial and tangential anisotropic directions. (Karsukovic et al. 1999, courtesy of Forest Product J)

Winistorfer (1990) emphasize the necessity of the alignment of the aperture with the anisotropic directions. Misalignment induces more measurement errors. Routine laboratory scanning for densitometric purposes is always performed in the radial direction and requires good alignment between the growth rings and the aperture. For wood species with large cell lumina, the porosity can also be determined by scanning in the radial direction (irradiation in the longitudinal direction) with a convenient size of the aperture. To avoid any underestimation of density and all the experimental artifacts, the authors proposed that the attenuation coefficient be measured in the tangential direction. To our knowledge, until today, no data have been reported in the literature related to the microdensitometric analysis of growth rings in the tangential direction because of the difficulty in observing the limit of the annual rings.

2.3.2.4 Beam Hardening

From a theoretical point of view, the X-ray attenuation coefficient is calculated for a monochromatic source having a single wavelength or energy. However,

Table 2.11. CT numbers of different materials. (Hattori and Kanagawa 1985; courtesy of The Japan Wood Research Society)

Material	Density (kg/m3)	CT number surrounding with granular sugar	CT number surrounding with air
Polyprophylene	905	−120	−87
Polyethylene	953	−71	−39
Nylon	1129	71	127
Acryl	1189	130	185
Polycarbonate	1191	97	139
Polyacetat	1405	342	418
Polyvinylchloride	1401	1315	1622

the scanner produces polychromatic X-rays. The attenuation of a polychromatic wave is not constant for all energies and is higher for the lower energy components of the X-ray beam (effect of beam hardening). This effect can significantly decrease the accuracy of quantitative results. However, it can be diminished by surrounding the sample with different compensatory materials, such as paraffin wax, granular sugar or acrylics, for small, clear specimens of wood (2 × 2 × 1 cm) as noted by Hattori and Kanagawa (1985). Beam hardening can also be compensated by reconstruction algorithms, as noted by Vinegar (1986). Calibration of scanners with attenuation standards is an essential step for calculation or comparisons of X-ray attenuation data and for enhancing of the accuracy of experimental measurements. As for classic radiographic analysis, calibration of the scanning system must be performed every day before starting the experiments to identify and characterize the drift of the X-ray source and the detectors. The reference material should exhibit characteristics close to those of the material to be analyzed. Correction must be made prior to calculation of density. Hattori and Kanagawa (1985) proposed the following materials as standard attenuators for wood specimens: polypropylene, polyethylene, nylon, acrylics, polycarbonate, polyacetal, and polyvinyl chloride. The CT numbers of these materials surrounded with air or with granular sugar are given in Table 2.11. CT numbers surrounded by air are inferior to those surrounded by granular sugar.

2.4 Applications

The field of applications of computed tomography is very large, including building inspection, wood preservation, stability of wooden construction elements, preservation of monuments, new constructions, pole inspection, inspection of wooden toys, tree examination and inspection, arboriculture, growth rate assessment, wood quality assessment, forest sciences, wood technology, wood biology, dendrochronology, dendroarcheology, climatology, geomorphology, phytopathology, tropical trees, plantation trees, and wood products (Rinn 1991). The choice of methods for specific inspections must be made on economic grounds. Using the great potential of this nondestructive method, it is possible to study the spatial distribution of matter on a scale from meter to micron in a very wide number of applications.

2.4.1 Examination and Inspection of Trees and Poles

X-ray computed tomography opened a new era for the nondestructive examination and inspection of trees and poles. The quantitative determination of decay, knots, checks, heartwood formation, ruptures, wet core, spreading of dry decayed tissue and other major defects as well as the distribution of moisture inside the stem of a tree is now possible. Fundamental differences among the images obtained for different species are easy to recognize.

Pioneering research results related to the development of portable instruments, suitable for outdoor use were published by several authors (Habermehl 1982a,b; Onoe et al. 1984; Habermehl and Ridder 1992a,b,1993, 1994, 1996; Niemz et al. 1998a,b). A description of these devices will be presented in the following pages.

For the examination of forest trees or of park and street trees, Habermehl and Ridder (1998) proposed two types of portable instruments, one with a parallel beam system (Fig. 2.14) and another with a fan beam system (Fig. 2.15), equipped with a γ-ray source. The source and detectors are sustained by the mechanical device shown in Fig. 2.16.

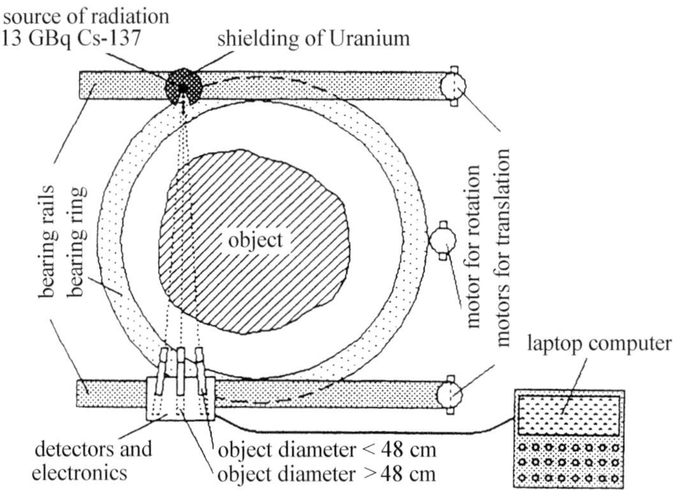

Fig. 2.14. Portable apparatus for the examination of tree with a parallel beam system. (Habermehl and Ridder 1998, with permission)

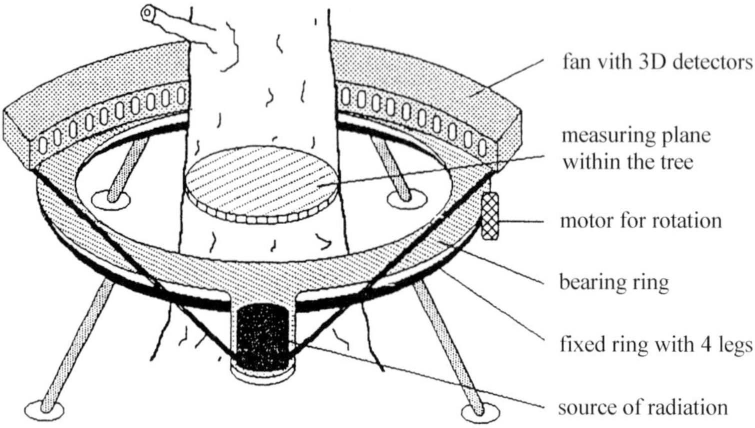

Fig. 2.15. Portable apparatus for the examination of trees with a fan beam system. (Habermehl and Ridder 1998, with permission)

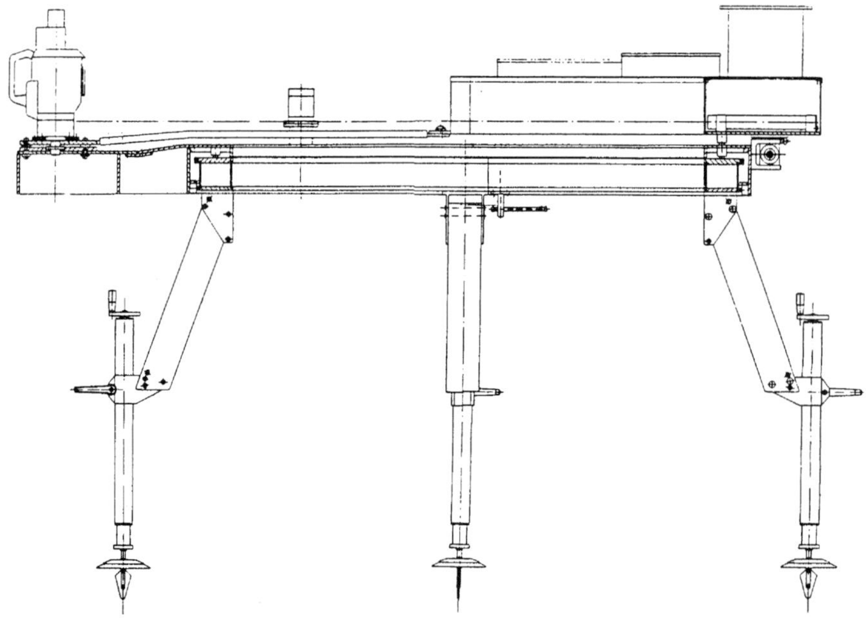

Fig. 2.16. Mechanical device for the sustaining of the source and detectors of the portable apparatus for the examination of trees. (Habermehl and Ridder 1992b, with permission)

The size and location of the defects that can be observed on a tomographed cross section are cavities and decay of the trunk, mechanical injuries, fissures, holes, zones destroyed by fungi, areas of lowered density with less moisture than in healthy zones, the presence of metallic inclusions in the trunk, like nails and bullets, zones of perfectly healthy wood, and the difference between sapwood and heartwood in some species because of the difference in their moisture content. The devices are very effective for the examination of park and street trees. The decision of the environmental authority to conserve trees or not after examination of tomographic images can be objective, and the safety of the traffic on the streets can be assured. For forest trees, the questions answered by the portable apparatus are related to the effect of liming and fertilization on the sapwood to heartwood ratio, to the water pathways above the cross section of the trunk, to the existence of frost cracks, and to the existence of wet cores and other specific internal structures that can affect the variation of the attenuation coefficient of the radiation and consequently of the density.

The source of radiation is ^{137}cesium having a half-life of 30 years and emitting γ-rays of a quantum energy of 662 keV. The parallel beam system is equipped with a source of radiation of 13 GBq and allows the translation and rotation of detectors and only translation for the source. The detector system is consists of three detectors with an angle of 3.75° between them. One translational displacement produces three simultaneous attenuation profiles at the same time. A total of 48 profiles are recorded in 16 rotational positions. The smallest diameter of the tree to be measured is 12 cm with a counting rate of 19,000 counts/s and the largest diameter of the tree is 72 cm. The tomograms reconstructed on the computer are performed in less than 2 min for one profile. The reconstruction process is based on the analysis of a matrix that contains values of the attenuation coefficient (cm^{-1}) of the examined slice. In the case of a tree, the reference value was considered to be 83 units for water and 0 for air. (Hounsfield units in medicine are −1000 for air and zero for water.) The spatial contrast for the standard adjustment is 10 mm for the full width at half-maximum (FWHM) of the signal. For trees with diameters up to 48 cm, the resolution must be 7.5 mm FWHM of the signal. The contrast resolution obtained was 2% and was related to the water attenuation.

The fan beam system is a more complex system and has 30 detectors. The source is 185 GBq and 8000 counts/s, mounted on a metallic ring with a 100 cm diameter. The complete data set produces 150 profiles, measured with a rotation of the source-detector system with a step of 2.4°. The fan system has the same spatial and contrast resolution as the parallel system. The algorithm reconstruction for the PC time is 4 min for one profile. For larger trees the total examination time is about 1 h. The tomograms can be coded by color or by a gray scale, as can be seen in Figs. 2.17 and 2.18.

2.4.1.1 Growth Rate Assessment

Using computed tomography Onoe et al. (1984) were the pioneers in the development of an automatic method for ring width measurement and ring counting of several species, including red pine, black pine, Douglas fir,

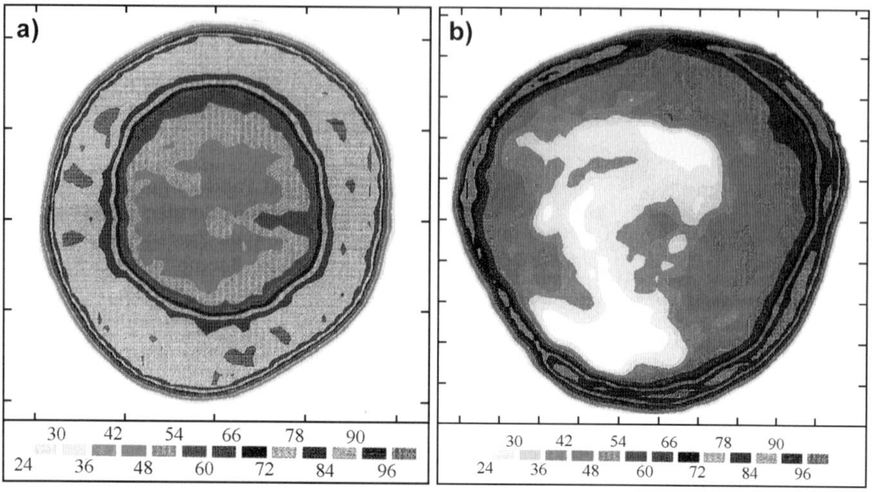

Fig. 2.17. Tomogram on a color scale of a transverse section of a healthy (**a**) and of a decayed (**b**) Norway spruce tree. (Habermehl and Ridder 1996, with permission)

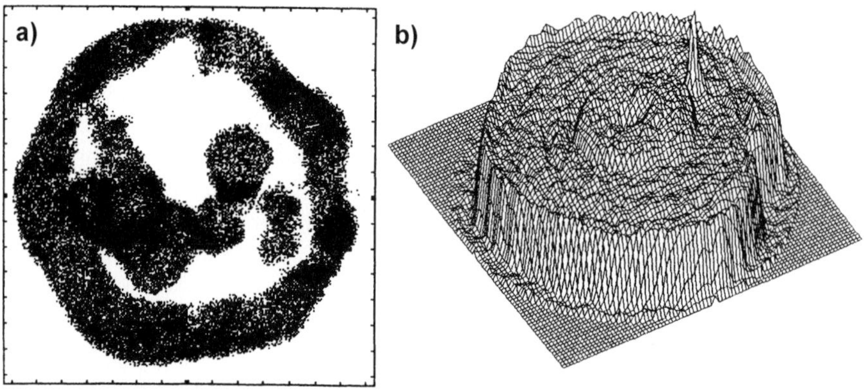

Fig. 2.18. Tomograms on a gray scale (Habermehl and Ridder 1996). Legend: a) a lime tree with internal decay and big holes b) an ash tree, showing a metallic inclusion (with a very high attenuation) (with permission)

Japanese cypress, birch, horse chestnut and maple. The details of the annual ring structure can be seen using high spatial resolution in the image reconstruction with computed tomography. A few tenths of millimeter resolution can be obtained with well-collimated, narrow X-ray beams and dense sampling. The device designed and used by Onoe et al.(1984) was portable, with an X-ray tube of 40–120 keV and three NaI scintillation counters. A fan beam algorithm was used for reconstruction, and 1200 samples of 16-bit projection data were taken at 2° intervals. The reconstruction images show very clearly the annual rings in Douglas fir as well as the heart wood central zone, having decay (Fig. 2.19).

Figure 2.20 gives the profile and the width of automatically measured rings. Firstly, the pith was determined by contour tracing of an inner ring. Secondly, several diameters were drawn through the pith. Thresholding after differentiation yields the number and the width of the rings. Hauffe and Mahler (2000) studied the possibility of evaluation of log quality using only the width of annual rings measured with a scanner on spruce logs from southern Germany. The technique was successful for 80% of the logs. Rinn (1991) proposed a mobile tomograph with 12 sensors for scanning growth rings in trees and poles.

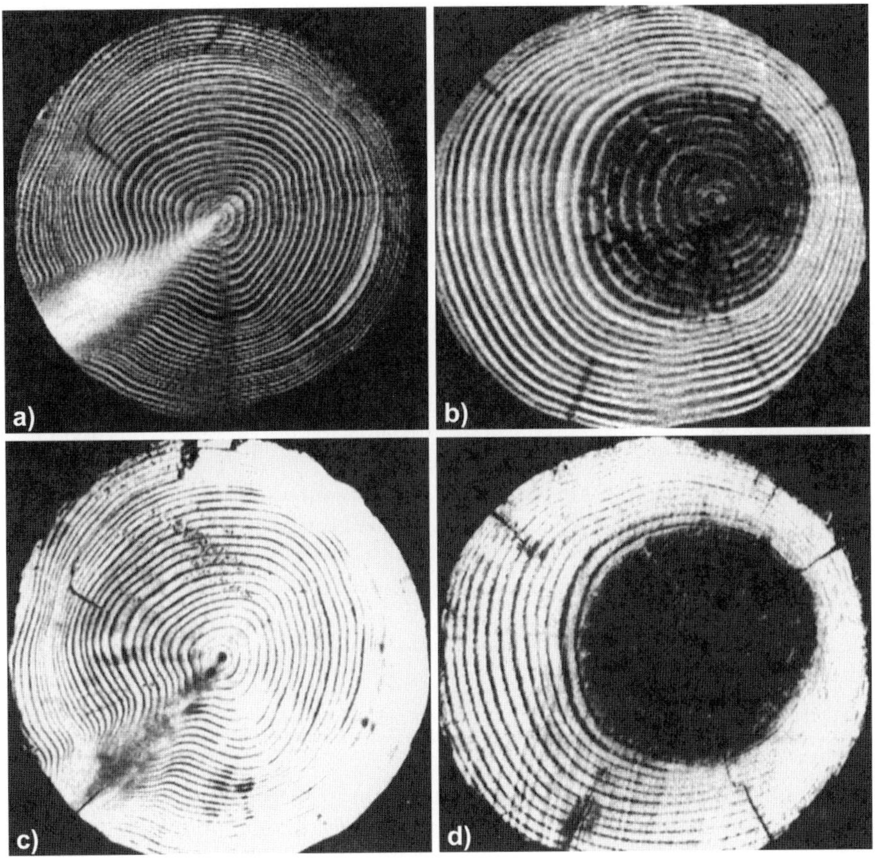

Fig. 2.19. Reconstructed images of the cross section of Douglas-fir pole compared with a photographic image of the same section (Onoe et al. 1984). **a** The reconstruction image of the wrapping pattern of the annual ring around a branch. **b** The reconstructed image of the heart decay in the central zone. **c** Photographic image of **a**. **d** Photographic image of **b**. (Reprinted from Nuclear Instruments and Methods in Physics Research, vol. 221, 1984, 213–220, 2001, with permission from Elsevier Science)

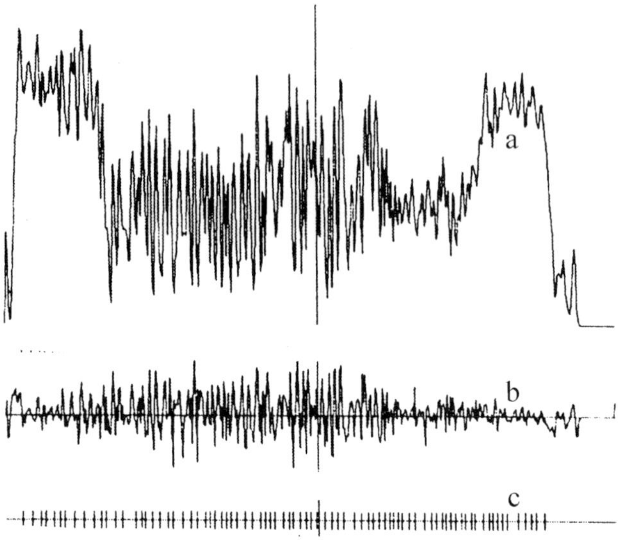

Fig. 2.20. Profile and width automatically measured annual rings (Onoe et al. 1984). (Reprinted from Nuclear Instruments and Methods in Physics Research, vol. 221, 1984, 213–220, 2001, with permission from Elsevier Science)

Another aspect of the examination of growth rate of trees where the application of X-ray computed tomography is very successful is related to the development of microtomography for 3D anatomic structure examination, as a very important step in the further development of wood microdensotometry in general. Davis et al.(1991) reported very well-defined images of transverse structures of wood of *Pinus sylvestris* and *Eucaliptus delegatensis*, as can be seen in Fig. 2.21. Details of the structure of the annual growth rings are clearly visible, and tracheids, rays and resin canals can be observed on the *Pinus sylvestris* specimen. For *Eucaliptus delegatensis*, a collapse check in the high-density zone is observed. The device for X-ray microtomography of wood was attached to an X-ray diffractometer. The schematic representation of this device for X-ray microtomography is shown in Fig. 2.22. The technical characteristics of this X-ray microtomograph are as follows:

- X-ray energy: 8 keV;
- beam characteristics: two beam collimation arrangements, one cylindrical with 20 μm and another rectangular of 5 μm width and 100 μm height;
- translation movement 255 stepwise, followed by 1.5° rotation repeated 120 times for 180° rotation of the specimen;
- reconstruction algorithm using a summation-convolution back-projection method;
- data collection time: 7 h for 20-μm scan step;
- computation time for image reconstruction: 5 min;

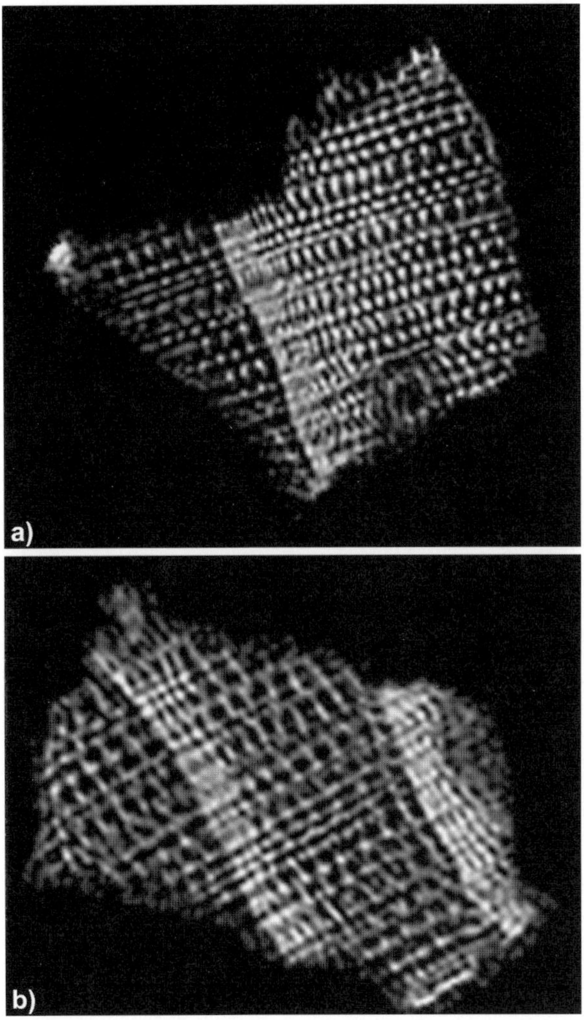

Fig. 2.21. Reconstructed images of the anatomic structure of *Pinus sylvestris* (Davis et al. 1991). **a** 5-μm spatial resolution of the transversal section (0.9 × 0.6 mm) of the annual ring with earlywood and latewood zones. **b** 5-μm spatial resolution of the transverse section (0.9 × 0.5 mm) in which the highest density is observed in the middle lamella between the cells. The images were obtained with the first generation scanner (8 keV, X-rays from the K_α peak of a Cu target tube). (with permission)

- pixel gray is zero (black) for the air and 255 (white region) for highest density in latewood;
- X-ray attenuation coefficient is defined with 5% accuracy;
- spatial resolution 1 μm;
- specimens section 2 × 2 mm of *Pinus sylvestris* and *Eucaliptus delegatensis*;
- annual ring width: 0.4 mm.

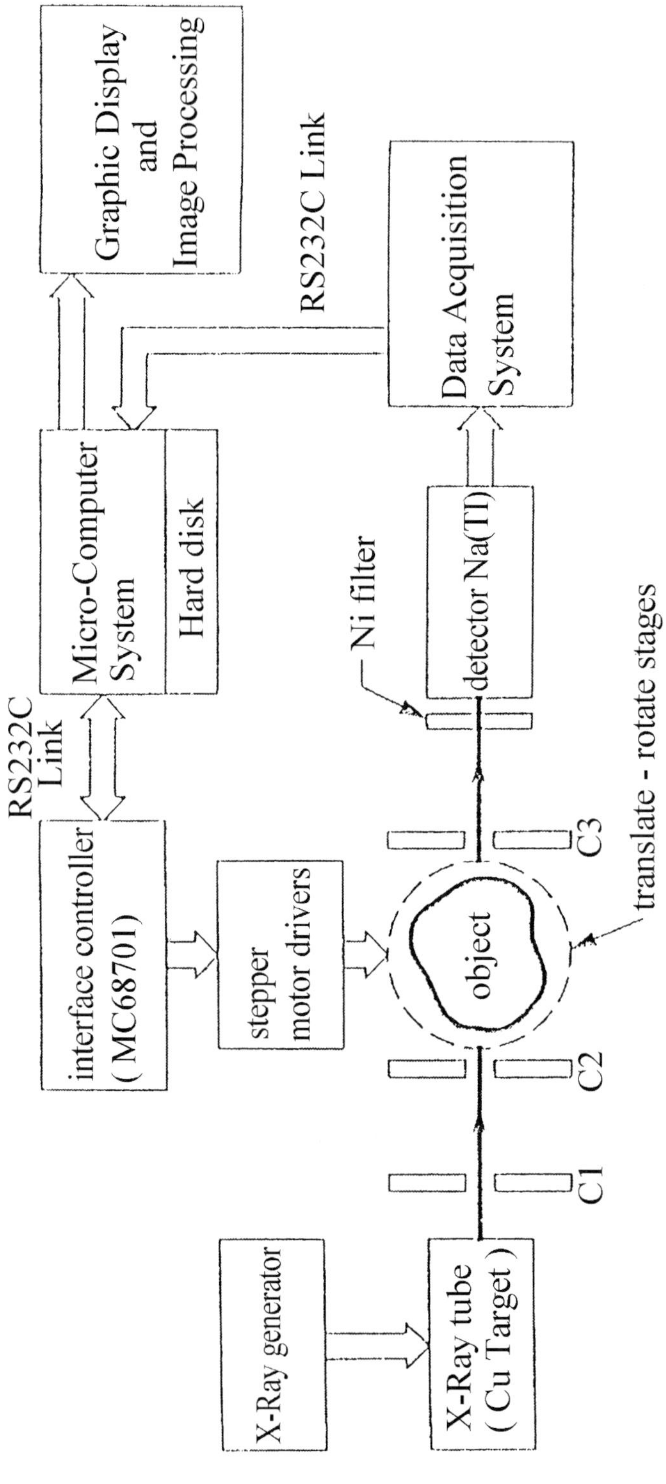

Fig. 2.22. X ray microtomographic device. (Davis et al. 1991, with permission)

Improvement of a miniature scanner allows (Sasov and van Dyck 1998) to produce images of biological specimens obtained on a X-ray microtomograph coupled with a corresponding microscope. The device fits in any laboratory environment and enables views of the structure in 3D, under ambient conditions, without any need for specimen preparation.

Badel (1999) produced 2D images of oak at a microscopic scale using an X-ray device with 70 kV and 95 μA. Vansteenkiste (2000) studied the ability of automatic analysis of X-ray images obtained in 2D for the detection of the annual ring limit, fiber and vessel zones and rays of oak.

2.4.1.2 Moisture Content in Trees

The moisture content of living trees is related to the viability of trees and is dependent on many factors such as climatic conditions, the moment of the day when the test of water flow is made, the sanitary conditions, etc. The forest decline observed today seems to be induced by the disturbance in water transport in trees.

The portable computer tomograph proposed by Habermehl and Ridder (1992a,b) and used by Szendrödi et al. (1994) permits mapping the distribution of water inside the trunk and observation of the mechanism of water transport in xylem. Raschi et al. (1995) and Tognetti et al. (1996) utilized a portable apparatus equipped with a parallel beam system and with a source of radiation of 13 GBq for measurement of the water content in the stems of sessile oak (*Quercus petraea*) and in Turkey oak (*Quercus ceris*) (Fig. 2.23) . For the portable computer used in this experiment, the CT numbers are on a scale corresponding to 83 units for pure water and 0 for the density of pure air, on a grid of 75 × 75 pixels for *Quercus ceris.* and of 63 × 63 pixels *Quercus petraea* with a reconstruction interval of 3–4 mm. The tomograms for *Quercus ceris* displayed green and blue colors, indicating a high density in the stem. The sapwood had a high density (green color) and the heartwood had slightly lower density (dark blue color). In *Quercus petraea,* the sapwood (blue and green rings) is well hydrated. The embolized zone is represented by the violet ring. The heartwood of highest density is represented by a dark blue color. These species are characterized by different susceptibilities to drought; the first one is a mesophytic type species while the second is xerophytic. The trees originated from a forest in Hungary at latitude 47°N, longitude 20° and altitude 320–340 m, with an annual precipitation mean of 600 mm. The trees from this forest had suffered a prolonged period of drought for many years before the experiments, which were performed in the spring. Characteristic anatomical structures of these species are schematically given in Fig. 2.24, where large vessels can be seen in the earlywood. The main function of large vessels is to increase the efficiency of sap flow. In this tissue, cavitation risk is always present. When the large vessels, malfunction, the sap flow is supplied by the

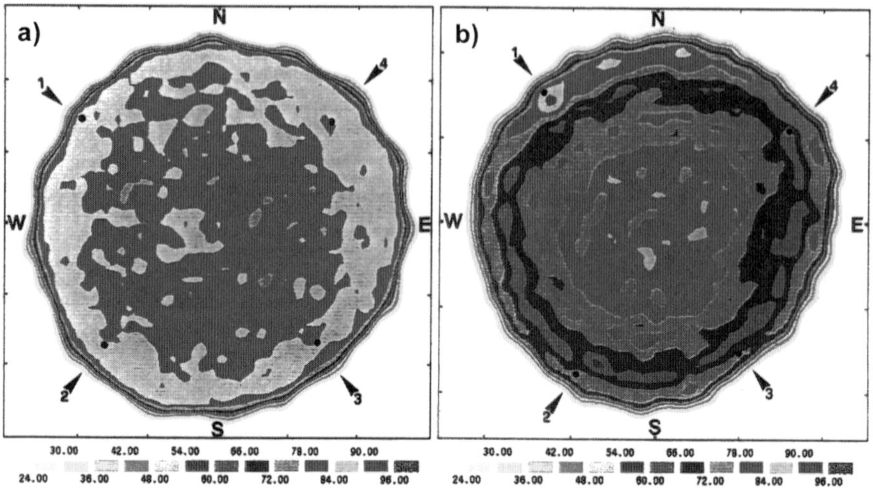

Fig. 2.23. Tomograms of *Quercus petraea* and *Quercus cerris* (Tognetti et al. 1996). **a** *Quercus cerris*; **b** *Quercus* petraea. *N*, *E*, *S*, *W* are the cardinal points; *1*, *2*, *3*, *4* are the position of measured points The density scale in CT numbers is represented by different colors (i.e., CT = 30 corresponds to dry condition and CT = 96 corresponds to high moisture content). (with permission)

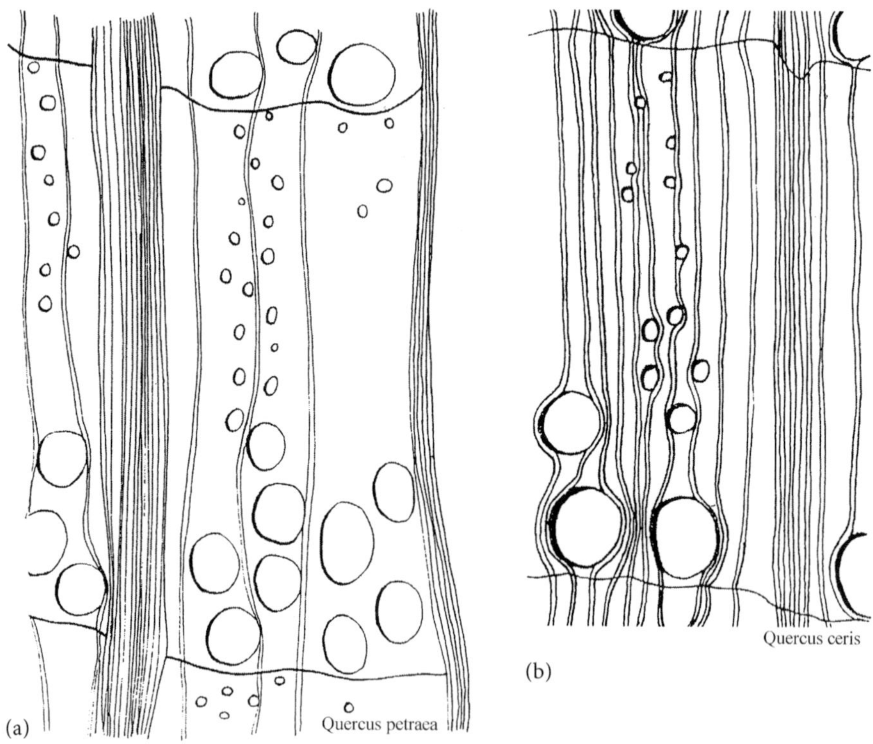

Fig. 2.24. Anatomic structure of **a** *Quercus petraea* and **b** *Quercus ceris*. (Jacquiot et al. 1973, with permission)

smaller latewood vessels and fiber-tracheids, which remain functional for years (Zimmermann 1983).

In both species, a narrow ring of low density surrounded the high-density ring which indicates that the outermost vessels buffer the sap flow rate (Granier et al. 1994). The sapwood area is influenced in different proportions by the sap flux as can be seen from the different colors displayed on the tomogram. The relative uniformity of the density in *Quercus ceris* compared with *Quercus petraea* could partially explain the better drought resistance of *Quercus ceris.* In *Quercus petraea* a high proportion of red and violet zones on the images indicate a relatively low wood density. The active sapwood appears in pale green and blue zones corresponding to 90 and to 72 units, respectively. The tree examined was not very healthy because of a prolonged dry period. However, it seems that the high capacity of water storage in the sessile oak vessels in the narrow external rings play an important role in tree survival during the periods of low soil water availability in the summer. The attenuation coefficient of the γ radiation of well-hydrated tissue is twice that in dry wood and a high correlation between the density of sessile oak sapwood and its water content was noted by Weibe (1992). It was also observed that the north-facing quadrant of the trees had a higher density. Tognetti et al. (1996) confirmed the previous statements of Wiede (1991) for sessile oak sapwood, namely, that in living trees the density is highly correlated with water content.

The key point in the interpretation of tomograms is related to the calibration of the pictures. For this purpose Beres and Fenyvesi (1994) as well as Raschi et al. (1995), Tognetti et al. (1996) suggested the use of several methods (ultrasonic acoustic emission, thermoelectric method, etc.) together with a computed tomography method. For quantitative interpretation of tomograms, the data deduced with these methods must agree with anatomic studies as well as with microdensitometric research.

2.4.1.3 Pollution Effects on Trees

Studies related to the pollution effect on trees (*Pinus sylvestris*) induced by ammonia produced by a large cattle farm were reported by Kätzel et al. (1997). Twenty-eight 73-year-old trees were investigated in situ, using mobile computed tomography. The nitrogen produced by the cattle farm and added to the soil over more than 20 years induced changes in the energetic, carbohydrate, amino acid and metabolisms of the tree needles up to a distance of 1 km from the source. The wood quality of the trees was modified because of long-term changes in metabolism. The ratio of sapwood to hardwood was modified, as can be seen from Table 2.12 and Fig. 2.25.

The tree cross stem area in the highly polluted zone is larger than that in the minimum polluted one. The corresponding ratio between the sapwood and heartwood areas is 1.5 for the maximum polluted area and 3.2 for the

Table 2.12. The influence of ammonia stress pollution of *Pinus sylvestris* on the sapwood/hardwood ratio and the moisture content of the trees. (Kätzel et al. 1997, with permission)

Site	Distance (m)	Pollution	Cross stem (area cm^2)	Sapwood (%)	Hardwood Ratio	Moisture Sapwood	Content Hardwood	CT number Ratio
1	200	Maximum	624	53.8	1.5	81	50	1.62
2	280	–	745	57.8	1.5	81	51	1.59
3	2900	Minimum	513	77.0	3.2	86	53	1.62

minimum polluted area. The diminishing of the sapwood area induced a reduction of sap flow in the tree and indirectly a reduction of water supply in the needles. The increasing heartwood area combined with the diminishing sapwood zone in the polluted tree can be understood as an accelerating aging process of the tree. The moisture content of both sapwood and heartwood zones in the polluted area was much reduced when compared with the unpolluted area. The survival prognosis of tree in the polluted zone is reduced. The results of this research demonstrated that computed tomography with a portable device is a very useful tool for nondestructive diagnosis of living trees.

2.4.2 Wood Quality Assessment

During the last 20 years, there has been a growing interest in the development of automatic methods for quality assessment of logs and lumber and in the determination of optimal edging and cutting procedures in the lumber industry (Birkeland 1985; Wagner and Taylor 1985).

The first commercial scanner developed in 1973 was relatively slow, requiring 4 min per image. The rotation of a significant mass around a heavy object like a log limited the rotation speed and increased the resulting scanning time. Recent technological developments permit the production of appropriate industrial devices able to detect small structural details. An equipment resolution of about 10 mm (or 1 cm^3 in volume) is of interest for the nondestructive control procedure. Wagner et al. (1989 a,b) described the first utilization of an ultra-fast computed tomographic medical scanner with 34 images/s for the detection of defects within hardwood logs. The fast scanning time was achieved by eliminating all moving parts of the device, using a fan detection system and a reconstruction matrix of 128×128 for a water oak log with a diameter of 38 cm and a length of 3.6 m. The tomograms were validated by visual comparison with real images of log cross sections, each 8 mm thick.

Chiorescu and Grönlund (2000) noted that the recent technological advancements in X-ray scanning techniques have a good potential for on-line implementation in softwood sawmills. It has been shown that the sawing

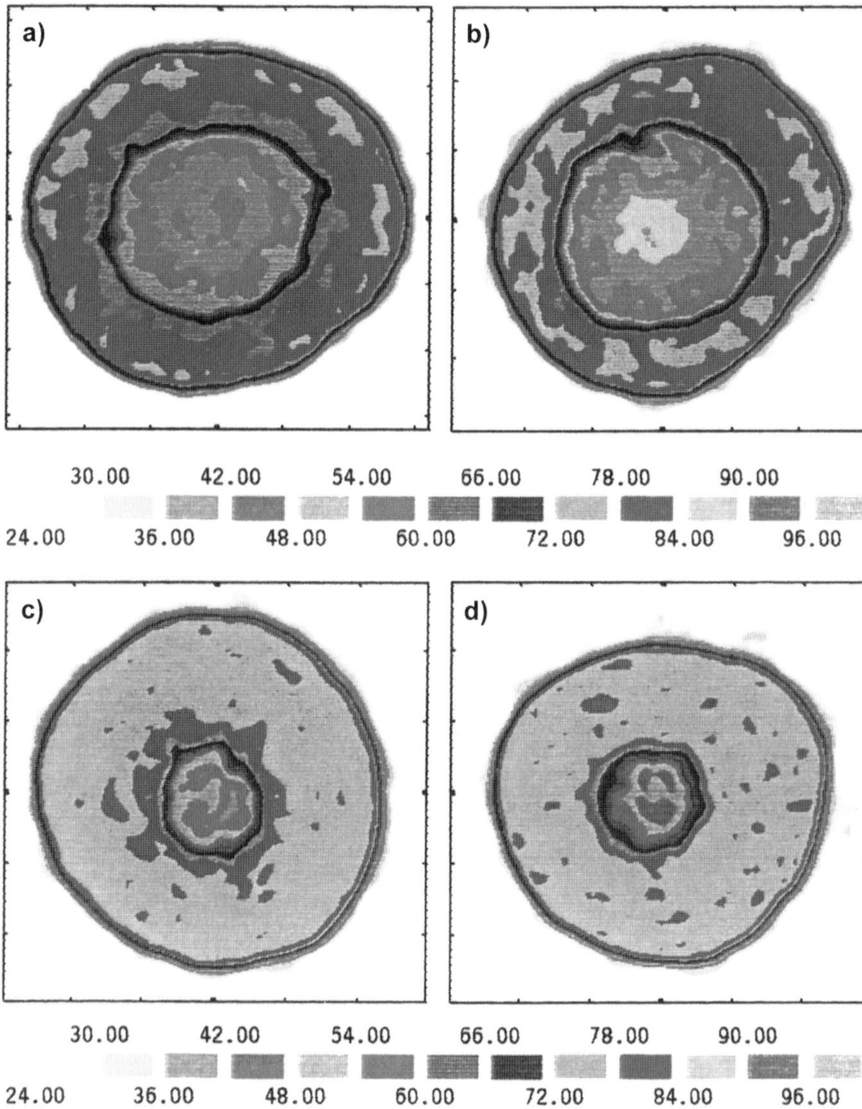

Fig. 2.25. The development of sapwood in Scots pine induced by ammonia pollution, compared with a normal tree at 100 cm stem height. Trees are located in a forest from the district Torgelow, Germany (Kätzel et al. 1997, with permission). **A** Tomogram for tree no. 1, with 624-cm^2 cross section located at site 1, 200 m from the farm (maximum pollution zone). **B** Tree no. 8, with 745-cm^2 cross section located at site 2, 280 m from the farm. **C** Tree no. 2 at site 3, with 513-cm^2 cross section located 2900 m from the farm (minimum pollution zone). **D** Tree no. 8 at site 3 located 2900 m from the farm (minimum pollution zone). The density scale in CT numbers is represented by different colors, CT = 30 corresponds to dry conditions and CT = 96 corresponds to a high moisture content. The tomograms were taken in September 1993 and 1994

process can be considerably improved by the utilization of software based on X-ray scanning data deduced from the scanning of 600 Scots pine *(Pinus sylvestris)* saw logs in Sweden (Gröndlund et al. 1994). For hardwood logs, the problems are even more complex because of the structural complexity and heterogeneity as well as the biological variability of their internal structure, both among and within species. Schmoldt et al.(1999) reviewed the progress that has occurred in scanning technology, image analysis, and data utilization in the sawing of for hardwood logs.

The final aim of the development of industrial X-ray computed scanning technology is to develop the scientific and technological foundation that is needed to make this nondestructive control method operational for sawmills.

2.4.2.1 Inspection of Logs for Lumber and Veneer

The main interest for the inspection of the internal structure of logs is related to the improvement of the grade and volume yield in the sawmill. Prior to studies for the development of commercial industrial tomographic log scanners for sawmill use, feasibility studies, and laboratory investigations were made. The stress was put on the detection and location of knots, decay zones, decay areas, cracks, internal holes, abnormal variation of density, abnormal growth rings, reaction wood, resin pockets, spiral grain, embedded metal, etc. (Taylor et al.1984; Funt and Bryant 1987; Swanson and Hailey 1987; Hodges et al. 1990; Maroc et al. 1996; Niemz et al 1997; Bahyl 1998; Karsulovic et al. 1999; Oja and Temnerud 1999; Nyström and Earl Kline 2000; Sepulveda and Grönlund 2000; Varga et al. 2000). The high moisture content in green logs makes the detection and location of defects especially difficult. Moreover, the scanning must be performed at a rate compatible with log conveyance speed in sawmills. During the transportation through the scanner, the logs must exhibit little side motion, rotation or vibration. The automatic interpretation of log tomograms started in the 1980s as feasibility studies with automated defect identification and labeling using different algorithms, first for softwood and later for hardwood logs. The main objective was to find a procedure for detection of defects able to operate at real time speed so that scanning, image reconstruction, interpretation, and display could be integrated into mill processing.

Scanning applications in sawmilling and wood manufacturing processes, as suggested by Johansson (1985), can be divided into three main domains:

- the control inspection, where production specimens are checked randomly;
- research and development for new technological procedures;
- on-line internal inspection before sawing at normal production speed.

The following is a review of the literature related to the internal inspection of logs before sawing. In softwood logs, the knots are the most common grade-reducing characteristic. As an example, the studies performed by Karsulovic et al. (1999) are chosen to demonstrate the ability of the linear attenuation coefficient of gamma radiation to detect the knots in *Pinus radiata*. The scanning devise is shown in Fig. 2.26. The irradiation of the samples (40 × 125 × 300 mm) was done with a gamma ray beam (^{241}Am) of 59.5 keV, with a 3-mm-diameter collimator for source and detector. The sample was 131 mm from the detector, and the distance from the source to the detector was 145 mm. The array for the irradiation was 5 × 5 mm. The variation of the linear attenuation coefficient in radial and tangential directions of the specimen was measured. Three zones of gray were used to distinguish knots from the surrounding zone of defect-free wood. A transition zone is clearly observed between clear wood and a knot, caused by a strong deviation of fibers from the natural anisotropic axes.

Using an industrial scanner with automatic image analysis, Taylor et al. (1984) presented the characteristic image of knots shown in Fig. 2.27. This slice was compared with the image analysis of the same section and with the photographic image of the cross section of a southern pine log to determine the correct location of knots, for which a good agreement was obtained.

In current industrial processing in choosing the best sawing pattern, the location of knots as well as the log geometry will be determined first. Skatter (1998) established the cross-sectional shape of softwood logs from three X-ray projections and modeled the cross section as an arbitrarily oriented ellipse. At the same time, the exterior shape, the shape under the bark, and the heartwood zone were determined. The comparison with experimental results on 40 Scots pine logs gave 1.64% accuracy in the prediction of the model. Secondly, the best log orientation will be selected to produce the lumber with the highest value for each log. Funt and Bryant (1987) developed computer algorithms to detect knots, decay, and cracks in hemlock logs with a medical scanner. The decay zone was the hardest feature to distinguish and was identified on the basis of a dark shade and a disruption in growth ring pattern. The program used a low ring uniformity pattern as a criterion for decay identification. The algorithm required 3 min/scan, which was very long compared with the speed of log conveyance in the sawmill. Wagner et al. (1989, a,b) reported the utilization of an ultra-fast medical scanning system for logs of 38 cm diameter, with 34 cross-sectional scans per second. They noted the necessity for the development of a specific industrial scanner for sawmills. The second-generation scanners with translate-rotate geometry required 20 s/scan. The third and fourth generation with rotate geometry only required 1 s/scan which is the maximum speed attainable today in rotating scanners. A short scanning time can be achieved by eliminating all mechanical movement parts, for example, by using an X-ray fan scanning system.

Schmoldt et al. (1993), and Schmoldt et al. (2000) emphasized the necessity for the development of specific computer algorithms for hardwood logs, which have a more complex internal structure than softwoods. An increase in the number of log defects increases the difficulty in making the best sawing decision. For automated interpretation of tomograms, data collection, and image

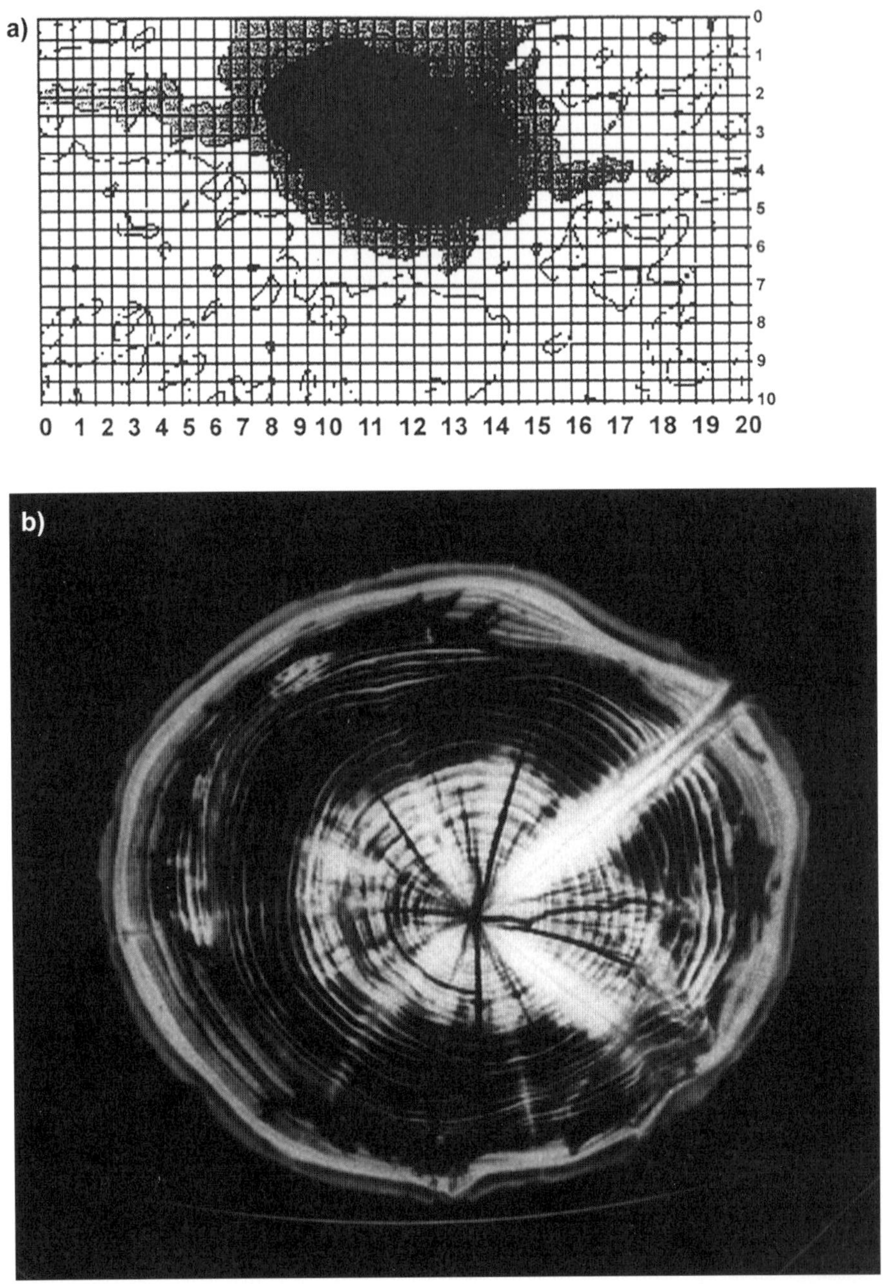

Fig. 2.26. Sound knots and surrounding area with different levels of gray. **a** Three levels of gray: *dark zone* for the knot, *white zone* for the sound wood, and *gray* for the zone around the knot (Krasulovic et al. 1999, courtesy of Forest Products J). **b** Tomographic image on a transverse section of fir, where the knots are visible as well as annual ring pattern, bark, sapwood, compression wood and cracks starting from the core. (Varga 2000, with permission)

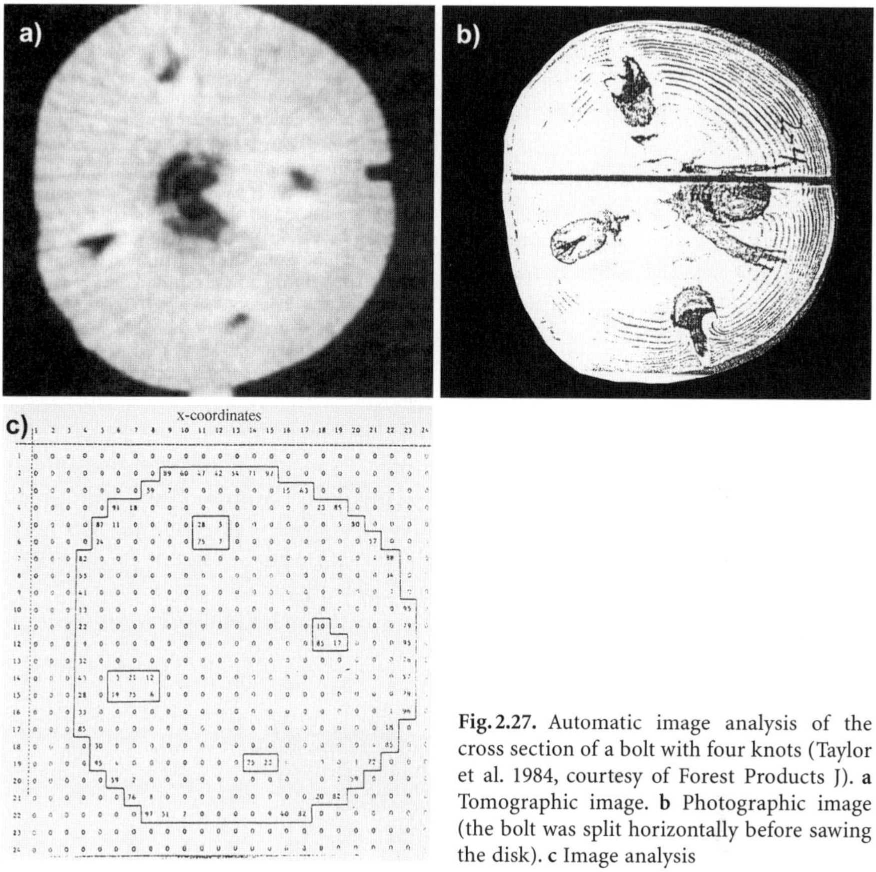

Fig. 2.27. Automatic image analysis of the cross section of a bolt with four knots (Taylor et al. 1984, courtesy of Forest Products J). **a** Tomographic image. **b** Photographic image (the bolt was split horizontally before sawing the disk). **c** Image analysis

analysis must be supported by robust methods of defect recognition developed previously with extensive data sets for different hardwood species and log configurations. The first decision in a sawmill is made for the processing of logs sorted as veneer logs or as high-quality saw logs. The computer simulation demonstrated an improvement of 21% of the value yield from red oak logs (Steele et al. 1990). Criticism of the simulated data can be made because of the ideal shapes of logs (cylinders or truncated cones), equations used to generate the defects, and low numbers of rotations.

Grundberg and Grönlund (1997) compared the accuracy of visual grading of Scots pine logs with simulated grading of the same X-ray scanned logs using an X-ray source of 160 keV. In the cross direction, the resolution was of 256 pixels, with a gray resolution of original signal of 4096 levels. Figure 2.28 shows the X-ray intensity variation across a log section. In this study, the logs were separated into three classes (butt, intermediate, and top logs) and processed with a normal cant sawing pattern All the boards were dried to 18% moisture content. The central boards were scanned on all four sides and then graded. The log classification was considered valid if at least one of the boards was graded the same as the predicted log grade. The agreement between the manual lumber grading and the simulated X-ray log scanner was between 71 and 89%. It was concluded that the simulation procedure and grading algorithm was a good predictor for the grades of logs and lumber.

A large scale validation test of computed tomography software against sawmill cutting was reported by Chiorescu and Grönlund (2000). This test was based on the Swedish Stem Bank X-ray computed tomography database (Grönlund et al. 1994) that acquired information on 600 selected Scots pine saw logs. The schematic path representation of the validation approach is presented in Fig. 2.29. The validation approach has two branches, the first corresponding to the simulated part of the study and the second to an actual medium-sized mill located in northern Sweden. The conversion process from the sawmill and the simulation is shown in Table 2.13. The volume recovery (48.08%) of the sawmill was very close to the simulation recovery (48.34%). A good prediction of the quality and length of boards was noted.

Schmoldt (1996) and Schmoldt et al. (1996a) and Schmoldt et al. (1997) developed specific X-ray computed tomography algorithms for hardwood logs. Specific scanning parameters are required for X-ray imaging of logs of differ-

Fig. 2.28. Real and simulated image of a log. (Grundberg and Grönlund 1997, with permission)

Table 2.13. The ability of the simulation software to predict board quality distribution. (Chiorescu and Grönlund 2000, with permission)

	Board Quality 1 (%)	quality Quality 2 (%)	Class Quality 3 (%)
Sawmill	27.00	38.80.	34.20
Simulation	28.15	39.74	32.11
Prediction	+1.15 (overestimation)	+0.094 (overestimation)	–2.09 (underestimation)

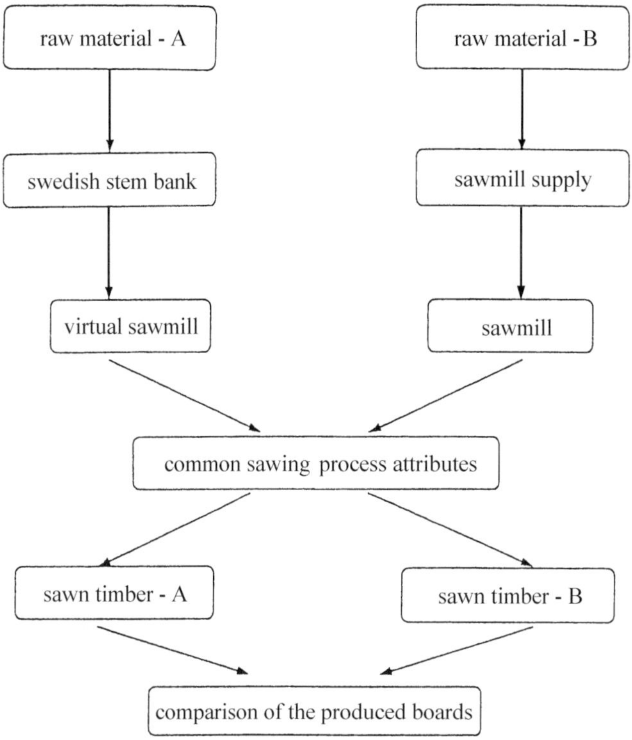

Fig. 2.29. Validation approach of the computed tomography software against sawmill cutting. (Chiorescu and Grönlund 2000, courtesy of Forest Products J)

ent hardwood species of different sizes, and with different image resolution and contrast. Data visualization and condensation must form a clear view of the geometry of the log and of its internal features. Only log defects that are important for subsequent processing need to be identified. Statistical labeling of accuracy was introduced, and segmentation and labeling into a single classification step were possible, largely by using local pixel neighborhoods. A neural network based on statistical classification of accuracy by using normalized pixel values as inputs to the classifier can formulate and apply aggregate features (average and standard deviation), as well as texture-related features. The recognition of different hardwood species was based on a histogram-based preprocessing step that normalized the CT numbers for density values prior to artificial neural network classification. Morphological postprocessing is used to refine the shapes of the detected image. The 3D identification of defects proposed by Schmolt et al. (1997) is a major step in the development of machine vision for wood species of very complex structure such as oak and yellow poplar. Initial results using a 2D classifier produced a lower classification accuracy than 3D data and it was suggested that the 2D classifier could be used successfully for hardwoods. Previous investigations published by McMillin (1982), Taylor et al. (1984), Funt and Bryant (1987), Zhu et al. (1991), Hagman and Grundberg (1995) were limited to 2D image analysis for softwoods only and no statistical approach was used.

2.4.2.2 Inspection of Poles

The large numbers of poles for telecommunications network in service in Australia, Japan, Canada, the United States (Miller 1988), and Europe (Rinn 1991; Habermehl and Ridder 1992a,b, 1993, 1994, 1995, 1996, 1998), explain the numerous projects that have been established to build and use field portable X-ray computed tomography scanners for the detection of the internal condition of wooden power poles. The need for the development of a nondestructive inspection method was supported by the high replacement costs. Major causes of internal deterioration of poles in the vicinity of the ground line are decay fungi and termite infestation.

The prototype device built in Australia (Cavies et al. 1990) satisfied several specifications: quantification of the internal structure of poles having the maximum diameter of 450 mm, production of tomograms from 300 mm below the ground line to 1000 mm above the ground line, relatively light (less than 50 kg), a spatial resolution of 5 mm or better ±50 kg/m^3 accuracy for the density of negligible radiation hazard while in operation, and production of a cross section scan within 5 min, an image of the cross section in 5min, and an image that is easily interpreted by the field operator.

The instrument operates with a translate–rotate (first-generation) mode for scanning poles of 450 mm diameter. The detector is an X-ray tube that is powered from a portable motor generator that can provide 160 keV at 18 mA. For each projection with 100 projections for 180° of rotation at a slice thickness of 5 mm, 99 ray sums are collected. The image is reconstructed from 99 × 99 pixels. The complete data acquisition phase is completed in 3 minutes, and the image is displayed as a 2D density map, with details on growth ring pattern and density histograms for immediate assessment by the operator. The spatial resolution is <5 mm. Data are stored on a floppy disk for further analysis if necessary. The total weight of the entire unit is 53 kg including the 8-kg X-ray tube.

2.4.2.3 Inspection of Lumber

Pioneering work on X-ray computed tomography inspection of the internal structure of lumber was done at the Forest Products Laboratory in Madison, USA (Suryoatmono et al. 1994). The main objective was to determine the density variation in a piece of lumber and to establish relationships between its structural performance and its mesostructure characterized by the presence of macroscopic defects like knots. The final goal of tomograms is to provide information for the optimization of the decisions in edging, trimming, and grading during the production of lumber. The characteristics of the scanning device were as follows:

- X-ray source of 50 keV, positioned 1810 mm from the image plate coated with photostimulable phosphor crystals, which act as energy traps when exposed to ionizing radiation;
- time of exposure 0.5 s at 200 mA;
- data readings with Philips Computed Radiography System to a grid of 0.9 × 0.9 mm. The readings are machine dependent numbers, ranging from 1 for the highest radiation exposure to 1024 for the lowest radiation exposure.
- calibration of the system with two small clear specimens of southern pine at 12% moisture content;
- constant mass attenuation factor assumed to be 0.2506 m^3/kg;
- southern pine lumber samples size: 520 × 38 × 140 mm. with a single wide-face knot ranging from 30 to 45 mm in diameter;
- high-density zones defined as having a density 20% greater than the specimen average.

The superposition of the real image on the graph of density variation along the length of the specimen shows the transition zone of increasing density within one knot diameter of the knot center point (Fig. 2.30). The knot density is twice the average density of the board. It was concluded that the transition zone density compensates for the mechanical weakening influence of the knot.

Successful lumber use for construction purposes is based on procedures to obtain strength grades. For sawmills to choose the proper raw material and predict the precise mechanical properties (modulus of rupture and modulus of elasticity in bending) of each board is an important step forward in the production of strength-graded lumber with special dimensions. Kliger et al. (1995) demonstrated that the density and the knot area ratio can be used to explain the variations in strength and stiffness of Norway spruce beams. Grundberg and Grönlund (1997) showed that the knot volume of logs can be measured with an X-ray log scanner. Wang et al. (1990) demonstrated a strong correlation between the X-ray tomographic green density of heartwood and the density of dry wood. Consequently, the hypothesis that relates density and knot volume of logs to strength and stiffness of the corresponding lumber was demonstrated by Oja et al (2000). For this purpose, eight Norway spruce (*Picea abies*) logs were scanned. The tomograms were used for the simulation of knot volume and green density of heartwood for a zone corresponding to four centerboards. After scanning the logs, four centerboards were sawn from each log, using a cant sawing pattern (45 × 145 mm). The boards were proof loaded and the bending strength and stiffness were determined using the Swedish norms, SS-EN 408. Statistical treatment of experimental data (multivariate models and partial least squares regressions) demonstrated reasonable correlations (R^2 ranging from 0.42 to 0.94) between knot volume and green density of

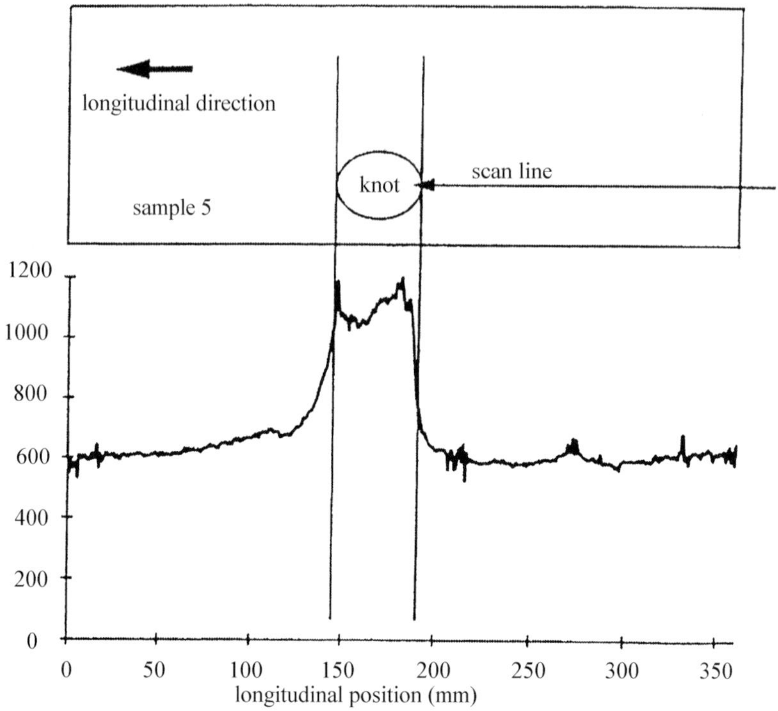
longitudinal direction
knot
scan line
sample 5
1200
1000
800
600
400
200
0
density (kg/cu.m)
0
50
100
150
200
250
300
350
longitudinal position (mm)

a)

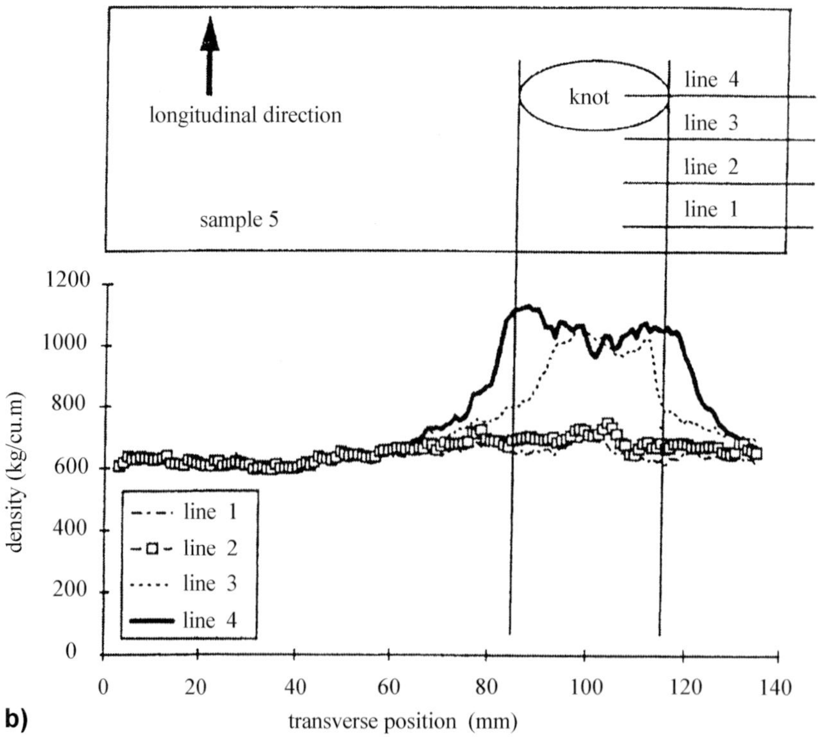
longitudinal direction
knot
line 4
line 3
line 2
line 1
sample 5
1200
1000
800
600
400
200
0
density (kg/cu.m)
line 1
line 2
line 3
line 4
0
20
40
60
80
100
120
140
transverse position (mm)

b)

heartwood deduced from tomographic inspection of logs, and the stiffness and strength measured on boards. It was concluded that the mechanical properties of each board are partly related to the sawing position of the corresponding board. From the information about the sawing position, it is therefore possible to predict the strength and stiffness of each board.

In the near future, this important advancement in lumber inspection needs to be confirmed with more studies with a larger number of specimens of the same species as well as with other softwood and hardwood species.

2.4.3 Wood Technology and Wood Processing

Today, the application of X-ray computed tomography in the wood industry is very limited because of the high costs of the device. Medical X-ray scanners have been used for laboratory tests performed for drying control of lumber and for the detection of voids in strand-based composites.

Specific scanners using gamma rays have also been designed for the measurement of the densitometric profile of wood composites. A continuous data profile of the board is provided by the device in a very short time, ranging from 2 to 5 min for a complete scanning of the board, depending of the sample thickness. The data are displayed in 2D, with the density as a function of the specimen thickness. No image of the internal structure of wood based composites was required. The main advantage of this scanner is the relative compactness of sources and detectors, without excessive shielding of the radiation source.

2.4.3.1 Control of Lumber Drying

Computed tomography during drying is an excellent nondestructive technique to study the moisture flux in lumber as a function of drying time. However, some difficulties related to the interpretation of tomograms have been identified:

- accuracy of local moisture content determination because of shrinkage;
- determination of the edge position and of the local moisture content at the wood surface with the conversion algorithm having an important density step at the transition between air and wood.

To elucidate these aspects, theoretical modeling and experimental approaches were proposed by Kanagawa and Hattori (1985), Kanagawa et al.

Fig. 2.30. Density variation observed during length scanning of a southern pine board of 350 mm length and 38 × 140 mm section, containing a knot (Suryatmono et al. 1993). **a** Longitudinal position; **b** transverse position. (Courtesy of Forest Products Society)

(1992), Pang et al. (1994), Pang (1996), Wiberg (1996), Arfvidsson et al. (1997), Pang and Wiberg (1998), and Wiberg and Moren (1999).

Kanagawa and Hattori (1985) used a medical scanner for the study of the distribution of moisture content during conventional kiln drying of 10.5 cm^2 lumber of hemlock (*Tsuga heterophylla*) and of almaciga (*Agatis* sp.) from green to 12% moisture content. The moisture distribution on sample sections was estimated following the procedure specified by Hattori and Kanagawa (1985) from the CT number of wet wood, oven-dry wood density, and volumetric shrinkage. A standard error of 1.1% was measured for the reference specimen with moisture content ranging between 2 and 27%. The errors were caused by the lack of uniformity of the images and lack of perfect linearity between the oven dry wood density, shrinkage and CT numbers.

The distribution of moisture on radial and tangential sections for both species are given in Fig. 2.31. In green *Agatis* sp. (50–70% moisture content) a high moisture content was observed at the limit of the annual ring and in ray tissue in the center of the sample. The moisture gradient in the tangential direction was more important than in the radial direction. In green hemlock

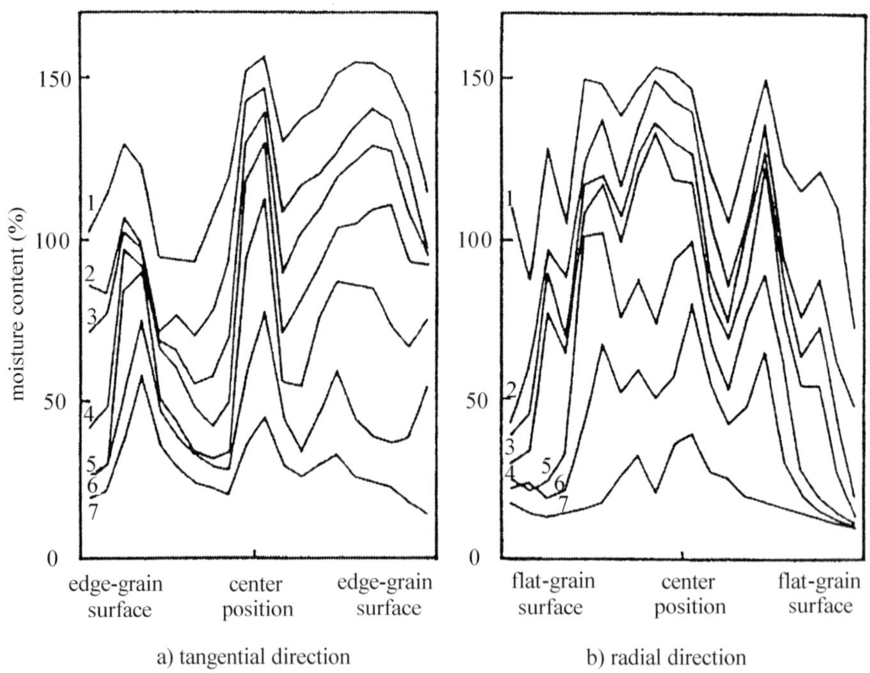

Fig. 2.31. Moisture content distribution during drying of a sample of 10.5 × 10.5 cm section of *Agatis* spp. (Kanagawa and Hattori 1985). The average moisture content is noted as: *1* 123%; *2* 98%; *3* 83%; *4* 73%; *5* 56%; *6* 42%; *7* 28%. (Courtesy of the Japan Wood Research Society)

(50–160% moisture content), moisture distribution in the tangential direction was relatively constant around the central position of the sample, while in radial direction a belt-shaped distribution was observed.

Pang (1996) and Pang and Wiberg (1998) studied the moisture distribution in *Pinus radiata* boards of 100 × 40 × 1200 mm during 30 h of drying (dry-bulb/wet-bulb temperatures were 90/62 °C). The moisture distribution and migration (Fig. 2.32) are shown in the wet board before drying (Fig. 2.32a), at an early stage of drying (after 9.6 h; Fig. 2.32b) and at a late stage of drying (i.e. after 30.4 h; Fig. 2.32c). A high moisture content is illustrated by the dark color. As can be seen from the tomograms, a dry shell is formed between the dry board surface and the wet front of wood. The thickness of the shell as predicted by Pang's model is between 0.5 and 2 mm. Pang's model simulated the mass transfer (moisture movement) and heat transfer following the width and the thickness of the board and predicted the moisture distribution, temperature profile and drying rate curves. The model proposed by Pang and Wiberg (1998) predicted density distribution (Fig. 2. 32d–f) and moisture content distribution (Fig. 2.32g–i). This model is more sophisticated than Pang's (1996) model, including inputs of wood basic density, within-ring variations of wood density, variation of moisture content, permeability of wood and permeability of latewood and earlywood, dimensions of lumber, drying conditions – temperature, humidity and velocity of the air – and actual sawing pattern.

In the early stage of drying (9.6 h), the moisture content near the surface decreases to below the fiber saturation point, inducing shrinkage of the board and corresponding drying stresses. However, the core of the board is still wet and drying stress can produce checks on the board surface. In the late stage of drying (30.4 h; Fig. 2.32c), when the average moisture content was 7.4%, the tomogram shows a high wet wood density in two latewood bands and in the core region of the board. The model predicted (Fig. 2.32f) a relatively high moisture content for the core zone, namely, 12 and 3.4% moisture content for the outer area. The model was not able to predict the moisture content in the two bands of latewood, probably due to the very high values of density in this zone, which was not included in the model. This is only of minor importance, and the validity of the model was confirmed by the drying curve expressed by the average moisture content of the board as a function of drying time (Fig. 2.33) which shows a very good agreement between experimental and predicted values. A small deviation of 6% was observed between 11 and 16 h of drying, corresponding to the decrease in moisture content from 60 to about 30%.

In view of these results, the modeling approach proposed by Pang and Wiberg (1998) is a major contribution in the field of detection of moisture distribution and validation of drying regimes using computed tomography. The causes of drying defects such as kiln brown stain and surface checking could be explained. At the same time, the model can be used to determine the influence of sawing pattern, wood density and permeability in earlywood and latewood, etc on the drying rate of lumber.

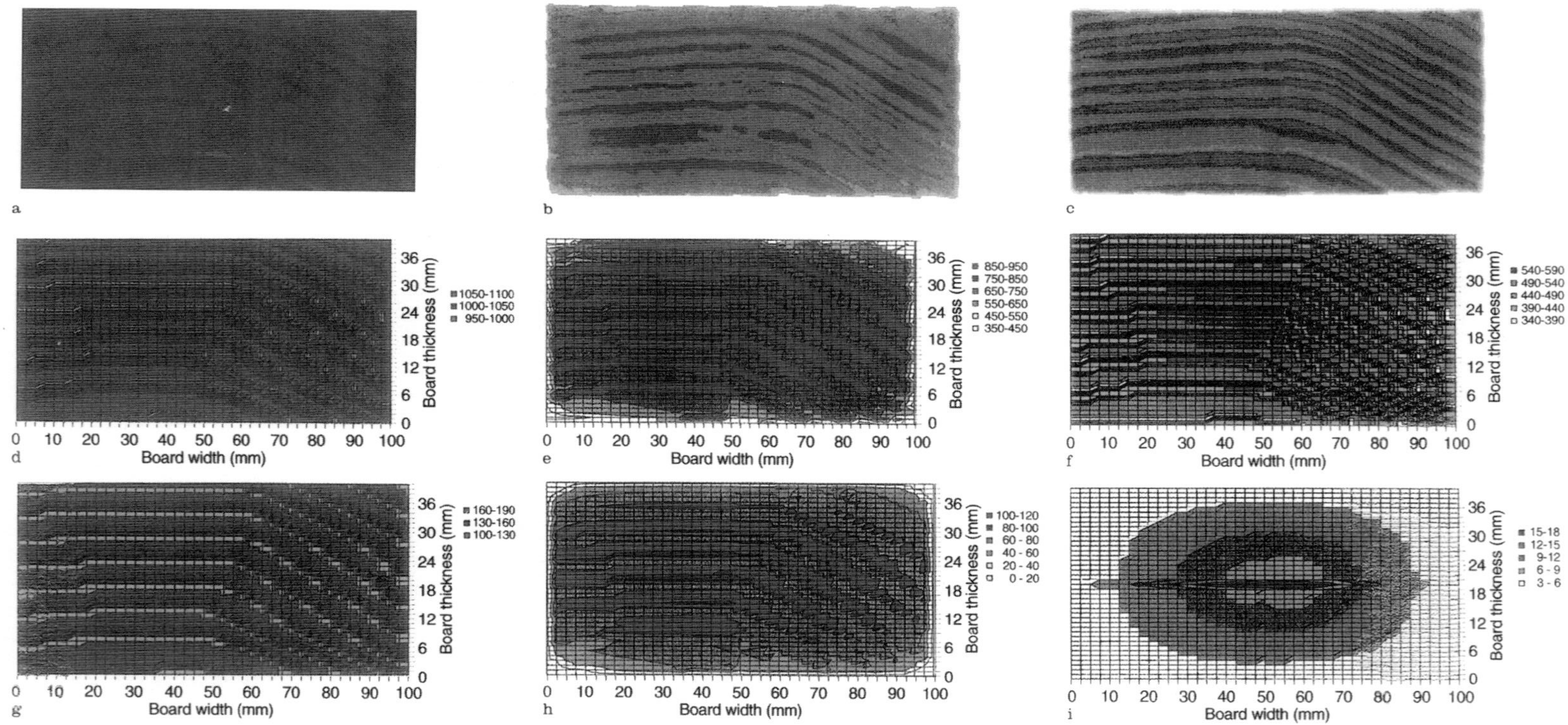

Fig. 2.32a–i. Dynamics of wood drying (Pang and Wiberg 1998). Experimental tomograms (with permission). **a** Scanning of the board before drying – corresponding to wet saturated wood. **b** Scanning of the board before drying – corresponding to wet wood after 9.6 hours of drying, and to the early stage of the drying process. **c** Scanning of the board before drying – corresponding to wet wood after 30.4 h of drying and to the late stage of the drying process. Predicted images for density distribution. **d** Wet wood density before drying. **e** Wood density after 9.6 h of drying. **f** Wood density after 30.4 h of drying. Predicted images for moisture content distribution. **g** Model of predicted wet wood density before drying. **h** Model of predicted wet wood density after 9.6 h of drying. **i** Model of predicted wood density after 30.4 h of drying

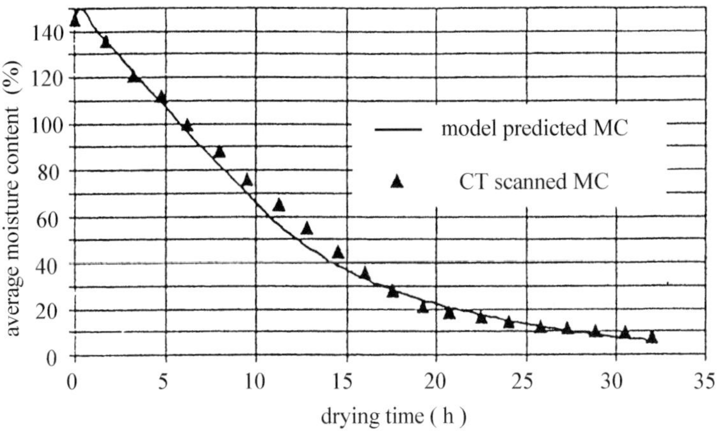

Fig. 2.33. Comparison between moisture content predicted by the model and the experimental data deduced from the tomograms of *Pinus radiata* board, expressed by the drying curve. (Pang and Wiberg 1998, with permission)

2.4.3.2 Control of Wood-Based Composites

The data concerning the utilization of X-ray computed tomography for the characterization of wood-based composites are very sparse. Sugimori and Lam (1999) reported the utilization of this technique for the study of macrovoid distribution in strand-based-wood composites of 630 kg/m^3 density. The macrovoids in strand-based composites, of various sizes and random distribution, are inherent to the forming process of boards. Their presence is due to the imperfect bonding between the adhesive and the wood strands.

The X-ray computed tomograms were obtained with a medical scanner, set at 120 kV, 200 mA, 10-mm slice thickness, and 0.86-mm pixel size and 0.74-mm^2 area, in a 512 × 512 pixel matrix. The image reconstruction was done every 3 s, and 130 slices were produced. The size of macrovoids was considered to be ≤2 pixels, i.e., 1.48 mm^2. The images obtained were converted to black and white with a conventional image-processing technique, and their size, number, position, and major and minor diameters were measured with the medical software. The variation along the length of a board of macrovoid area as well as their distribution is shown in Fig. 2.34. The authors noted that the experimental size measurements are in agreement with their mathematical prediction.

Gamma ray scanning densitometry (Feraz 1976; Ranta and May 1978) has been more frequently used for the measurement of the densitometric profile of wood-based composites. Winstorfer et al. (1986) measured the densitometric profile of particleboard (12, 18 and 44 mm thickness), waferboard (13 mm thickness), fiberboard (6 and 12 mm thickness) and medium density fiberboard (10, 19 mm thickness). A gamma source of 60 keV was mounted in a collimating holder, and the sample was placed on a carriage that held it stationary

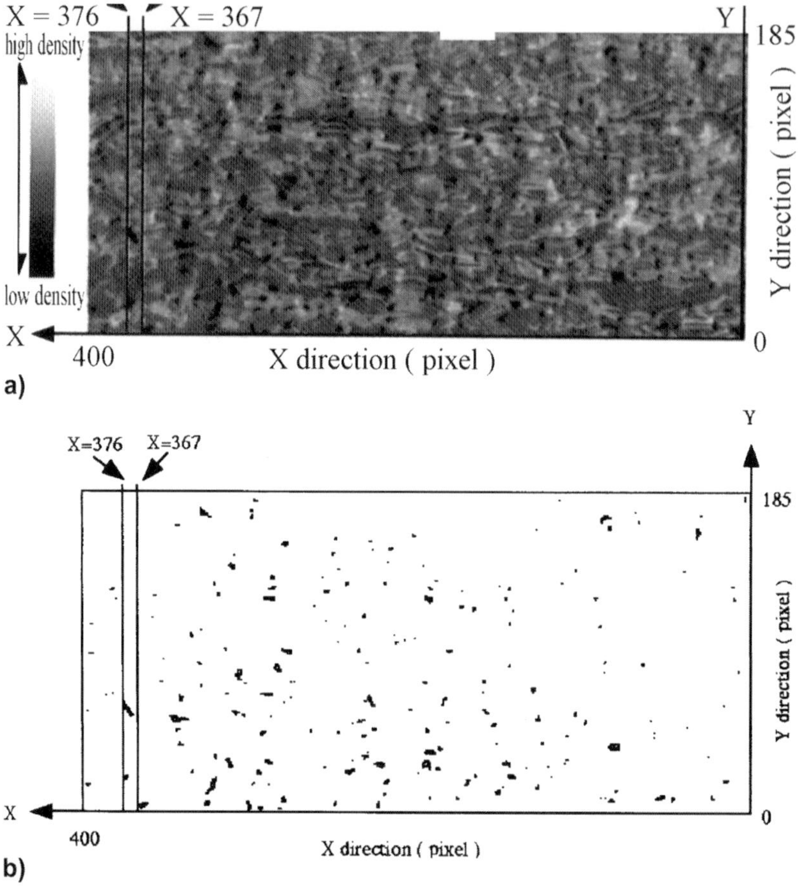

Fig. 2.34. Macrovoid distribution in particleboard of 630 kg/m^3 (Sugimori and Lam 1999). **a** X-ray image; **b** the same image after conversion with a 270-threshold value. (Courtesy of the Japan Wood Research Society)

during counting and then automatically moved to the next step. At each step, which can vary from 15 μm to 30 cm, the counter is automatically read, reset and restarted. Counts at each step were taken every 10 s. The average counting time for a complete scanning of a board depends on the thickness, and is between 2 and 5 min. The authors suggested using this as a standard method for nondestructive evaluation with the accuracy and precision of density profiles for wood panel products.

Laufenberg (1986) also studied the use of gamma radiation to measure the density gradient in reconstituted wood products. The mass attenuation coefficients were determined for different resins: phenol-formaldehyde (0.181 cm^2/g), urea-formaldehyde (0.182 cm^2/g) and isocyanate (0.178 cm^2/g) as well as water (0.196 cm^2/g) and wood–red oak and Douglas fir (for both

$0.183\,cm^2/g$). He noted that the large differences between the mass attenuation coefficient of resins, wood, and water can induce errors in the prediction of the wood density of the board. For this reason, the boards must be well conditioned at a minimum 12% moisture content prior to density measurement. In this case, the expected error of the density prediction is <1%.

Elbez (1990) reported the utilization of computed tomography with a medical scanner for the control and characterization of cured adhesives and glue joints in balsa, spruce, poplar, oak and two tropical species, moabi and ipe. The adhesives were urea-formaldehyde plasticized with furfurylic alcohol, polyvinyl acetate with metallic salt and resorcinol phenolic formaldehyde. The tomograms clearly showed structural details of the glue joints and of the adjacent zone composed of adhesive and wood. The density gradient was observed with a maximum in the center of the joint and with a minimum in the proximity of the wood. The minimum density is probably due to the lower degree of condensation in urea formaldehyde glue or to the lower polymerization in the case of the polyvinyl glue. Lindberg et al. (1996), using X-ray CT, measured water diffusion through acrylate latex paint films with 0.1- and 0.4-μm particle sizes. It was proven that the techniques employed were successful for the measurement of water diffusion in films. As expected, a 0.1-μm particle film had a much lower water diffusion coefficient than an 0.4-μm film.

2.4.4 Dendroarcheology, Dendrochronology, and Dendroclimatology

The physical and mechanical parameters of archeological buried wood, saturated with water, were studied by Schniewind (1990). Compared with fresh wood of the same species, the mechanical properties of archeological wood are diminished because of abiotic deterioration.

Passialis (1997) noted that changes in chemical composition are due to a slow hydrolysis of polysaccharides, with a notable increase in lignin content. Morphological deterioration of cellular tissue and separation of cells in the growth rings are found. Density is also reduced because of the losses of chemical constituents and of deterioration of cell integrity. Kim and Singh (2000) emphasized that all species can survive in a fully saturated environment for thousands of years. The slow degradation of polysaccharides, lignin and extractives can be caused by biological agents such as fungi and bacteria that are tolerant to oxygen limiting conditions. Wood deterioration can be caused by both microbial and chemical agents. Unger et al. (1988) reported the use of X-ray computed tomography for the inspection of archeological oak wood with different amounts of water, ranging between less from 185 to more than 400% moisture contents. The images clearly show the advanced degradation zones.

Klein and Vogel (1993) noted the interest of using X-ray CT for the identification of wood species in dendrochronology. Madsen (1994) and Divos et al.

(1994) employed a gamma backscattering technique for inspection of ceilings and roof structures of old buildings. Hoag and McKimmy (1988) proposed direct scanning densitometry for dendroclimatological studies. A significant advancement was observed in nondestructive inspection of fine art objects, historic musical instruments, etc., using computed tomography. An example is given by Fioravanti and Ricci (1991) for the study of the densitometric profiles of a top plate of a violin. Onoe et al. (1984), and Schirone and Lo Monaco (1988) also suggested the use of computed tomography in archeology and dendrochronology. Unfortunately, computed tomography has not been extensively used in dendrochronology, dendroarcheology, or dendroclimatology.

2.5 Summary

Ionizing radiation (X-ray or gamma ray) computed tomography in wood science is a cornerstone in the field of the imaging of solid wood and wood-based composites. High resolution X-ray imaging is a valuable tool for nondestructive evaluation and inspection of trees, logs, poles, boards, etc. and for nondestructive control in wood processing. The ability of this method to identify and quantify the internal inhomogeneities and defects in wood products, the distribution of moisture inside the trunk, etc. explains its importance and increasing implementation in wood science and technology.

The aim of computed tomography is to create a precise map of the internal inhomogeneities of a sample that can be a trunk, piece of lumber, laboratory testing specimen, etc. This map is obtained through the determination of the attenuation coefficient of ionizing radiation in a single thin layer of the sample. The images produced are reconstructed from projections using advanced computational and mathematical procedures underlying the data collection, image reconstruction and image display. The spatial information in 2D or in 3D is able to discern zones of low attenuation contrast. First generation scanners produced tomograms with one source and one detector acting in parallel by translation. The sample was rotated by a 1°-step angle and the whole image obtained for 180° of collected data. Second-generation images were also obtained by translation, using an array of detectors that made simultaneous measurements through different angles in a single transverse inspection. The sample was rotated by the array beam angle. The third generation of tomograms were produced by a fanning movement, with a scanner having many detectors located on an arc focused on the radiation source. The fourth scanner generation has also a fan system of detection, and the detection array is located on a circle that surrounds the source and the sample.

In inhomogeneous materials like wood, the attenuation coefficient depends on both the quantum energy of the ionizing radiation and on the chemical composition of the sample. Because of the high quantum energy of radiation and the low atomic number of the chemical constituents of wood, the attenuation phenomenon is caused mainly by the Compton effect and, consequently, is proportional to the mass density of wood. Each zone of the specimen having the same density is evaluated statistically and is represented by one color, or by a gray scale.

Worldwide, during the last 20 years, stress has been put on understanding of the basic aspects of ionizing computed tomography in view of the development of specific industrial scanners for routine quality control of wood products. The main advantages of the ionizing radiation computed tomography compared with conventional radiography are:

- elimination of the intermediate steps involving photographic film and optical densitometry;
- data available in real time;
- improvement of the calibration procedure, since the mass attenuation coefficient can be determined directly, using a scintillation detector with standard radiation pulse shaping and counting equipment that allows the user to select the energy range counter;
- large-volume material can be inspected quickly and implementation of scanning technology in sawmills and other mills for wood will have an important payback for wood processors.

The scanners for ionizing radiation computed tomography can be fixed or portable. Medical scanners were used in the 1980s for imaging wood specimens, because the physical properties of wood do not differ greatly from that of the human body. The development of non-medical scanners, initially experimental and later commercial, occurred in the mid-1970s. The industrial systems in wood industry require: high energy X-ray sources, accommodation for a wide range of specimen sizes and densities and operations in industrial environments with a speed consistent with efficient wood processing.

Fixed and portable equipment are used in wood science and technology for different purposes. If the object can be brought to the scanner as in the case of logs and boards, fixed equipment is recommended. In this case, an X-ray source is commonly used. Portable equipment was designed to be used for in situ inspection of trees, poles, and building elements. Gamma rays were used in this case because of many advantages with respect to the wide range of energies available, portability and cost. In spite of these superior characteristics, gamma ray sources are of low intensity compared with X-rays.

The factors affecting the quality of the image are: beam path, spatial and contrast resolution, anisotropic direction of wood and beam hardening. X-rays travel in a straight line through materials. Two main parameters are related to the beam path: range of transmission probability and combined effects. The range of transmission is defined as the radiation resolution of a given wood X-ray experiment, in reference to the separability of transmission probabilities resulting from density variations. The combined effects are a global factor taking into consideration the combination of X-ray energy, wood specimen thickness, and heterogeneities induced by the anatomical structure.

The maximum radiation resolution is determined by density variations in wood specimens, energy and intensity of X-rays, spatial and contrast resolution and beam hardening. Because X-ray attenuation is highly dependent on the energy of the X-rays, it is always necessary to provide this information.

The main applications of ionizing radiation computed tomography discussed in this chapter are:

- examination and inspection of trees, related to the growth rate assessment, water content in trees and pollution effects;
- wood quality assessment, in the inspection of logs and poles;
- processing of wood for control of lumber drying and the quality of wood-based composites;
- in dendrochronology and related fields, such as dendroarcheology and dendroclimatology.

2.6 Annexes

Annex A. Measured and calculated CT numbers for wood of different densities and 0% moisture content. (Lindgren 1991a)

Density (kg/m^3)	Measured CT number	Calculated CT number
415	–603	–606
419	–601	–602
496	–534	–529
499	–529	–526
619	–414	–412
612	–419	–418
352	–670	–666
468	–554	–555
400	–621	–620
456	–570	–567
506	–518	–519

Annexe B. Density and CT number for different species. (Klein and Vogel 1993)

Species	Density (kg/m^3)	CT number (H)
Lophira alata	1030	62.6
Millettia laurentii	750	210.1
Afzelia spp.	700	178.9
Quercus robur	650	483.9

Species	Density (kg/m^3)	CT number (H)
Fraxinus excelsior	650	347.8
Pinus palustris	630	441.6
Salix spp.	520	666.6
Tilia grandifolia	490	510.4
Tilia spp.	490	640.3
Populus nigra	450	649.0
Pinus cembra	450	641.7
Picea sitchensis	430	616.4
Picea abies	430	607.6
Bombax spp.	400	874.2

3 Thermal Imaging

3.1 Introduction

Thermography is a generic term for a variety of techniques used to visualize the temperature at the surface of objects and refers to imaging of a full-field temperature distribution. Active heating or passive heating procedures can be used for thermographic imaging of solids (Liezers et al. 1985; Puttick 1987; Vavilov 1992). Thermography provides a map of surface temperature distribution. It can be said that thermography is a noncontact and nondestructive method that gives information about the thermal properties of a sample. The main thermographic methods for solids fall into three major classes: pulse thermography (Lau et al. 1991; Thomas et al. 1995), heating up thermography (Reynolds and Wells 1984), and lock-in thermography (Busse et al. 1992b; Busse 2001). The main differences between the thermographic methods are determined by the type of heating and by evaluation of the thermal properties of the samples. The pulse thermography system records the transient thermal response of an object to a pulse of thermal excitation, provided by a flash tube.

Heating-up thermography utilizes heat sources such as: commercial photographic lighting devices, which deliver energy up to 5 J/cm^2 in 1 μs at high repetition rate, hot liquids, and air dryers of the high power "heat gun" type. As an alternative to heating, cooling can be used with an aerosol freezer spray. Video cameras in visible light wavelengths record the images in a rapid corresponding time. The transient character of the temperature field demands a dynamic tuning of the thermo-vision system. Sometimes, the large local nonuniformity of heating can induce noisy signals that influence the image quality obtained throughout the whole scanned area. The development of infrared video cameras has extended the wavelength range of visible light video cameras to the thermal infrared range, between 3 and 12 μm. Images obtained with infrared cameras correspond to a mixture of emitted and reflected radiation of the object and of its surroundings and this test is particularly severe in the range of 8–12 μm wavelength, which corresponds to the peak of blackbody radiation at room temperature. For this reason, a calibration test must always be performed. In conventional thermal wave-imaging techniques, this problem is overcome by the use of a lock-in analyzer, synchronized to the source of the thermal wave.

Lock-in thermography can provide three types of images in a very short time: thermographic, phase and magnitude images. The development of lock-in vibro-thermography was an important step in thermal wave imaging which allows the utilization of thermographic equipment to image the average temperature distribution of oscillating components. The thermal waves are generated by a periodical stress at very low frequency (0.03 Hz), which do not induce overheating of the surface of the sample. Vibro-thermography is a very useful method for inspection of solid wood and wood-based composites because of the high emissivity and low thermal conductivity of these materials.

In wood science, thermography is a relatively new field of study. Thermal imaging of subsurface temperature distribution on wood-based composites to ascertain the integrity of the subsurface structure was introduced less than 10 years ago. We note here the pioneering work by Busse (1988, 1994, 2001) and coworkers at Stuttgart University in Germany who developed these techniques for polymers and wood-based composites.

Heat propagation phenomena in solids are influenced by the thermal impedance of the materials, which varies as a function of the interaction of thermal waves with the structure. Since these interactions are associated with a modification of the thermal properties, it is possible to obtain a map of the sample characteristics and consequently to obtain an image.

The active heating procedure, or stress-generated thermal field, was used in early work on wood thermographic imaging (Masuda et al. 1995; Masuda and Takahashi 1998). This makes use of the fact that under fatigue cycling loading in bending or compression, the temperature rise is determined by the increase in energy in the hysteresis loop. The material is cyclically stressed while its surface temperature is monitored thermographically. Heat dissipation results in a temperature distribution on the material surface that is related to the stress distribution in the material. Loading of the specimen can be done either by direct fatigue or by vibrational means.

The advantage of this active heat generation technique is that the influence of defects on the mechanical properties of solids can be easily and rapidly examined. The quantitative evaluation of a stress field can be obtained with a highly sensitive infrared camera capable of detecting changes of 10^{-3} K in regions of less than 1 mm. The main disadvantage of the cycling loading of the material is related to the possibility of inducing new damage in the sample when trying to locate existing defects. The use of a low-amplitude vibrational excitation can be an alternative and can alleviate the damage propagation. With pulse thermography the response of the sample is recorded after its warm-up period with a short energy pulse (light, ultrasonic pulse) and an infrared image is obtained. This image indicates the inhomogeneities and defects at the surface of the sample or at different depths.

With lock-in vibrational thermography, which is a combination of thermal wave technique and thermography, a large sample depth range can be inspected, and no overheating occurs at the surface. The sinusoidal thermal wave is excited on the whole surface of the sample and then monitored, using both the thermographic technique and lock-in data analysis which very rapidly provide a phase angle image. This remote method was used for detection of flaws such as delaminations in wood composites (Wu and Busse 1995, 1996; Wu et al. 1996, 1997b) because of its sensitivity to subsurface defects and its imaging capability in short time.

In the case of the so called passive method, the material is subjected to heating from an external source. The thermal image obtained is the result of a very complex interaction between the heating source, the material and the presence or absence of defects. The rate of heat application and the mode of heating by contact or by radiation are factors of major importance for obtaining good images with correct resolution. The thermal gradient and not the temperature itself makes it possible to obtain information about the integrity of the structure. The thermal gradient is calculated from the temperature distribution. The heat application may be at a single point or over a wide area. This passive mode produces images that are transient and therefore difficult to capture for a material that is not subjected to any mechanical stress. The elimination of the loading fixture makes the rapid testing of objects of various geometries in situ possible . The time interval of interest for passive thermal testing is a few seconds. During this time the surface temperature gradients reach the maximum and yield the greatest resolution of the thermal image.

An advantage of the passive heating method over the active heating procedure is its ability to produce a temperature distribution without resorting to mechanical loading of the material. The thermal stress or the thermal shock is too low to damage the material. The disadvantage of the passive method is that the thermal images are transient and require a fast recording system to capture the most interesting images during the test.

Electronic imaging systems or chemical imaging systems can be used for the visualization of the temperature distribution on surfaces. The most widely used system is an electronic system with a scanning infrared camera, which produces an image of an object through electronic detection of infrared radiation emitted from the object.

Chemical imaging is obtained with liquid crystals applied to the surface of the object. They selectively scatter the light according to their temperature. The thermal resolution is of 0.007 °C and the spatial resolution of 1 μm (Wilson and Charles 1981). A photographic recording system is attached to capture liquid-crystal images during the test.

Catena (1992) reported the use of an infrared passive technique for the detection of cavities and for phytopathological diagnosis of ornamental trees. The infrared camera scanning technique was also used for the detection of different defects, such as delamination in wood composites (Berglind and Dillenz 2000) or locating knots and voids in lumber (Tanaka 1994; Quin et al. 1998; Steele et al. 2000a,b; Tanaka and Divos 2000). In wood technology, the scanning infrared imaging technique seems to be one of the most convenient remote sensing methods available today for thermographic inspection of trees, wood and wood-based composites in situ.

The aim of this chapter is to provide a general description of thermal imaging techniques for solid wood and wood-based composites The methods described here are divided into two main groups: the active and the passive methods, for imaging defects in trees and in lumber, delaminations in wood-based composites and in different types of joints for lumber connection.

3.2 Basic Aspects

Significant progress in the understanding of thermal properties of solids was made with the understanding of the physics of diffusion waves. Diffusion waves are neither electromagnetic nor elastic but thermal and their mathematical formalism describes how a temperature modulation propagates in a medium. In the 19th century, Fourier, Ångström and Bell were the pioneers in investigating thermal waves. Only 100 years later, the advance of laser technology made possible the coherent explanation of the optoacoustic or photoacoustic effect (Rosencwaig 1983) induced by the propagation of thermal waves.

Mandelis (2000), in an article devoted to diffusion waves and their uses, noted that "diffusion waves lack wave front, can't be beamed and don't travel very far, yet they form the basis of several new and revolutionary measurement technologies." Diffusion waves have a complex wave vector and their transport takes the form of a spatial diffusion gradient, which obeys a linear law. In the imaging application of thermal waves, the map of the characteristics (magnitude or phase angle) of a locally produced signal shows the thermal parameters of the structure under test. Busse (1988) noted that the image generated with "locally observed phase shift is usually more significant" than that obtained with the magnitude of the signal.

Thermal waves can be generated by any modulated energy. Imaging with thermal waves is based on the dynamic effect of the modulation of the temperature by the structure.

When a static effect is used, the image is obtained with a thermal infrared emission that shows the static temperature distribution weighted by a local infrared emission coefficient. In both cases, physical contact with the samples is not necessary.

Heat flow in a solid is governed by Fourier Eq. (3.1)

$$k\, d^2T/dx^2 = \rho\, C\, dT/dt \tag{3.1}$$

where T is the temperature at depth x beneath the surface at time t, k is the thermal conductivity, ρ is the density, and C is the specific heat capacity of the solid.

If the angular frequency of the intensity-modulated laser heat source is ϖ, the solution of Eq. (3.1) has the form

$$T(x,t) = T_O e^{-x/\mu} \cos(2\,\pi\, x/\lambda - \omega t) \tag{3.2}$$

with thermal waves, the sample undergoes periodic heating when it absorbs intensity-modulated laser radiation. If diffusion is periodical, produced by absorption of radiation, the thermal diffusion length is defined as the distance where the temperature decrease due to conduction is e^{-1} times that at the origin of the heating at the surface of the sample. The signal waveform phase lag relative to the excitation waveform increases by 57 degrees.

The thermal diffusion length μ is given by Eq. (3.3)

$$\mu = [2\,k/\omega\,\rho\,C]^{1/2} \tag{3.3}$$

where ω is the radial frequency at which the light is intensity modulated. In metals μ is about 1 mm. For other materials, the thermal diffusion length is smaller. At the same frequency, the thermal wavelength $2\pi\mu$ is much smaller than the acoustic wavelength.

The phase velocity is the speed with which the temperature modulation moves along the z direction of the specimen and is given by:

$$v = [2\,\omega k/\rho\,C]^{1/2} \tag{3.4}$$

It depends on frequency, thermal conductivity, density and the specific heat of the sample. The dependence of the phase velocity on frequency undergoes a strong dispersion. On the other hand, the wave velocity is not dependent on the geometry of the sample in which the wave propagates.

The thermal wave length is given by:

$$\lambda = 2\,\pi\mu \tag{3.5}$$

Thus, high frequency thermal waves propagate at high velocities but because of their short diffusion lengths, they decay rapidly. In contrast, low frequency thermal waves penetrate deeply into the subsurface with a low velocity.

It has been demonstrated that thermal wave imaging phenomena can be explained with an extreme near-field diffraction theory, and standard scattering techniques can be employed for the calculation of thermal wave imaging (Rosencwaig 1983). Measurement of the thermal diffusion length and of thermal conductivity can give information about the thickness of the sample and can be used as a routine technique for depth profiling of a subsurface structure. Most applications of relevance to quality control can be imagined for depth profiling of layered wood-based composites or for measuring the thickness of paint or varnish layers as demonstrated by Wu and Busse (1996). Moreover, for thermal waves, the real and imaginary parts of wave numbers are equal. This characteristic allows thermally driven waves to be observed at any frequency with a very small wave vector magnitude, which permits identification of subsurface features in the submillimeter range.

The spatial damped nature of thermal waves has two major advantages over other tomographic techniques: the high near-surface feature resolution and at the same time less susceptibility than other waves to the appearance of spurious signals from a remote interface. For these reasons, tomographic techniques were developed for clinical applications and for nondestructive evaluation of critical and valuable manufacturing components.

A laser beam that has a high spectral selectivity, spatial coherence and resolution is usually the light source used for generation of thermal waves. The choice of laser characteristics (wavelength and power) depends on the thermal properties of the sample. He-Ne laser and Ar+ laser were used for a power level of about 20 W. They have an acceptable ratio of signal to noise and cause no optical damage to the sample (Busse 1988). Tsuchikawa et al. (2000) proposed the utilization of a semiconductor near-infrared laser of 830 nm wavelength and 48 mW output power for the study of the effect of physical characteristics of a specimen (sample thickness, fiber direction, surface roughness, moisture content) and the geometric conditions of illumination on the transmitted output power.

The relationship between the sample thickness and transmitted output power for Sitka spruce and beech specimens (Fig. 3.1) shows an exponential reduction of power for both species. This graph argued in favor of the development of a nondestructive technique for measurement of veneer thickness. The thickness of the sample can be controlled by varying the modulation frequency of the incident radiation and hence the thermal wavelength.

The combined effect of light irradiation and of the anisotropic directions of wood is shown in Fig. 3.2. It can be noted that the variation of output power is strictly related to the geometric conditions of the experiment. The effect of

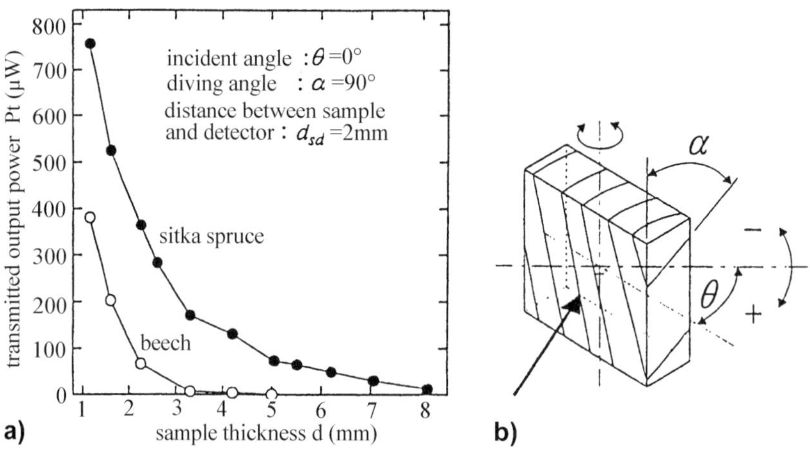

Fig. 3.1a,b. Transmitted output power under air-dried conditions as a function of sample thickness (Tsuchikawa et al. 2000). **a** Relationships for beech and Sitka spruce. **b** Specimen geometry and parameters. (Courtesy of Forest Products J)

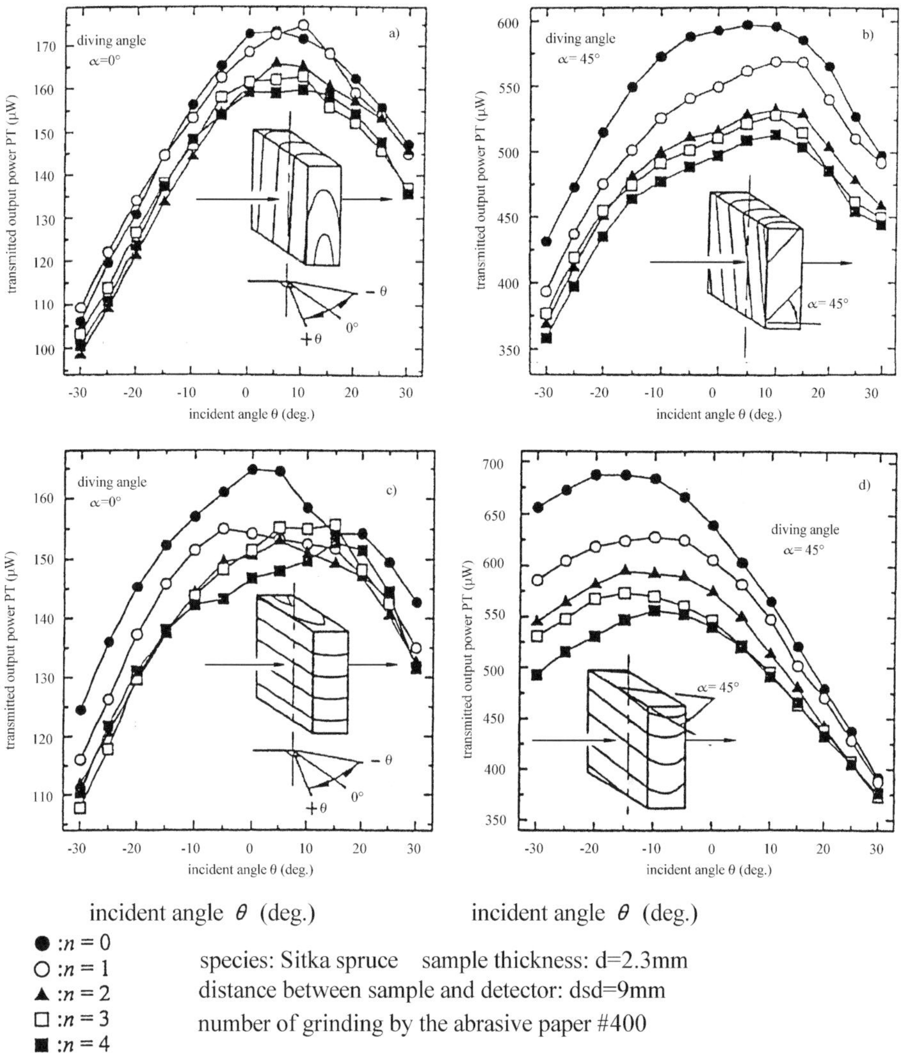

Fig. 3.2a–d. The combined effects of light irradiation, anisotropic directions and surface roughness of Sitka spruce specimens on transmitted output power (Tsuchikawa et al. 2000). **a** Diving angle $\alpha = 0$; light irradiation in LR plane and propagation in T direction. **b** Diving angle $\alpha = 45°$; light irradiation in LR plane and propagation at 45° in T direction. **c** Diving angle $\alpha = 0$; light irradiation in LR plane and propagation in R direction. **d** Diving angle $\alpha = 45°$; light irradiation in LR plane and propagation at 45° in R direction. (Courtesy of Forest Products J)

moisture content was related to the reduced transmitted output power, which corresponds to the distance between the sample and the detector of 1 mm for Sitka spruce and beech specimens of different thicknesses. For both species and for all thicknesses, the reduced transmitted output power increases with increasing moisture content. This means that the output power increases with the presence of free water in cell lumina and influences the conditions of light reflection and scattering.

The second part of this section is devoted to an analysis of some particular aspects related to thermal conductivity of wood. Because of its anisotropic and porous nature, the thermal behavior of wood must be related to the parallel and perpendicular direction of the fibers and to the hygroscopic level, below and below fiber saturation point. The relationships between the thermal conductivity of wood and some physical parameters such as the density or specific gravity and the porosity were studied by Siau (1995). The model proposed by Siau (1995) shown in Fig. 3.3 was used for the derivation of the thermal conductivity equations, for which three elements were considered: the thermal conductivity of the cell wall substance, of bound water, and of air.

Below the fiber saturation point, when the heat flux is applied in the direction of the fiber axis, the thermal conductivity is given in Eq. (3.6):

$$k_L = k_L{}^w (1 - v_a) + k_a\, v_a \tag{3.6}$$

where k_L is the thermal conductivity in the longitudinal direction, $k_L{}^w$ is the thermal conductivity of the cell wall in the longitudinal direction, v_a is the porosity of the wood, k_a is the thermal conductivity of the dead air.

When the heat flux is across the cell walls, the transverse thermal conductivity k_T can be deduced in a very simplified form by Eq. (3.6) for which the thermal conductivity of cell-wall substance was 0.318 and 0.401 W/mK for bound water and 0.024 W/mK for dead air in the cell lumina.

$$k_T = 0.50 - 0.46a \tag{3.7}$$

where a x a is the lumen size that corresponds to the porosity v_a = a x a

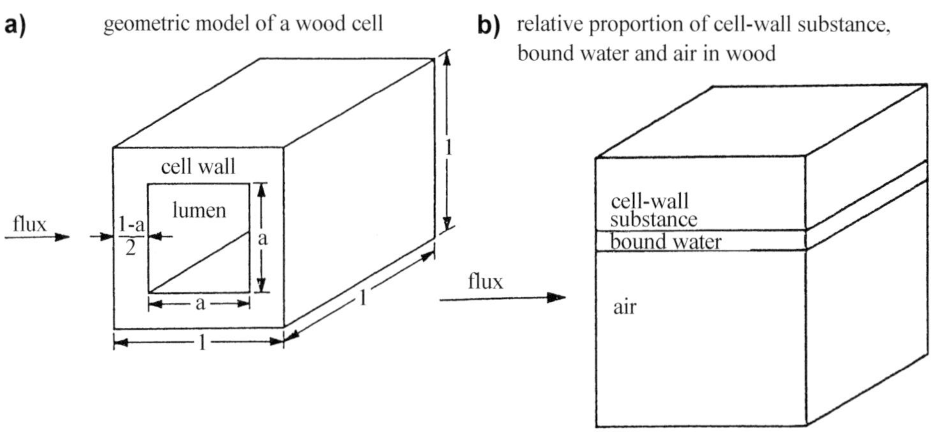

Fig. 3.3a,b. Simplified model of a cell wall and relative proportions of cell-wall substances, air and bound water in wood (Siau 1995). **a** Geometric model of wood cell. **b** Relative proportion of cell-wall substance, bound water and air in wood. (Courtesy of Wood Fiber Science)

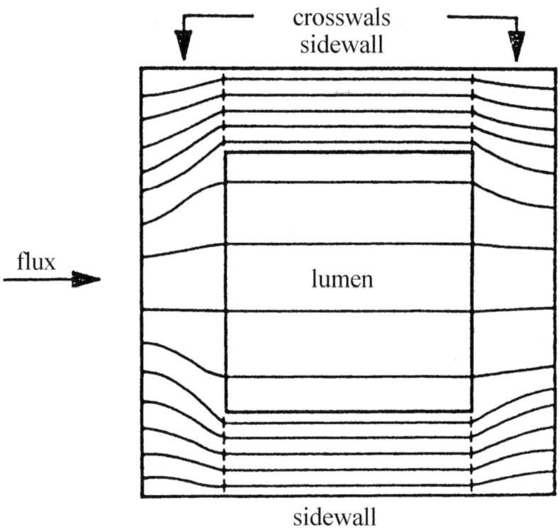

Fig. 3.4. The flux of thermal energy in lumen and in cross walls and in side walls (Siau 1995; courtesy of Wood Fiber Science)

The relative flux concentration through the cross walls and side walls is given in Fig. 3.4. It should be noted that because of the nonuniformity of the flux, the entire width of the wall is not available for conduction.

Above the fiber saturation point, the empirical equation for the determination of k_T as a function of the specific gravity of the wood (G) and of the moisture content (MC) is given in Eq. (3.8)

$$k_T = G(0.27 + 0.0055\ MC) + 0.024\, v_a \tag{3.8}$$

The relationship between the longitudinal and transverse thermal conductivity is given by Eq. (3.9):

$$k_L = (2.25 \ldots 2.75) k_T \tag{3.9}$$

Thermal properties of wood and wood-based composites have been reported in several reference books and articles (Weatherwax and Stamm 1947; Kollmann and Côté 1968; Giordano 1971; USDA Forest Prod. Laboratory 1972; Siau 1995; Olek et al. 2000) Some values of thermal parameters are given in Tables 3.1 and 3.2.

3.3 Equipment for Thermal Imaging

In this section, we describe the imaging equipment used for thermal imaging of solid wood and of wood-based composites. The most popular device for thermal imaging is the infrared scanning camera, which is a thermal wave detector in the infrared domain and which can be used for both field and laboratory measurements. The popular infrared camera can be used for all thermal wave-type imaging. The thermal wave describes how a temperature

Table 3.1. Some values of thermal conductivity of wood and other materials in the SI international system (Siau 1995).(courtesy of Wood Fiber Science)

No.	Material	Thermal conductivity (W/mK)
1	Wood, 450 kg/m^3; 12% MC, perpendicular to fibers	0.13
2	Wood, 450 kg/m^3; 12% MC, parallel to fibers	0.31
3	Wood, 700 kg/m^3; 12% MC, perpendicular to fibers	0.18
4	Wood, 700 kg/m^3; 12% MC, parallel to fibers	0.44
5	Cell wall substance perpendicular	0.44
6	Cell wall substance parallel	0.88
7	Water, free	0.59
8	Air, dead	0.024
9	Douglas fir plywood	0.12
10	Hardboard, medium density	0.10
11	Particleboard, medium density	0.13
12	Cement mortar, common brick	0.72
13	Concrete	0.93
14	Stone, lime, sand	1.80
15	Glass 1.05	1.05
16	Fiberglass	0.039
17	Silver	419
18	Cooper	396
19	Aluminum	202
20	Stainless steel	16.3

Table 3.2. Thermal diffusivity of wood as a function of moisture content and density. (Data from Kollmann and Côté 1968)

Moisture content (%)	Diffusivity (10^{-7} m^2/s) for different wood density			
	Density (200 kg/m^3)	Density (400 kg/m^3)	Density (600 kg/m^3)	Density (800 kg/m^3)
10	1.89	1.56	1.42	1.36
20	1.75	1.47	1.36	1.33
30	1.64	1.39	1.31	1.28
50	1.50	1.31	1.22	1.19
100	1.44	1.14	1.06	–

modulation propagates if a sequence of thermographic images is recorded during one or more modulation cycles of excitation. If a Fourier transform of the time-dependent signal is performed at each pixel, the whole sequence is compressed into an amplitude image and a phase angle image. The latter is more relevant since it has a larger depth range and is less dependent on surface topography and optical surface features (Busse 1979, 2001; Rosencwaig and Busse 1980; Lehto et al. 1981; Bennet and Patty 1982). This kind of combining thermography with imaging based on periodical excitation and thermal wave analysis is named "lock-in thermography" (Carlomagno and Berardi 1976; Beaudoin et al. 1985; Kuo et al. 1987; Busse et al. 1992b; Busse 2001). In previous sections, we have seen that the discontinuities in objects scanned with infrared equipment can be observed on thermographic images since a tissue discontinuity produces a difference in thermal conductivity that yields a non-homogeneous distribution of surface temperatures.

The video infrared camera converts the electromagnetic thermal energy radiated from the object under test into electronic video signals. Furthermore, the electronic signals are amplified and transmitted to a monitor that displays the image on a screen. The presence of internal cavities or damaged zones induces a surface temperature different from that of sound wood.

As noted by Catena and Catena (2000), the infrared scanning camera used for the detection of cavities in trees (Fig. 3.5) is a light, mobile piece of equipment (3 kg). Only objects larger than the instantaneous field of view can be

Fig. 3.5. Infrared device for the detection of cavities in trees. (Catena and Catena 2000, with permission)

seen by the infrared camera (Fig. 3.6). The equipment used by Catena and Catena (2000) for imaging of pathological attacks and internal cavities in trees is composed of the following subsystems:

- a set of infrared lenses that focus the electromagnetic energy radiated from the object.
- matrix of 320 × 240 microbolometers pick up the temperature at the tree surface
- an electronic device that displays the image on the screen

For the inspection of trees, it is possible to use any infrared camera with high thermal sensitivity and geometric resolution. The images obtained by Catena and Catena (2000) had a thermal sensitivity of 0.1 °C at ambient temperature and a geometric resolution of 1.3 mm at 1-m distance from the tree. The camera is mounted on a tripod and can be stationed at a variable distance of up to 20 m from the tree. The average time for the inspection of one tree is about 2 min. The equipment has the advantage of being noncontact and very easy to use without the need for additional personnel. Table 3.3 gives the characteristics of the lens for the instantaneous field of view of the infrared camera used by Catena et al. (1990) for the detection of large internal cavities of trees.

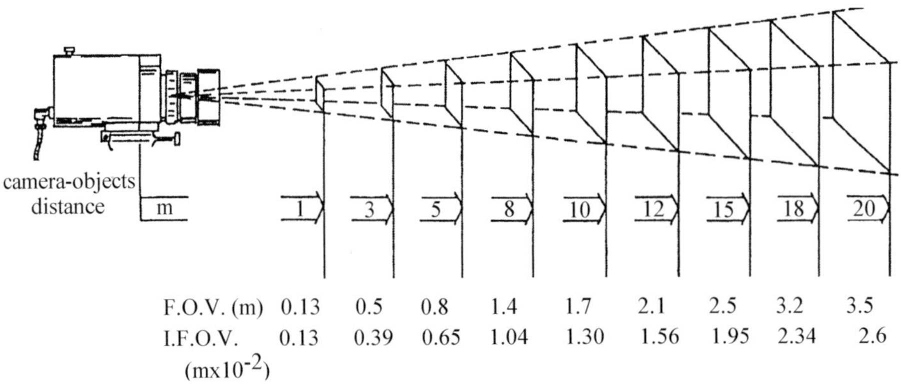

Fig. 3.6. Geometric description of camera field of view and of the instantaneous field of view of the sensing element equipped with a 10° × 10° lens. (Catena 1993a, with permission)

Table 3.3. Thermal scanner lens for the instantaneous field of view of sensing elements and the corresponding field of view of the camera. (Catena et al. 1990, with permission)

Lens	Instantaneous field of view of scanning element (cm)	The square width side of the field of view of camera (m)
10° × 10°	1.3	1.75
25° × 25°	2.5	4.45
45° × 45°	3.5	8.16

The infrared camera used by Catena (1993a,b) was used in the ambient temperature range from −15 to +55 °C. The sensitivity was 0.1 °C for +30 °C object temperature.

For the nondestructive evaluation of the mechanical behavior of wood under loading, Luong (1996) used mobile equipment (AGA 782 SW) that comprises a set of infrared lenses that focus the electromagnetic energy radiated by the object, an electro-optical mechanism that determines the field of view with a scanning rate of 25 fields/s, a set of relay optics containing a selectable aperture unit and a filter cassette unit, a photo-voltaic shortwave infrared detector that delivers a signal proportional to the radiation from the object with a spectral response between 3.5 and 5.6 μm, a liquid nitrogen Dewar chamber and a control electronics unit that produces a videosignal on the screen. Figure 3.7 shows the utilization of an infrared camera for laboratory measurements for thermographic detection of artificial defects and delaminations in plywood. The specimens were illuminated by halogen lamps, and the temperature was measured in defective and clear zones.

Transient thermography, for which the heat flow is generated by a short light pulse, is able to give more detailed information about the inspected object, by a most sophisticated analysis of the temperature differential. Lock-in thermography combines conventional thermography with modulated thermal waves. The block diagram of the experimental device for thermal wave imaging with lock-in thermography is shown in Fig. 3.8.

The sample is exposed to a sinusoidal illumination of a heat beam. A computer controls the sinusoidal illumination power produced by a conventional halogen lamp. An infrared filter is used to avoid the emission of infrared radiation. During each modulation period, successive scans are performed to obtain the image from temperature differentials from the magnitude and phase of the signals. The lock-in thermographic equipment used for the detection of delaminations in wood-based composites is shown in Fig. 3.9. The phase angle

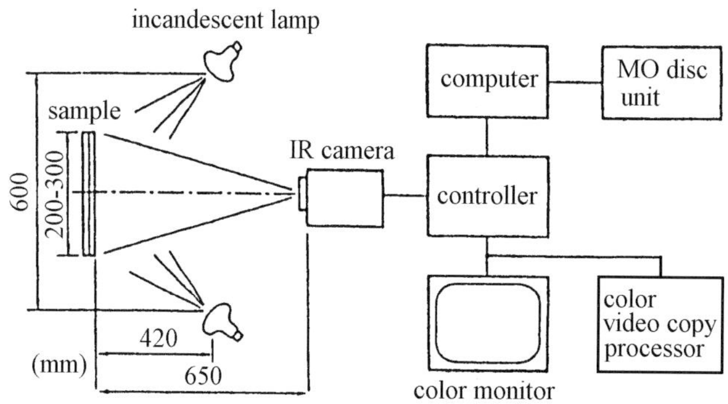

Fig. 3.7. Infrared camera for laboratory measurements, for thermographic detection of artificial defects and delaminations in plywood (Xu et al. 1994; courtesy of Forest Products Society)

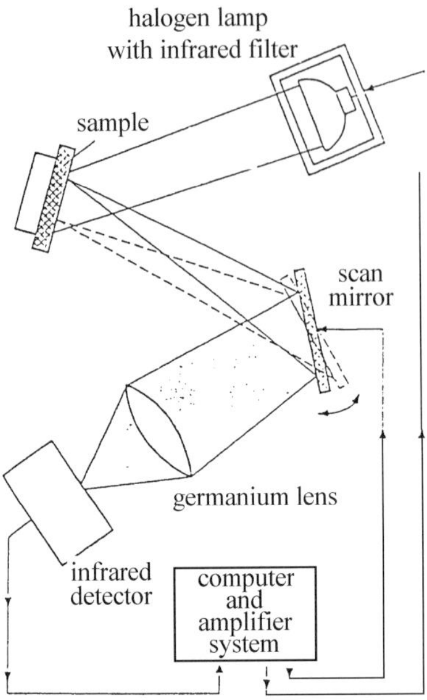

Fig. 3.8. Block diagram of the experimental device for thermal wave imaging with lock-in thermography. (Wu et al. 1992, with permission)

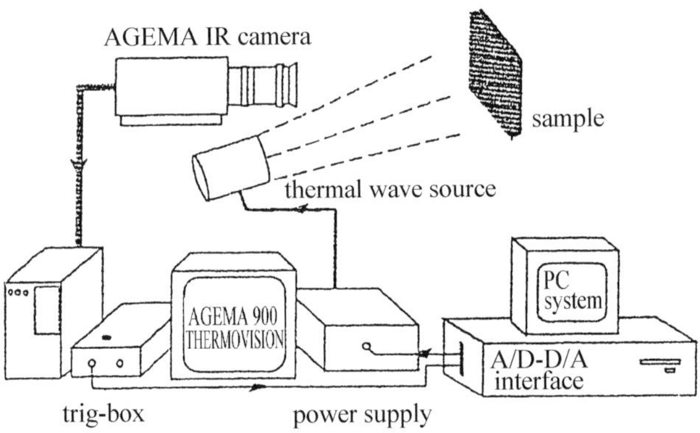

Fig. 3.9. Equipment used for the detection of delamination in wood composites samples. (Wu 1994, with permission)

images were taken with 0.03, 0.06 and 0.12 Hz modulation frequencies. Three minutes are necessary to obtain an image.

Lastly, we wish to cite one of the most interesting techniques, which is the lock-in vibro-thermal method (Busse et al. 1992a), for which the thermal waves are generated by a periodical stress inside the sample and not only at the surface. The phase of thermal waves with respect to the mechanical excitation provides information about the integrity of the samples. Figure 3.10 shows the experimental set-up used by Rantala et al. (1996) for polymer materials.

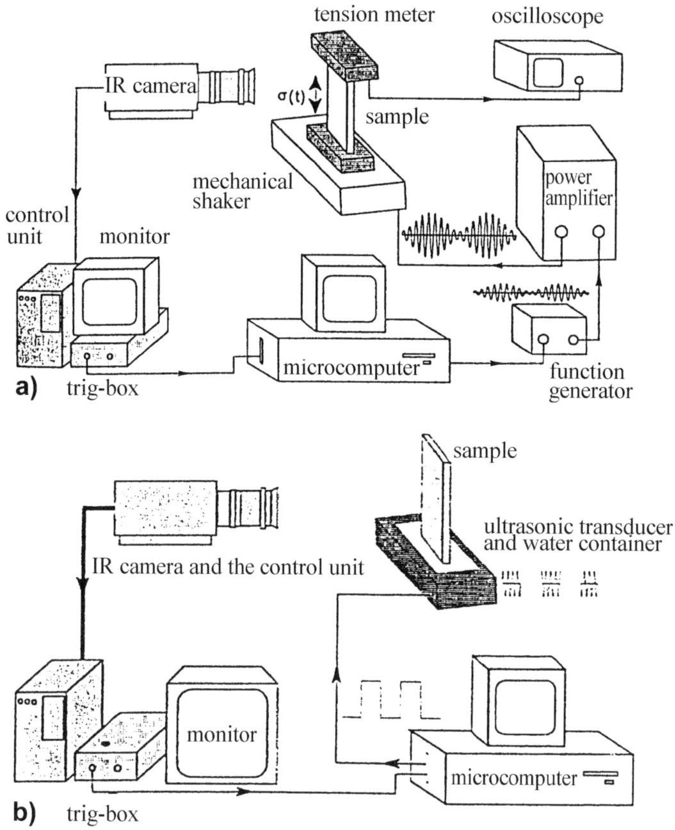

Fig. 3.10a,b. Lock-in vibro-thermography setup. **a** Amplitude modulated frequency in tension. **b** Modulated ultrasonic excitation. (Rantala et al. 1996, with permission)

The amplitude modulation can be obtained with a mechanical shaker or with ultrasonic excitation. To our knowledge, this method has not yet been used in wood science, but its high efficiency for imaging of polymers encouraged us to present some details at the end of this chapter with the hope that one day it can be a source of inspiration for further research. Ultrasound lock-in thermography is a very interesting research field as was noted by Hennecke et al. (1979) and Salerno et al. (1997, 1998). Ultrasound burst phase thermography combines the advantages of lock-in and pulse thermography as pointed out by Dillenz et al. (2000b).

3.4 Applications

The field of application of thermal wave nondestructive testing in forestry science is very large, ranging from forest fire surveillance with live infrared imaging superimposed on visible terrain view (Tice and Euskirchen 1976) to mapping forests (Catena 1991; Castagnoli et al. 1997), recording thermography on forest environments (Wickmann 1985), locating potential frost pockets in the forest (Laurence and Banner 1980), establishing the effect of acid rain on trees (Catalano et al 1986), inspection of the sanitary state of urban green zones (Catena et al. 1995) or making decisions regarding the possibility of saving trees of historical interest (Catena 1989, 1997).

Thermal infrared imaging has also been used to study the consequences of induced embolism in the path of sap flow in living trees (Anfodillo et al. 1993). The field of infrared thermography is endless (Nicolotti and Miglietta 1998). An interesting application was reported by Wisniewski et al. (1997) who studied the thermal response of plants during freezing. This technique could be extended to a study of frost crack propagation in standing trees. Knowledge of ice nucleation in wood is important in devising strategies for frost control.

In the following pages, we will consider some applications of thermal imaging related to specific problems of wood in living trees, in timber and in wood-based composites such as imaging of cavities in standing ornamental trees, imaging of the internal structure of lumber at the macroscopic level, imaging of moisture distribution in lumber, imaging of different stages of solid wood rupture phenomena, and imaging of defects in wood-based composites and in lumber joints.

3.4.1 Imaging of Defects in Trees

The assessment of the stability of weakened ornamental trees in parks and gardens is today a public safety hazard problem. We have to emphasize here Catena's pioneering activity in this field in his laboratory in Rome (Istituto Superiore di Sanita). Thermal detection of cavities and internal decayed tissue in trees has been reported in several publications since 1986 by Catena and coworkers. (Catena from 1989, 1991, 1992, 1993a,b, 1997, 2000). The validity of the thermographic method to observe temperature differentials with an infrared camera was demonstrated for softwood and hardwood trees.

Figure 3.11 shows the photographic and infrared image of an old *Acer platanoides* tree. On the photographic image, a big cavity is observed, which corresponds to an old pruned branch. The question to be answered is whether this cavity has ramifications within branches. The presence of internal cavities induces a variation of the surface temperature, which will appear on a monitoring screen as a darker or lighter area, as can be seen from Fig. 3.11. The size and the precise location of the cavities can be easily observed on the infrared image. The sound wood is represented in green. The blue color defines well the presence of cavities in trunk and branches. The cavity observed in the stem is extended into the bifurcation of the branches. The cavities are smaller in the branches than in the stem. Catena and Catena (2000) also demonstrated

Fig. 3.11a,b. Cavities in standing trees. **a** photographic image. **b** thermographic image of cavity (in blue) in trunk and in branches. (Catena and Catena 2000, with permission)

that pathological attack on the roots of trees can be observed with thermal imaging. For the nondestructive estimation of the sanitary state of trees in parks and public gardens, the authors suggested different periods for the infrared inspections, ranging from 1 year for trees having advanced deterioration to 3 years for trees with 'medium deterioration' and up to 8 years for healthy trees.

3.4.2 Timber Imaging

Thermographic images of timber have been obtained with active and passive methods. The active methods generate the required temperature field in the volume of the component under test, which acts as a distributor of the heat source caused by hysteresis or by thermoelastic effects. One of the major interests in the development of a thermal imaging technique for the lumber industry is related to the improvement of the accuracy of stress grading. Masuda and Takahashi (1999, 2000) studied the feasibility of this methodology using western hemlock. Steele et al. (1998, 2000a,b) used passive methods for detection of knots in softwoods and hardwoods. Sadoh and Murata (1993) and Murata and Sadoh (1994), Tanaka (1994) as well as Naito et al. (2000) reported the thermographic measurements of the slope of grain in lumber. In this section, both approaches for timber imaging will be discussed.

3.4.2.1 Imaging of Knots

For the detection of knots, Masuda and Takahashi (1999) reported measurements on western hemlock lumber under cycling loading in bending and compression. For bending loading, the measurements were performed on lumber with knots. The specimens had a 45 × 105 mm cross section and a length of 600 mm. The lumber was tested under cycling bending at four points, starting with 5 Hz. This frequency decreased with increasing bending load because of experimental equipment limitations. The tension and compression sides of the specimens were inspected. Two experimental configurations are discussed, the first one with a specimen including an edge knot at the tension side and the second one with a specimen including a central knot, for which the variation of the temperature distribution as a function of loading are shown. For the bending loading, a temperature rise was observed from 1000 N loading, corresponding to 120 N/cm^2. The ultimate failure load was observed at 2973 N, corresponding to 360 N/cm^2. Thermographic detection of defects was started with one third of ultimate strength. Improvement of defect detection can be obtained with an increase in repetition loading to more than 50 times. The hysteresis effect favored the low thermal conductivity of the defective zone. Imaging of wood structure and knots was possible under high speed cycling loading. Clearly defined annual rings were observed at 6000 N (Fig. 3.12).

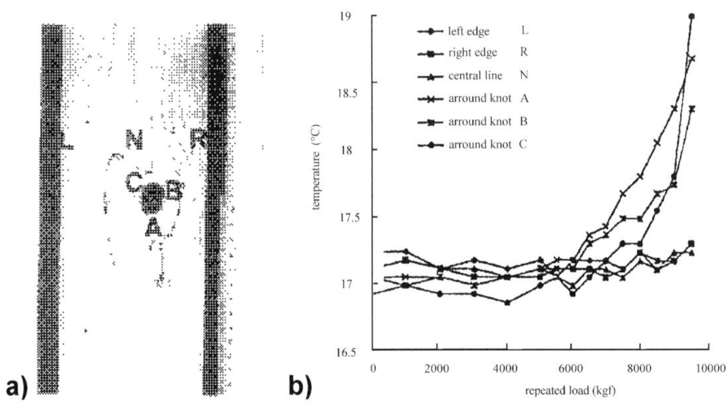

Fig. 3.12a,b. Imaging of wood structure and knots at 6000 N cyclic loading of lumber. **a** Image of annual rings around different knots at 6000 N. **b** *L* left edge knot; *R* right edge knot; *N* central line of the sample; *A*, *B*, *C* knots in the central part of the sample. (Masuda and Takahashi 1999, with permission)

The relationships between the temperature and the hysteresis curves of deformation and of energy loss by a hysteresis loop are shown in Fig. 3.13. Temperature rises in compression were much higher than that in bending. In compression loading, high speed cycling loading induced larger hysteresis heat than did bending, but only small differences were observed between static and dynamic compression.

For improvement of the accuracy of stress grading of lumber it was suggested that measurements of the static modulus of elasticity should be

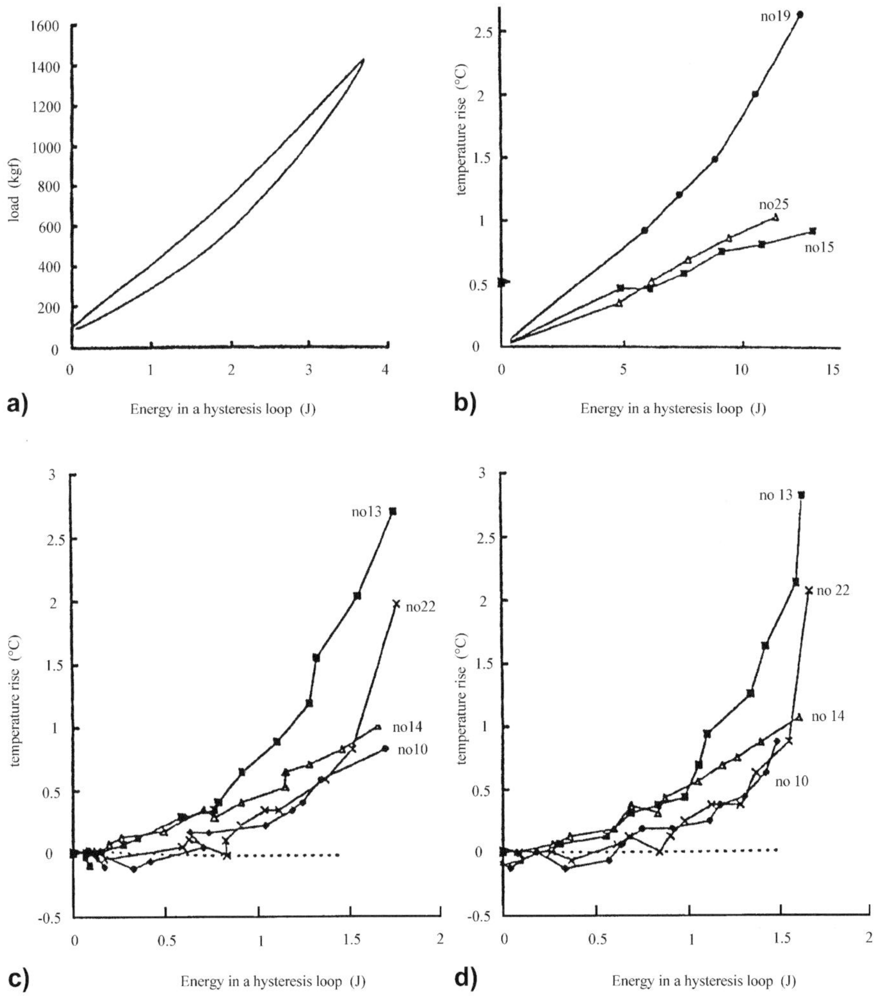

Fig. 3.13a–d. Relationship between temperature rise and energy loss with hysteresis. **a** Hysteresis loop on load-deflection curve. **b** Static bending. **c** Static compression. **d** Dynamic compression. The number of the specimen is referred to on each curve. (Masuda and Takahashi 1999, with permission)

associated with thermal imaging of the temperature rise induced by the energy loss in the hysteresis loop during cycling compression loading. Passive methods for detecting the differential thermal response of knots and of clear wood were reported by Steele et al. (2000a). The thermographic images were produced by two methods: by heating with a radio frequency heating system during 10 s at 35 kHz, and by heating with infrared quartz lamps of 6000 W at in-line lumber production speed. Temperature differences and differences between several physical parameters of wood are shown in Tables 3.4 and 3.5. For softwoods, the difference between the physical parameters of knots and that of clear wood was more important than in hardwoods.

It was concluded that for softwoods the temperature difference of 2.07 °C between knots and clear wood was significantly large enough to allow the detection of knots with thermographic imaging methods. For hardwoods, the

Table 3.4. Temperature differences between clear wood and knots in lumber as a function of species and of the heating system. (Data from Steele et al. 2000a, with permission)

Species	Heating Radio frequency system Temperature difference (°C)	System Infrared system Temperature difference (°C)
Softwoods		
Loblolly pine	3.29	0.63
Eastern white pine	2.10	2.33
White spruce	0.80	1.19
Ponderosa pine	2.66	1.22
Douglas fir	2.07	2.35
Eastern red cedar	1.49	0.65
Hardwoods		
Red oak	0.40	0.75
Yellow poplar	0.10	0.40
Black walnut	0.19	0.12
Basswood	0.12	0.77
Black cherry	0.56	0.66
White ash	0.49	0.95

Table 3.5. Difference between the mean values of physical parameters of clear wood and of knots in different softwoods and hardwoods. (Data from Steele et al. 1998, 2000a, with permission)

Parameters	Softwoods	Hardwoods	Ratio
Specific gravity (kgm^3)	400	180	2.22
Moisture content (%)	1.12	0.51	2.24
Extractive conrtent (%)	12.11	0.58	20.87
Fibril angle (°)	14.11	6.22	2.27
Temperature with radio-frequency system	2.04	0.31	6.58
Temperature with infrared system	1.40	0.61	2.29

difference is very small and only 0.31 °C, which makes the differentiation between clear wood and knots quite difficult. Rapid heating with infrared lamps is less efficient than radio frequency heating for 10 s. The on-line validity of the infrared heating system is strongly dependent on the technological development of new and more sensitive infrared cameras.

The experimental data reported by Steele et al. (2000a,b) were in agreement with data obtained by Murata and Sadoh (1994) on heat absorption by knots of hinoki *(Chamaecyparis obtusa)* and sugi (*Cryptomeria japopnica*) lumber. The rise in temperature of the lumber surface irradiated with incandescent lamps indicated an increasing heat absorbency by the knots of about 1.6 times that of the clear surrounding wood.

Quin et al. (1998) employed an infrared camera to determine the influence of the radiant heating periods and the subsequent cooling times on temperature differences between knots and clear wood in southern yellow pines. The study demonstrated that knots can be detected regardless of their diameter in heated lumber, depending upon the length of heating period and of elapsed cooling time.

For in situ detection of knots, Tanaka and Divos (2000) proposed measurements of temperature differences between knots, clear wood and decayed wood in sugi samples of 10 × 10 × 80 mm, using the daily natural temperature variations ranging between 26.3 and 35.7 °C. Figure 3.14 shows the temperature differences between knots and decayed areas and the temperature of the air during daytime. The air temperature increases from 5 a.m. to 4 p.m. The evolution of temperature differences for knots is opposite to that for decayed areas. A good image resolution was achieved with a temperature difference of 0.7 °C, which implies that the method can be successfully used for the detection of defects in situ. The detection of knots on wood boards with ultrasound thermography was also studied. The amplitude image was obtained with

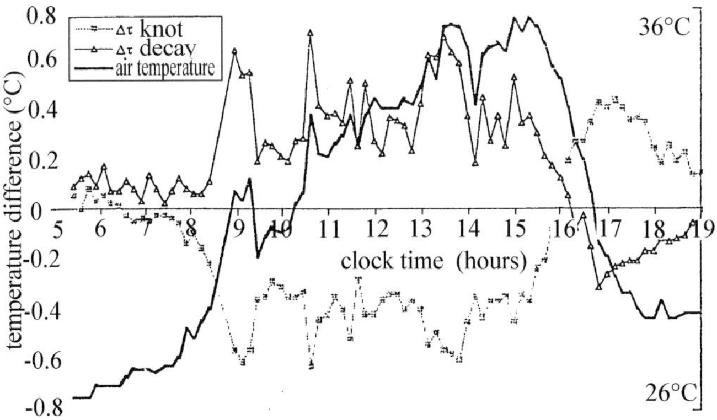

Fig. 3.14. Detection of knots and surrounding area. Temperature difference of knot and decay areas compared with the temperature of the air during daytime. (Tanaka and Divos 2000, with permission)

ultrasound lock-in thermography. This method works, but unfortunately the input coupling of high power ultrasound is very critical. The damage threshold can also be exceeded in wood.

3.4.2.2 Imaging of the Slope of Grain

Modulation of the temperature is affected by the anisotropic nature of wood, which creates an elliptical pattern of temperature rise around the heat source. The spatial dependence of the temperature modulation is affected by the thermal structure of the sample. Consequently, the determination of the slope of grain with a thermal method is an inverse problem approach.

Thermal conductivity measured by the laser flash method as a function of the grain angle is shown in Fig. 3.15. The relationship between the thermal conductivity and the grain angle deduced by Murata and Sadoh (1994) is given for hinoki and sugi in Eqs. (3.10) and (3.11).

$$\lambda = 0.154\ \rho\ \sin^2\theta + 0.476\ \rho\ \cos^2\theta + 0.025 \tag{3.10}$$

$$\lambda = 0.146\ \rho\ \sin^2\theta + 0.471\ \rho\ \cos^2\theta + 0.025 \tag{3.11}$$

where λ is the thermal conductivity (kcal/mh°C), ρ is the density (g/cm^3) and θ is the grain angle (degree).

The correlation between the slope of grain estimated by thermography and by visual inspection for three different species, sugi, western hemlock and buna, is significant and is between 0.71 and 0.87. Bernard et al. (1996) measured the infrared birefringence and determined the slope of grain

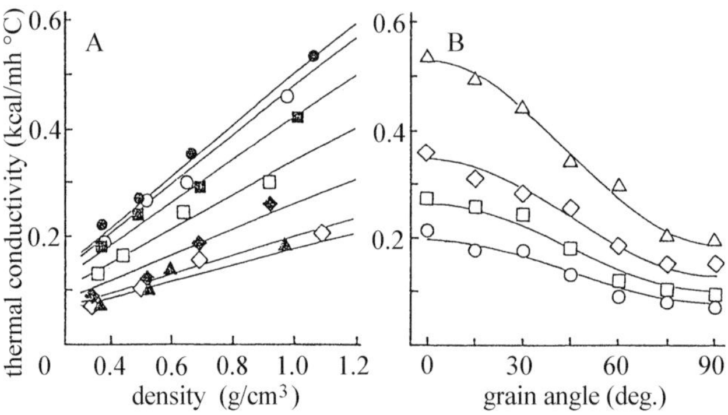

Fig. 3.15A,B. Relationship between thermal conductivity, density and grain angle. **A** Thermal conductivity and density at constant grain angle. *Filled circles* The angle of heat flow is 0°; *open circles* 15°; *filled squares* 30°; *open squares* 45°; *filled diamonds* 60°; *open diamonds* 75°; *filled triangles* 90°. **B** Thermal conductivity and grain angle at constant density, symbolized as wood with density ranging between 360 and 680 kg/m^3. *Open circles* The density for 360 g/m^3; *open squares* for 500 kg/m^3; *open diamonds* 680 kg/m^3: *open triangeles* also for 680 kg/m^3. (Murata and Sadoh 1994; courtesy of the Japan Wood Research Society)

and noted that this technique is insensitive to wood moisture content and temperature.

Naito et al. (2000) verified the feasibility of thermographic imaging, using the finite element method to demonstrate the transient temperature on the wood surface. The experimental device is shown in Fig. 3.16. The finite element is seen in Fig 3.17. The variation of the temperature distribution with heating time, rising from 10 to 60 s is shown in Fig. 3.18. These results were compared with thermographic images. Good agreement was observed between finite

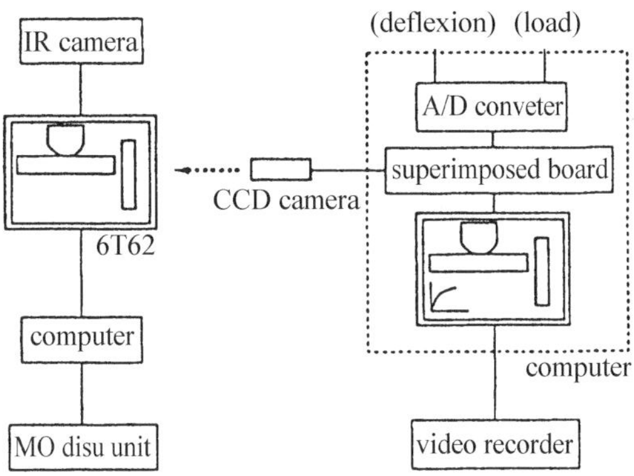

Fig. 3.16. Experimental device. (Naito et al. 1998, with permission)

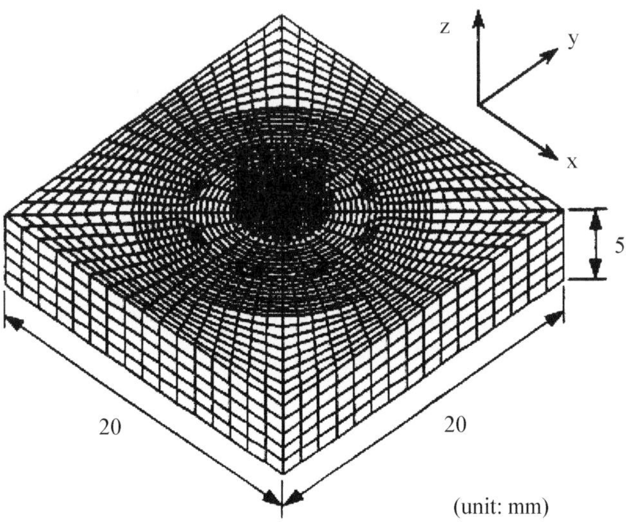

Fig. 3.17. Thermographic measurements of the slope of grain. Finite element model. The heat flux is represented by the *vertical arrows* in the central region. (Naito et al. 2000; courtesy of the Japan Wood Research Society).

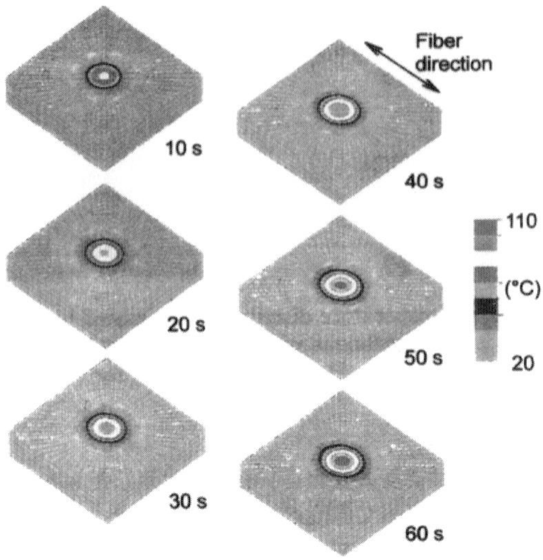

Fig. 3.18. Temperature distribution as a function of heating and slope of grain deduced with the finite element method time obtained with the finite element method. (Naito et al. 2000; courtesy of the Japan Wood Research Society)

elements simulation and the real experimental images. This allows the conclusion that the slope of grain can be measured with an infrared camera.

3.4.2.3 Imaging of Wood Density

Density variation mapping of the annual ring with high resolution (100 μm) can be achieved with the far-infrared spectroscopy method in the range of T Hz. The experimental device as described by Koch et al. (1998) is presented in Fig. 3.19. Density imaging is obtained with coherent far-infrared pulses which are generated and detected by small photoconductive dipole antennae gated by ultra-short laser pulses in the near-infrared. The pulses are "guided by off-axis parabolic mirrors to form an intermediate focus of sub-millimeter diameter through which the wood samples are scanned in a raster pattern. At each sample point a complete transmitted waveform is acquired and subsequently Fourier transformed. The transmitted intensity integrated over a given window is translated into a gray value to finally obtain a pixel. The data acquisition rate is 20–50 Hz and the time required to generate a two-dimensional image is typically 10 min, depending on sample size and resolution".

Figure 3.20 shows the density mapping of beech and balsa obtained by the transmission method through a 1.7-mm-thick microscopic section. In beech, the dark blue and red areas are zones of low transmission in latewood and in medullary rays, ranging between 700 and 1400 kg/m^3. It is evident that the large

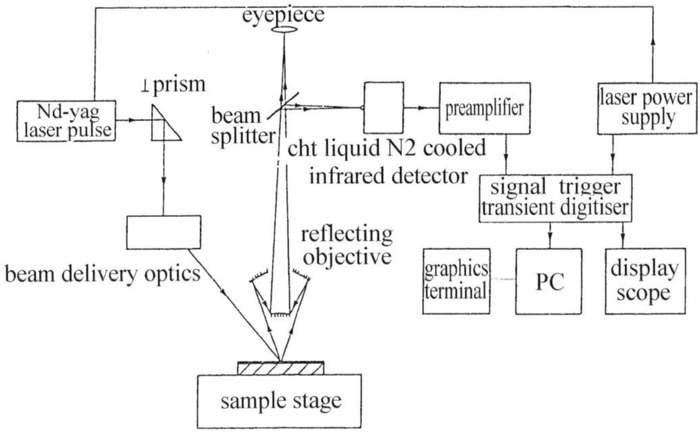

Fig. 3.19. Experimental device for T Hz imaging of density variations in the annual ring. (Koch et al. 1998, courtesy of Wood Science and Technology)

rays have a higher density than the latewood. This is in agreement with data obtained by Keller and Thiercelin (1975) with an X-ray technique. In earlywood containing vessels of about 250-µm diameter, the density is much lower than in latewood, ranging between 500 and 600 kg/m^3.

It is well known that in contrast to beech, in balsa only the vessels and fibers can be identified. In Fig. 3.20, the vessels are very well identified as large yellow zones. The large dominant red zones are produced by the presence of fibers with a density of about 470 kg/m^3, while the radially oriented blue zones correspond to the parenchyma ray cells with a relatively low density of about 300 kg/m^3 and where starch is a major component.

Concluding this study, it can be said that the false color image of density variation in transverse sections of wood gives us a better understanding of the anatomical structure of wood.

Since the aim of our analysis is to show the way in which thermal techniques can illuminate wood structure, it is important to note the pioneering work of by Moksin (1993), who measured the decay time of opto-thermal signals observed across a pine sample scanned with a pulsed laser source with 100-µm step size. The opto-thermal decay time τ was calculated from Eq. (3.12)

$$\tau = \alpha^{-2} D^{-1} \tag{3.12}$$

where α is the absorption coefficient and D is the thermal diffusivity of the sample.

Figure 3.21 shows the decay curve profile of growth rings pine scanned with a step size of 100 µm, which is consistent with the earlywood and latewood zones of the annual ring. The minima represent the thinner and dark zones of latewood, which are more efficient in heat dissipation than the clear earlywood

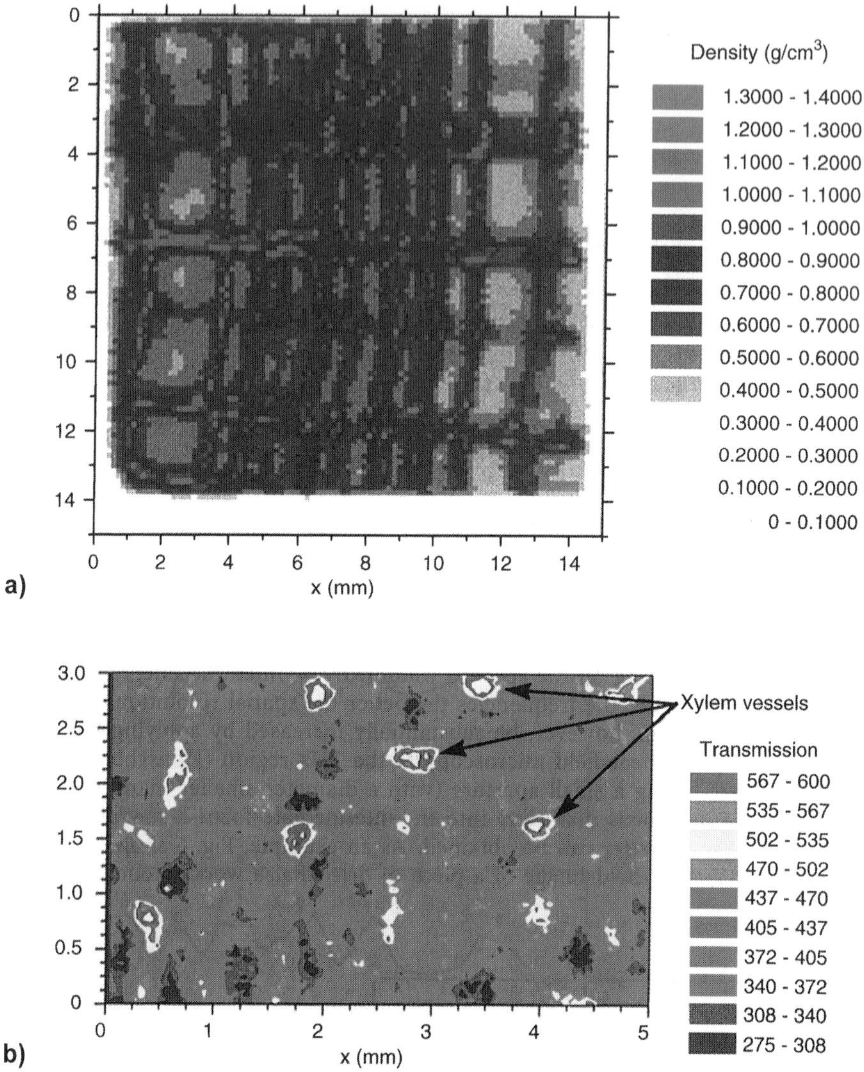

Fig. 3.20. Density mapping of a transverse section of two species: **a** beech and **b** balsa. The size of the specimen was 14 × 14 × 1.7 mm. (Koch et al. 1998; courtesy of Wood Science Technology)

areas. The distance between two minima corresponds to an annual ring width of 4.4 mm.

For the future, we suggest that this technique be utilized for the measurement of the density profile of wood. This opto-thermal method can be validated with the X-ray densitometric method.

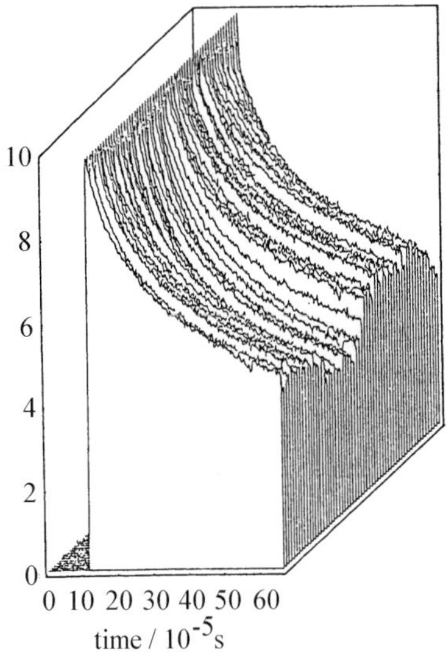

Fig. 3.21. Decay curves of an opto-thermal signal in pine. (Moksin 1993, courtesy of Wood Science Technology)

3.4.2.4 Imaging of Moisture Content Distribution

Indirect evaluation of moisture content in wood and its distribution can be performed with photothermal techniques as demonstrated by various authors (Cielo et al. 1988; Bernatowicz 1994). The photothermal parameters related to wood moisture content are: thermal conductivity, K, thermal diffusivity, α, thermal effusivity *e*, wood specific heat, C, wood density, ρ, and surface losses. The specific heat for wood is between 0.25 and 0.35 cal/g K for wood densities ranging from 300 to 900 kg/m^3. The specific heat of wood is lower than that of water which is 1 cal/g K.

Measurements of the photothermal parameters of a wood sample can be performed on the front or back surfaces. The front surface measurements can be carried on a solid wood specimen of normal thickness, on which the surface temperature measurements need shorter observation times than for that on the back surface. The back surface measurements are convenient because they are not affected by reflected light noise. The value of the thermal diffusivity can be evaluated from the shape of the thermograms without any absolute calibration of the detector. Cielo et al. (1988) noted that surface temperature measurements are more sensitive to lumber moisture content than thermal diffusivity measurements.

In one-dimensional approximation, the surface temperature rise ΔT induced by a constant power flux Q absorbed at the front surface of a semi-infinite irradiated material during a time period τ is expressed by Eq. (3.13), if the time $t < \tau$

$$\Delta T = 2\,Q(t)^{1/2}(\pi K \rho C)^{-1/2} - P(T) \qquad (3.13)$$

and if the time $t > \tau$:

$$\Delta T = 2\,Q(\Pi\,K\,\rho\,C)^{-1/2}\left[(t)^{1/2} - (t-\tau)^{1/2}\right] - P(T) \tag{3.14}$$

where P(T) is the temperature-dependent heat losses due to air convection, radiation and phase changes.

The thermal diffusivity of the material, in this case of spruce submitted to laser irradiation for 5 s, can be calculated from:

$$\alpha = K/\rho\,C = 0.139\,d^2/t^{-1/2} \tag{3.15}$$

where d is the sample thickness and *t* is the half-rise time.

The thermal effusivity *e* can be calculated from Eq. (3.16) if the absorbed power Q and the heat losses are known:

$$e = (K\,\rho C)^{1/2} \tag{3.16}$$

The anisotropic character of solid wood was demonstrated also for its thermal properties as can be seen from Table 3.6 and Fig. 3.22. The data in Table 3.6 were obtained with a flash thermometry analysis, the front of the 3.4-mm veneer sample was heated for 1 s, and the back surface temperature was recorded with an infrared detector. The thermal diffusivity in the longitudinal direction is twice that in the radial or tangential directions. The isotherms are quite circular in the radial tangential plane. In the longitudinal-tangential plane the fast axis is in the fiber direction.

The effect of the moisture content in the range 0–160%, on ΔT and α deduced with Eq. (3.13) are shown in Fig. 3.23 for softwood veneer, red cedar and spruce. Both of the thermal parameters studied decreased with increasing moisture content. The variation as a function of moisture content of the ratio between the thermal characteristics (conductivity, effusivity and temperature rise) at a given moisture content and under oven-dry conditions is given in Fig. 3.24. An increase in all characteristics was observed with the increase in moisture content. The contribution of these three variables to thermal phenomena is of the same order of magnitude. Above 30–40% moisture content, the increase in the temperature rise ratio is much more important than the conductivity and effusivity ratios. Referring to Eq. (3.13), it can be noted that at low moisture contents surface losses have a relatively small effect on the temperature rise. At high moisture contents, corresponding to the presence of free

Table 3.6. Thermal diffusivities in red cedar at 30 percent moisture content as a function of wood anisotropic direction. (Cielo et al. 1988, with permission)

Thermal propagation direction	Thermal diffusivity (mm^2/s)
Longitudinal	0.41
Radial	0.15
Tangential	0.16

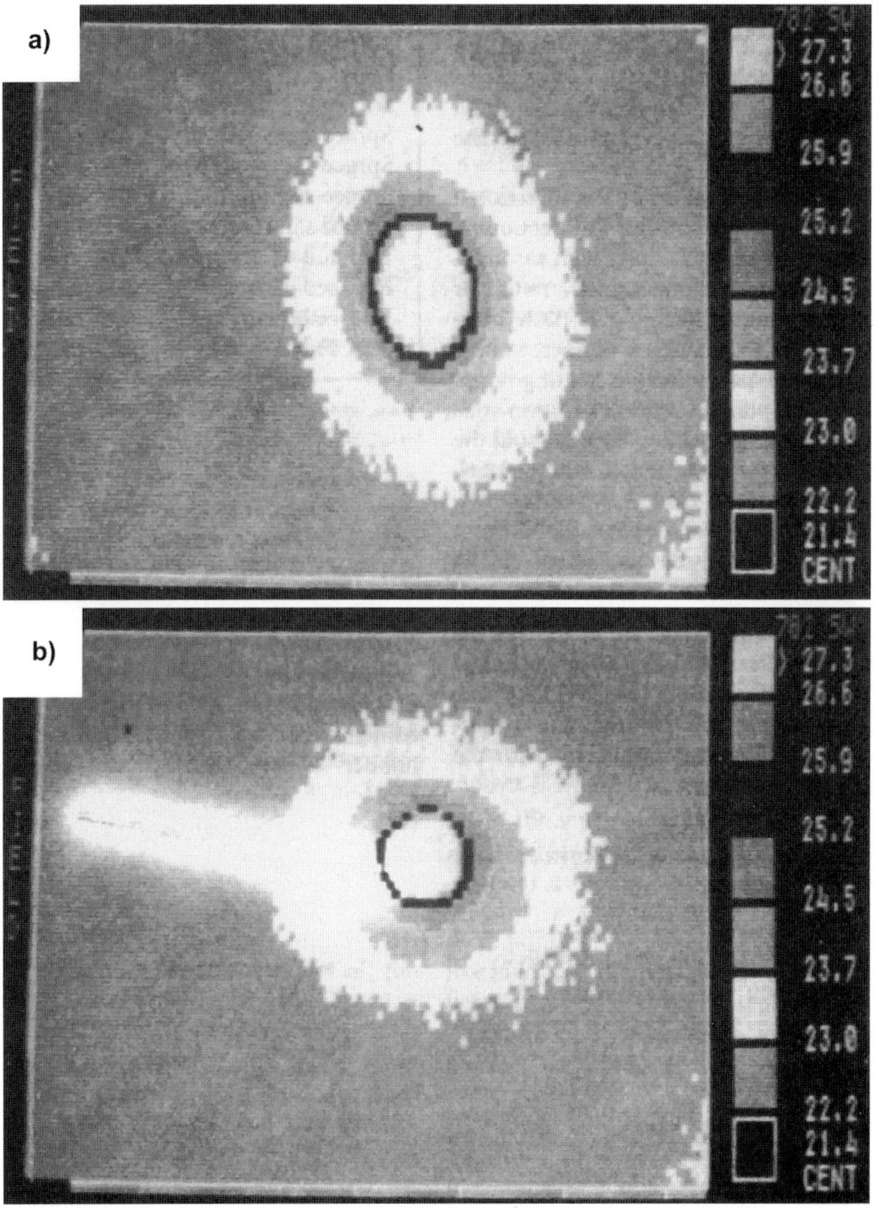

Fig. 3.22a,b. Thermal anisotropy of a 3.4-mm softwood veneer sheet at 40% moisture content in **a** radial tangential plane; **b** longitudinal tangential plane. The temperature distribution is displayed over a 2×3 cm area. (Cielo et al. 1988, with permission)

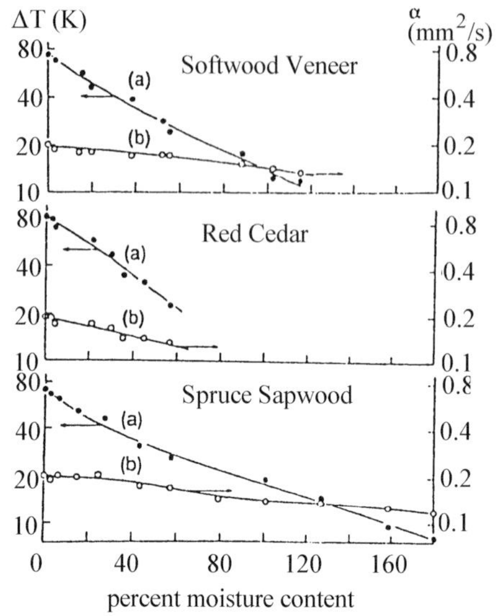

Fig. 3.23. Thermal diffusivity (α) and temperature differential (ΔT) as a function of moisture content on the front surface of the specimen, on which the temperature increases after 5 W/cm^2 laser irradiation for 4 s. (Cielo et al. 1988, with permission)

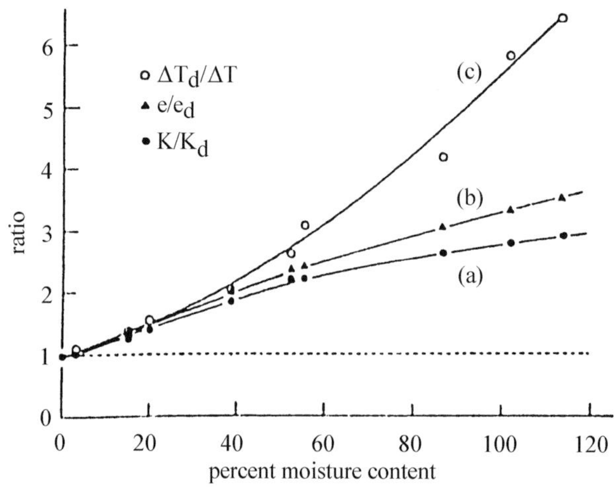

Fig. 3.24. Relationships between moisture content and the ratios of thermal conductivity; *a Filled circles* Thermal effusivity: *b filled triangles* front surface temperature variation; and *c open circles* under laser irradiation for a softwood veneer sample. (Cielo et al. 1988, with permission)

water, the effect of the rise in temperature is predominant, probably because of the losses due to surface evaporation.

Figure 3.25 shows a thermal image of a veneer sheet with zones of different moisture contents, which was artificially moistened on the left side and was heated at 10 K above ambient and finally allowed to cool down by vaporization. The knots dry faster than the surrounding clear wood because

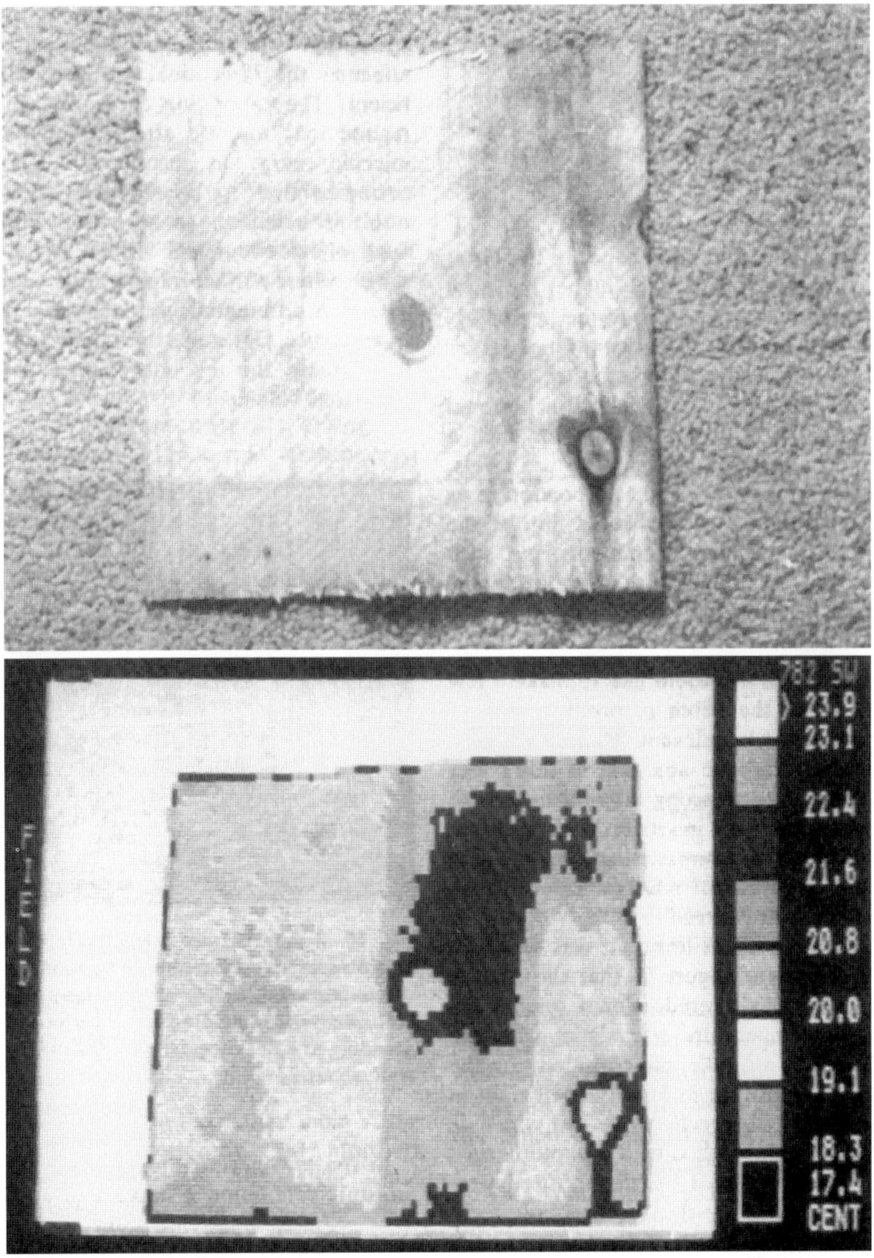

Fig. 3.25. Thermal image of a veneer sheet of 30 × 30 × 0.4 cm with zones of different temperatures and moisture content. (Cielo et al. 1988, with permission)

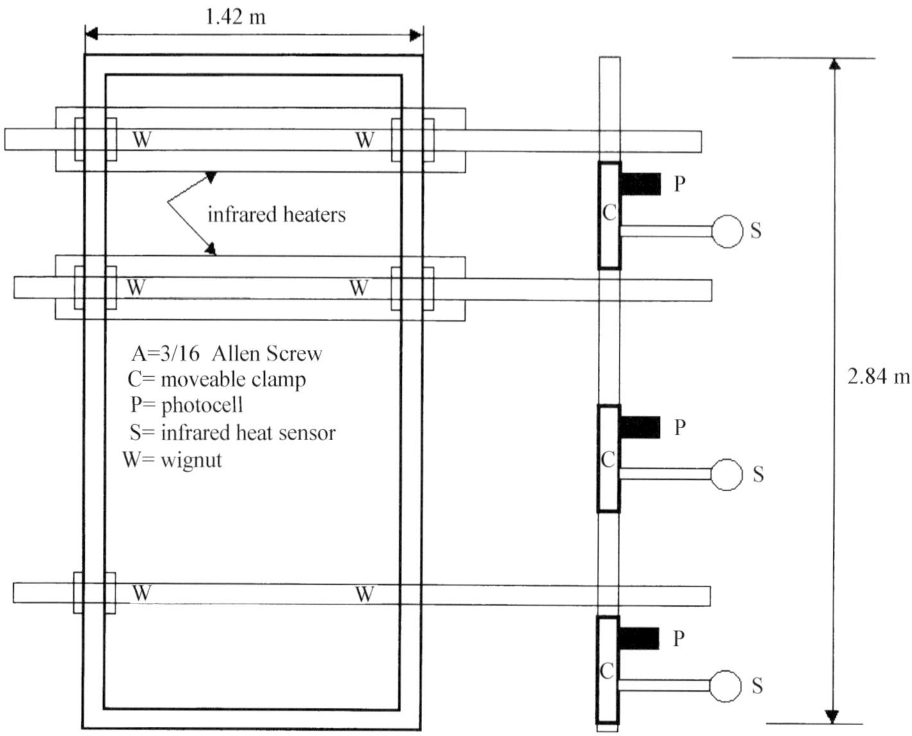

Fig. 3.26. Infrared measuring device for green veneer and lumber *S* Infrared heat sensor; *P* photocell; *W* wingnut. (Troughton and Clarke 1987; courtesy of Forest Products J)

of their particular structure. The moisture content of such a specimen containing various defects exhibits important local variations. The moisture content of the whole surface can be expressed by an average value with a wide distribution.

For in-line measurements of the moisture content in unseasoned veneer and lumber Troughton and Clarke (1987) used the system shown in Fig. 3.26. The device designed for a mill trial is equipped with two 5000-W heaters and three infrared sensors connected with a computer and used for unseasoned 60-mm lumber with a speed of 13 m/min. The relationships between the moisture content and the surface temperature for veneer and lumber are shown in Fig. 3.27.

3.4.2.5 Imaging of Wood Rupture Phenomena

Thermal wave generation in wood under stress has opened new directions for studies of wood rupture phenomena. The propagation of thermal waves in specimens under stress can be monitored by a noncontact infrared technique,

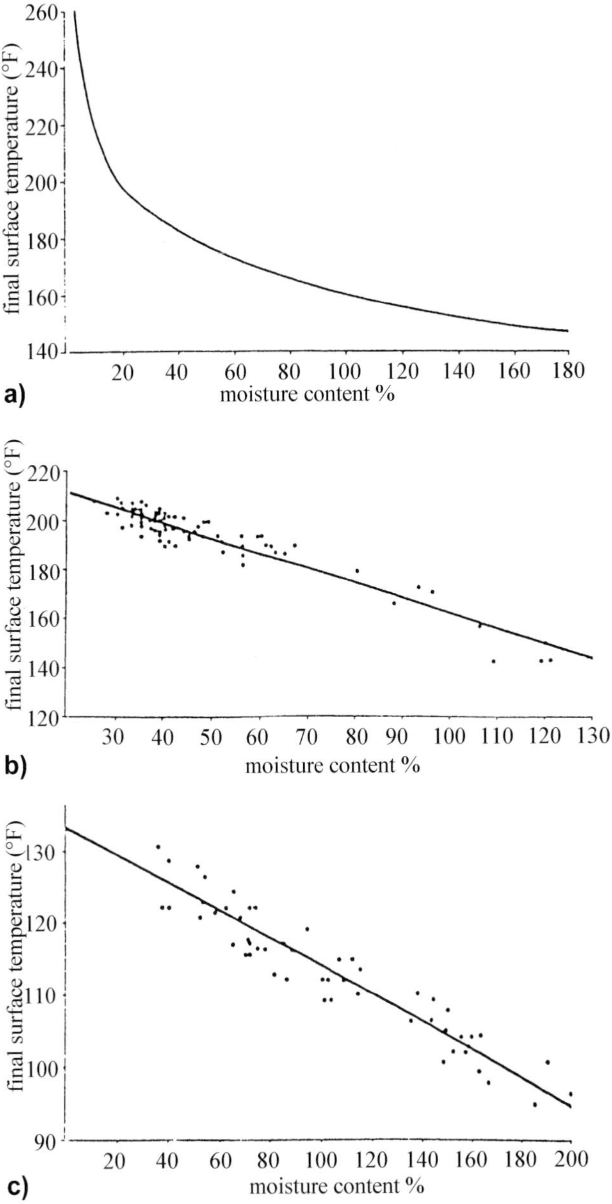

Fig. 3.27a–c. Relationships between moisture content and surface temperature. **a** Relationship between veneer surface temperature and moisture content. **b** Relationship between final surface temperature and moisture content (%) for unseasoned 2 × 6-in. lumber. **c** Relationship between final surface temperature and moisture content (%) for 1/8-in. spruce veneer. (Troughton and Clarke 1987; courtesy of Forest Products J)

as demonstrated by Luong (1995, 1996), Okumura et al (1996) and Naito et al. (1998).

Infrared imaging on pine specimens under an unconfined compression test is shown in Fig. 3.28. The effect of anisotropy was demonstrated using three different specimens having the major axes in longitudinal (L), radial (R) and tangential (T) directions. Analysis of the thermal images was performed "by discriminating the intrinsic dissipation from thermal noise by subtracting the thermal image at a reference time from the thermal image taken at a given applied stress level." It was observed that for each anisotropic direction the rupture mechanism is different, and typical modes of intrinsic dissipation localization describe in colors the damage process.

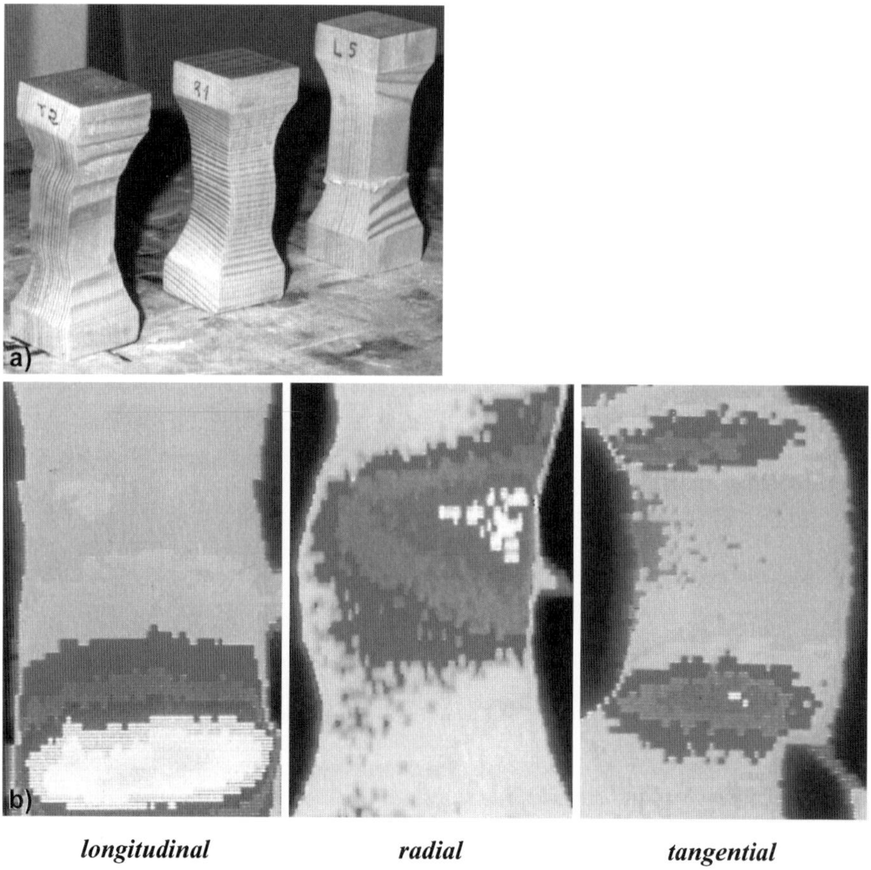

Fig. 3.28a,b. Infrared imaging of pine specimens under compression. **a** Specimens with major axis in longitudinal, radial and tangential directions. **b** Localization of intrinsic dissipation in wood specimens as a function of anisotropy (each color hue corresponds to 0.2 °C) in longitudinal, radial and tangential directions. (Luong 1996, with permission)

Okumura et al. (1996) studied the behavior of three tree species under static unconfined compression. The profile information of the temperature distribution was related to strain measurements obtained on an array of 66 points (Fig. 3.29). With specimens under longitudinal compression, a rapid increase in temperature from 19.6 to 22 °C along the fracture plane was observed (Fig. 3.30 at point d). The temperature increase in specimens under radial and tangential compression (Fig. 3.31) was between 20 and 21.2 °C. In both cases, the highest temperature corresponds to the fracture line. As can be seen, the most important increase in the temperature was observed in specimens under compression in the longitudinal direction.

Narito et al. (1998) studied the temperature distribution in yellow cedar specimens under static bending test in four configurations. In this section, we present the thermographic images obtained for specimens under three-point bending (Fig. 3.32), with a 5-mm-diameter hole drilled in the compressed zone of the specimen. At a very low loading level, the temperature increased 0.6 °C under the loading point and the presence on the hole is not observable. Increase in the loading resulted in a better image of the hole. The rise in local temperature around the hole is similar to that in the fractured zone of the specimen.

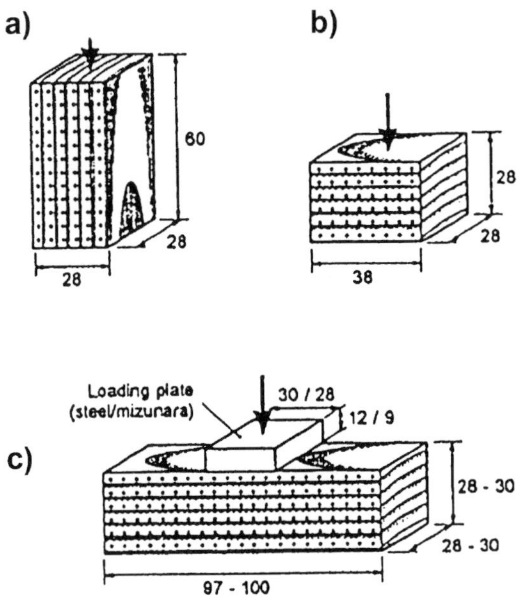

Fig. 3.29a–c. Specimens for testing under compression loading. **a** Longitudinal loading on hinoki specimen; **b** radial loading on hinoki specimen; **c** loading on hinoki, sugi and white seraya composite specimens. Note that for strain measurements strain gauges were used as a lattice represented by the lattice dots. (Okumura et al. 1996; courtesy of Wood Research Institute, Kyoto University)

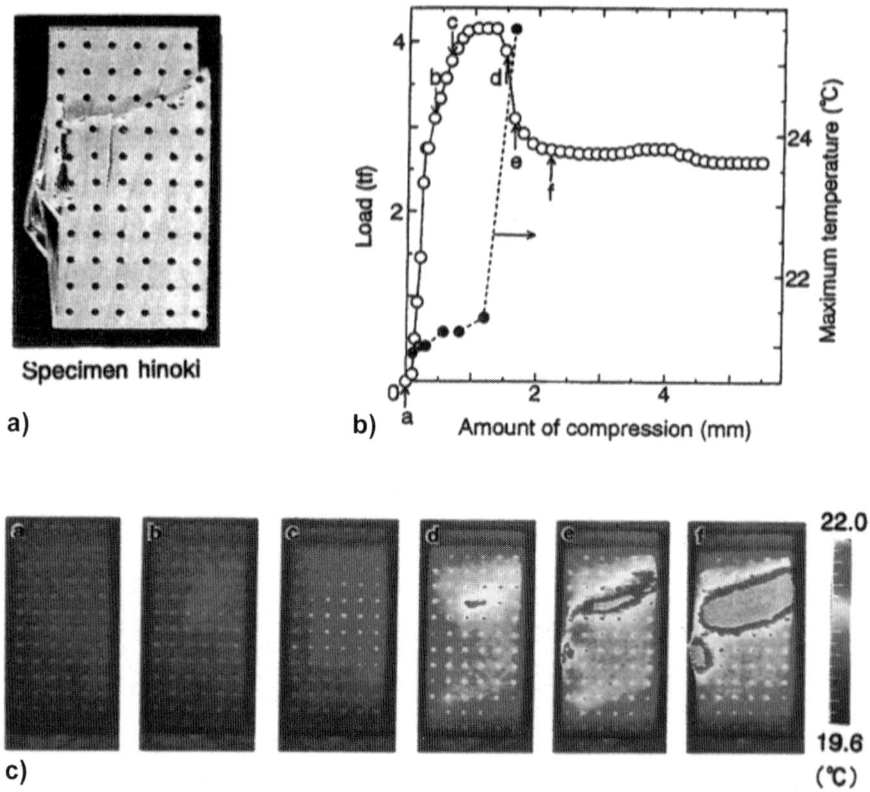

Fig. 3.30a–c. Infrared imaging of specimens under compression loading related to stress distribution in the longitudinal direction. **a** specimen; **b** load strain relationship; **c** thermographic images corresponding to the points *a, b, c, d, e, f* from **b**. (Okumura et al. 1996; courtesy of Wood Research Institute, Kyoto University)

3.4.3 Imaging of Delaminations in Wood-Based Composites

Detection of delaminations in wood-based composites with noncontact and nondestructive methods is one of the most important aims of industrial quality inspection. Thermographic methods detect the delamination areas in composites, using a heat flow that is disturbed by the presence of the defect. It also causes a variation in temperature at the surface of the inspected object. Heat can be applied in the front or to the back surface of the laminate. In the second case, the front surface temperature is observed. In this way, the heat flow is conducted from the heated back surface toward the cooler front surface. This heat conduction phenomenon is very complex and is determined by the anisotropic and heterogeneous nature of the laminate. The defect present between two laminae is a barrier to the heat flow. The heat flow, which produces an image on the front of the sample, "incorporates" the presence of the

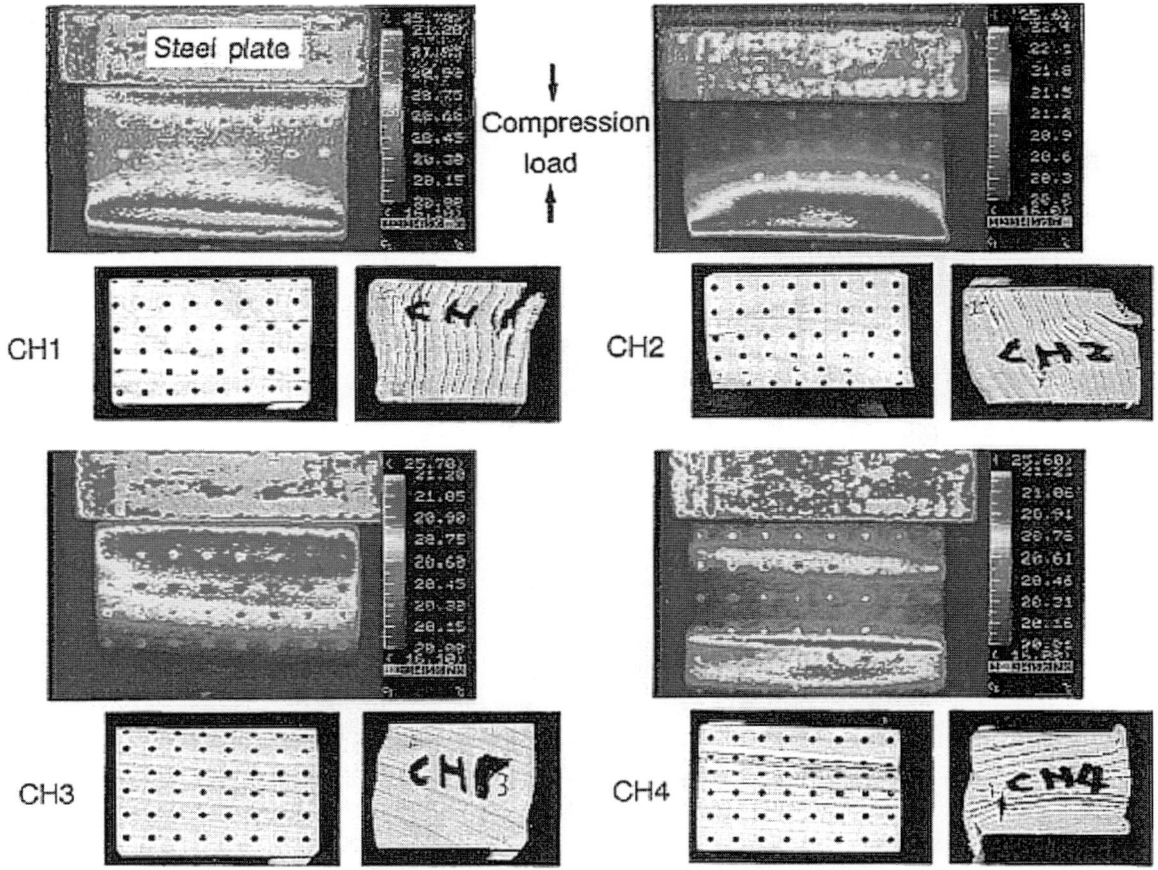

Fig. 3.31. Infrared imaging of specimens under compression loading related to the stress distribution in the transverse plane. Images of specimens at rupture under different compression loads applied in radial (*CH4*) tangential (*CH1*) and inclined directions vs. the annual rings (*CH3* inclined vs. T and *CH2* inclined vs. R). (Okumura et al. 1996; courtesy of Wood Research Institute, Kyoto University)

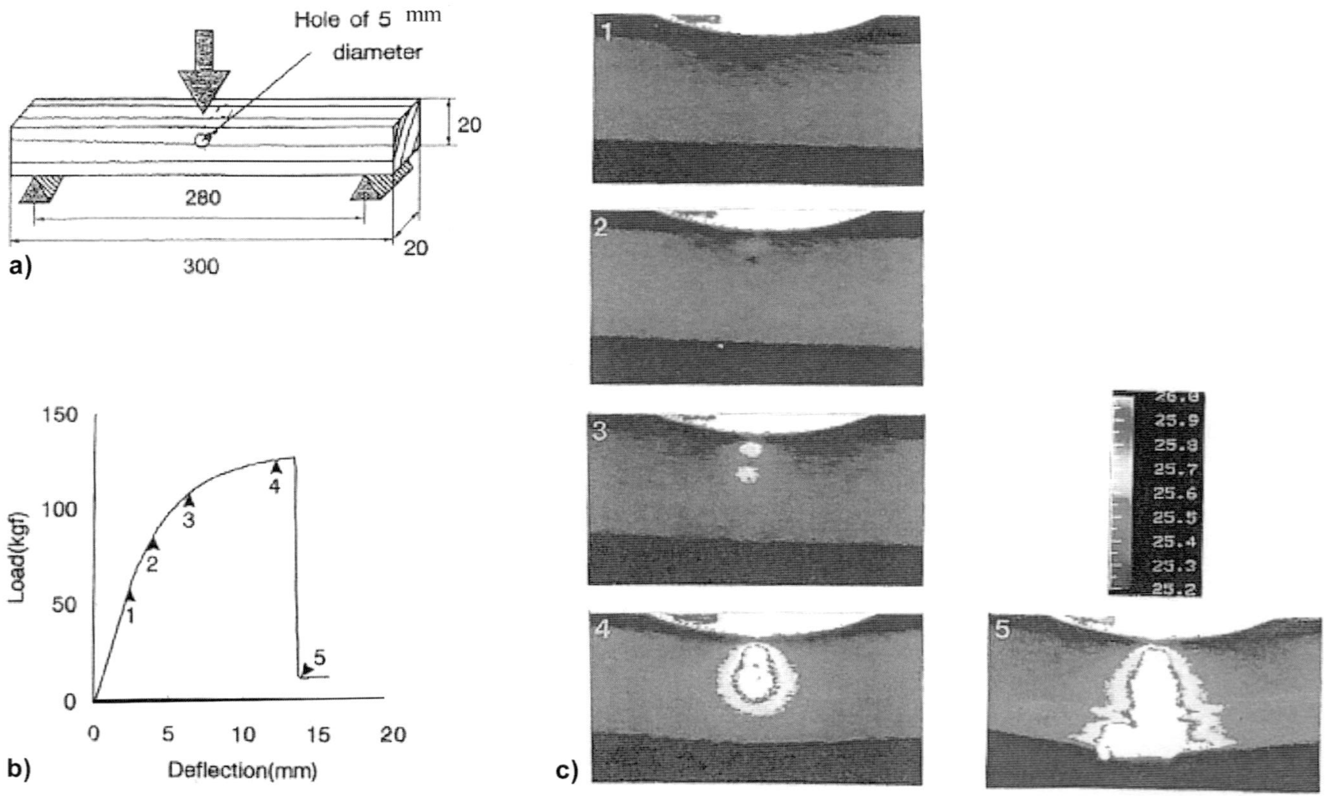

Fig. 3.32a–c. Infrared imaging of specimen under static bending loading. **a** Specimen with a central hole under static bending. **b** Stress strain curve. **c** Temperature distribution as a function of loading and strain, corresponding to points *1*, *2*, *3*, and *4*. (Naito et al.1998; courtesy of Wood Research Institute Kyoto University)

defect. The thermal gradient on the surface is one of the most important parameters to by studied, when the temperature distribution is known.

During the last 20 years, a large number of important publications have been devoted to the detection of delaminations in different materials, such as those used in the aerospace industry, the composites for a large variety of applications (Henneke and Jones 1979; Charles and Wilson 1981; Kou et al. 1987; Busse et al. 1992a; Wu 1992; Wu et al. 1992, 1997a; Bauer et al. 1992; Lebowitz et al. 1996; Dillenz et al. 2000) for ceramics (Lyamshev et al. 1995), for materials with microinclusions (Willis et al. 1997, 2001), for electromagnetic materials (Balageas et al. 1996), etc. The techniques developed for the wood industries are an extension of technological progress achieved in this field for different wood-based composite materials (Schulte et al. 1995). The purpose of this section is to describe the thermographic methods used for imaging of delaminations in a large variety of wood-based composites such as: glued laminated lumber, plywood, medium-density fiber boards and chipboards.

3.4.3.1 Detection of Adhesion Defects in Laminated Wood Composites

The methods for the detection of delaminations in glued laminated wood discussed in this section were reported by Xu et al. (1993, 1994), Masuda and Takahashi (2000) and Berglind and Dillenz (2000).

Xu et al. (1994) used an infrared camera to detect the delamination defects between two veneer laminae of different thicknesses (1.3, 2 and 3.1 mm). The surface of the artificially produced defects varied from 10×10 to $50 \times 50\,\mathrm{mm}^2$. The specimens were heated with incandescent lamps. Thermal images clearly show the area of the implanted defect surrounded by a sound zone, in which the adhesion between the two laminae is correct.

For the detection of delamination under cyclic bending, Masuda and Takahashi (2000) used an active laboratory infrared thermographic technique. The specimens were composed of two laminae to which were glued several zones of glue discontinuities ranging from 4 to 16 cm in size as can be seen in Fig. 3.33. They were tested between 63 and 72% cyclic loading of the ultimate failure load. The thermographic image for the specimen with two 16 delamination zones is given in Fig. 3.34 in which the temperature rise was observed at the glued end of each zone and at the central loading point, corresponding to tension and compression zones, respectively.

The mechanical performance of the specimens and the corresponding parameters for thermally detected defects are shown in Table 3.7. The higher failure load was observed on the specimen with two zones of 8-cm delamination, which also corresponds to the higher Young's modulus and failure stress. It was concluded that the proposed technique can be used for the detection of delamination of specimens.

For industrial applications, Berglind and Dillenz (2000) studied the feasibility of a heating-up thermography method for the detection of delamina-

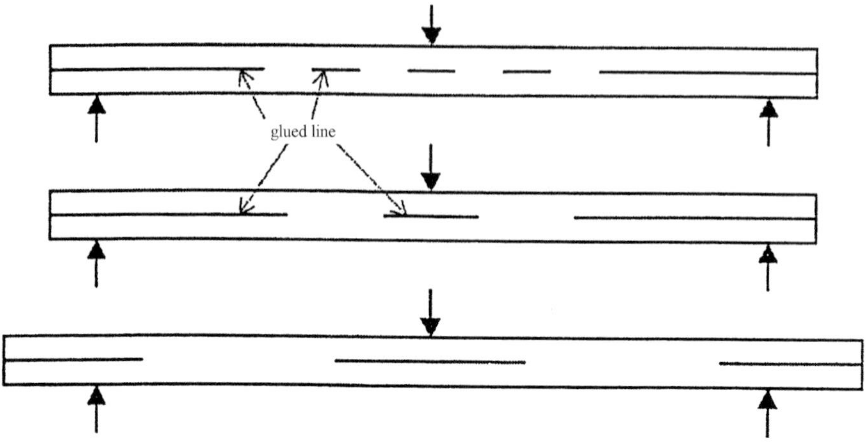

Fig. 3.33. Glued laminated specimen with different zones of discontinuities in glue layer. (Masuda and Takahashi 2000, with permission)

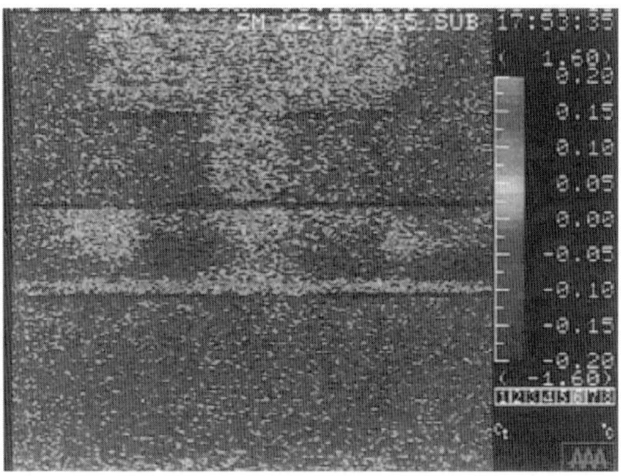

Fig. 3.34. The thermographic image for a specimen with two 16-cm delamination zones. (Masuda and Takahashi 2000, with permission)

Table 3.7. Mechanical performance of the specimens and the corresponding parameters for thermally detected defects. (Masuda and Takahashi 2000, with permission)

Mechanical parameters	Delamination length in four zones of 4 cm each	Delamination length in two zones of 8 cm each	Delamination length in two zones of 16 cm each
Detected load, P (kN)	10.3	12.1	5.9
Failure load, P_r (kN)	15.3	16.8	9.03
Ratio P/P_r (%)	67	72	63
Density (kg/m^3)	509	551	458
Bending Young's modulus, three points (GPa)	9.16	11.17	7.49
Failure stress, σ_{max} (MPa)	83.9	91.1	48.4

Fig. 3.35. The geometry of specimens used for the detection of the delaminations between wood and urea-formaldehyde glue. (Berglind and Dillenz 2000, with permission)

tions, the defect depth and in plane size of a specimen, as shown in Fig. 3.35. The heat camera operated with infrared radiation between 8 and 12 μm and had a temperature resolution of 0.08 K. The heat source was produced by two halogen lamps of 1000 W each, which were placed 1.7 m in front of the test piece.

The image contrast was calculated as the difference between the signal parameter corresponding to the defect and to the sound area, as can be seen in Fig. 3.36. The multiple correlation equation (Eq. 3.17) between the contrast, *C*, the defect width, *D*, and the defect depth, *L*, below the surface layer of the exotic wood merbau had a coefficient of determination of $R^2 = 0.90$, which means that the variability of the population not explained by this regression is only 10%.

$$C = 1.3 + 0.28\,D - 0.91\,L \qquad (3.17)$$

The relationship between the contrast and the thermally determined defect size is shown in Fig. 3.37, in which large defects result in a higher contrast compared to smaller defects, and defects below 3 mm cannot be detected.

Several factors such as inhomogeneous lighting of the specimen surface and natural variability of wood species, can influence the temperature distribution and the corresponding thermal image contrast. The effect of wood species on the detection of subsurface defects of 24-mm width and 2-mm depth is shown in Fig. 3. 37 in which four species were analyzed namely: oak, merbau, pine and alder. A higher contrast was observed for merbau, a lower for alder, depending on the degree of "darkness" of each species. "Dark" species have a high absorption that gives rise to a greater heat flow that propagates into the specimen, reflects again the defect, returns to the surface and is detected with a corresponding contrast by the infrared camera. The sources of errors that can influence the surface temperature of the specimen are: inhomogeneous lighting of the surface and the natural variability of wood, differences between earlywood and latewood, zones of vessels and fibers, the presence of knots and of reaction wood. The penetration of thermal waves was between 2 and 4 mm, depending of wood species.

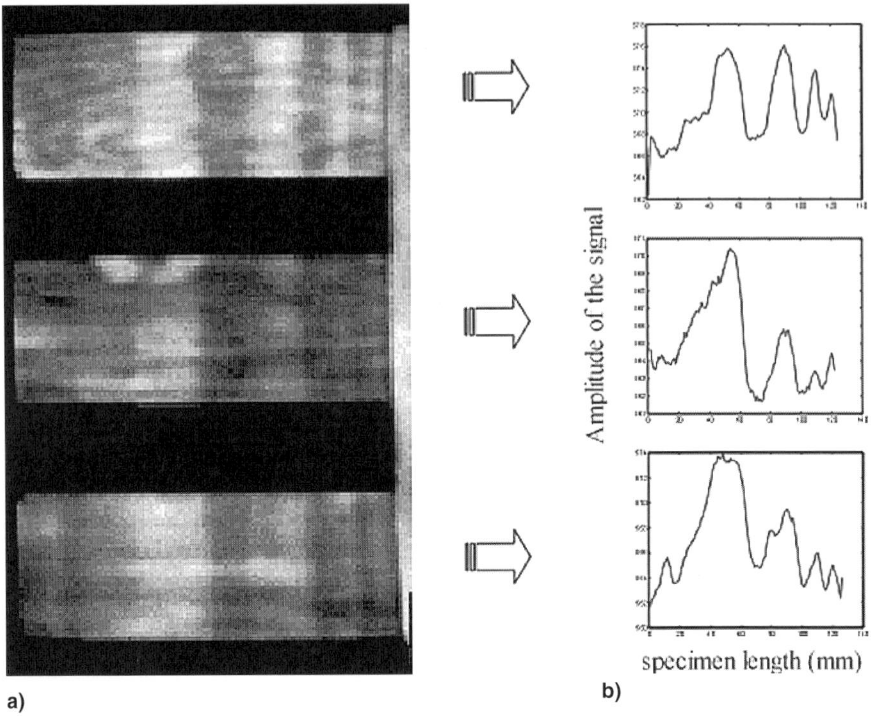

Fig. 3.36a,b. Infrared image of specimens with defective glue zones and the corresponding signal as a function of specimen length. **a** Infrared image where the bright zones corresponds to the defective areas. The maximum contrast also corresponds to defective zones. **b** The signal profile vs. the length of the specimen. (Berglind and Dillenz 2000, with permission)

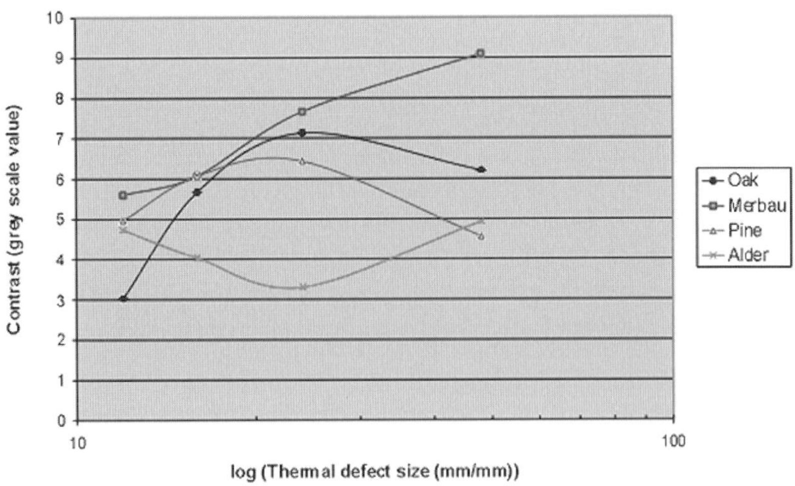

Fig. 3.37. Contrast in gray scale between a defective glued area with delaminations and sound area vs. the logarithm of the defect size for different species. (Berglind and Dillenz 2000, with permission)

3.4.3.2 Detection of Subsurface Defects Under a Veneer Lamina

The plywood composed of veneer laminae is probably the most ancient wood composite known from antiquity. The decorative value of fine veneer is highly prized. The veneer is produced from solid wood as thin sheets, usually cut on a rotary lathe. The thickness of veneer sheets that compose the plywood can be below 1 mm and the plywood thickness ranges between several millimeters to several centimeters. The laminae are alternated with grain at right angles. Today, plywood has a wide field of utilization, from furniture to material for building framing and other structural applications. The development of synthetic adhesives and the improvement of quality control contributed to the rapid expansion of the plywood industry, despite the declining size and quality of the timber resource.

For the inspection of defects in wood composites under a veneer lamina, Wu and Busse (1995) suggested the utilization of the lock-in thermography method. The feasibility of this technique was demonstrated, using a sample composed of several veneer laminae of different thickness, glued on chipboard or on a solid wood block of a different tree species. The phase images were taken at 0.03 Hz. The following items were studied:

- the detection of artificially induced defects in the chipboard. through the surface veneer sheets of various thicknesses. For this purpose the veneer thickness was varied in five steps from 0.5 to 2.5 mm. At the same time, under the surface, holes of different diameters were made in the chipboard, as can be seen in Fig. 3.38. Holes with a diameter greater than 4 mm were detected.
- the detection of natural defects such as knots present on solid wood blocks covered with veneer sheets (Fig. 3.39). The detection of knots was possible until about 2 mm underneath the surface layer.
- the detection of the nature of the solid wood blocks (oak and maple) glued under the plywood composed of different veneer sheets (Fig. 3.40). The differences between the thermal properties of maple and oak can be observed for all thicknesses of plywood composed of different laminae.

3.4.3.3 Detection of Defects in Particleboards

Quality monitoring is one of the most important steps in present particleboard production, with more and more complex equipment, within the framework of a decrease in quality of raw materials (small diameter logs, chips, wood waste of different species). To reach the goal of constant quality, the producer must monitor all the parameters that can effect particleboard fabrication, using advanced on-line nondestructive testing methods for quality control of the panels.

The measuring tasks of the on-line production of particleboard are presented in Table 3.8. In front of each measuring task are the corresponding

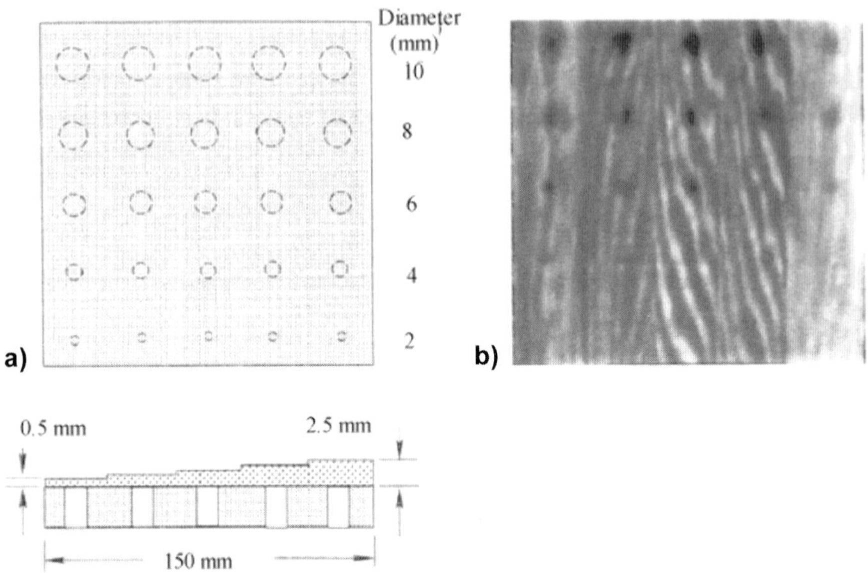

Fig. 3.38a,b. Detection of artificially induced defects in chipboard through the surface veneer sheets of various thicknesses. **a** Geometry of the specimen. **b** Thermal image. (Wu and Busse 1996, with permission)

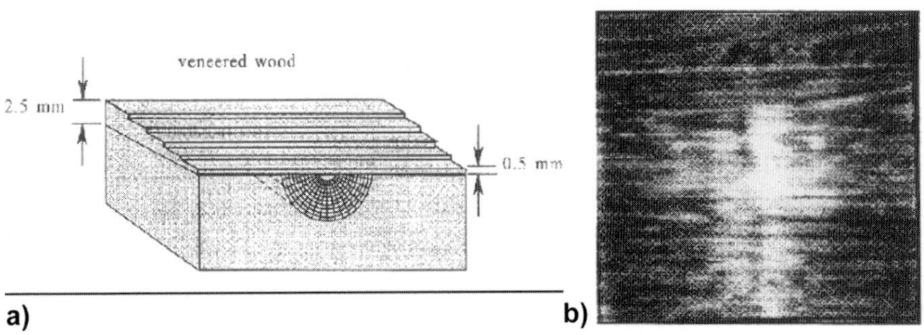

Fig. 3.39a,b. Detection of knots present in solid wood covered with veneer sheets. **a** Geometry of the specimen. **b** Thermal image. (Wu and Busse 1995, with permission)

nondestructive methods used today for quality control. The quality control operations among which thermal methods have an important role are:

- drying, for noncontact infrared reflectance spectroscopy measurement of the moisture content of wet and dry chips
- gluing, for the measurement of glue distribution on chips with near infrared spectroscopy as described by Niemz and Sander (1989)
- panel thickness and its surface properties could be detected by phase image. The thickness accuracy could be of the order of several μm. (Wu 1994; Wu and Busse 1995)

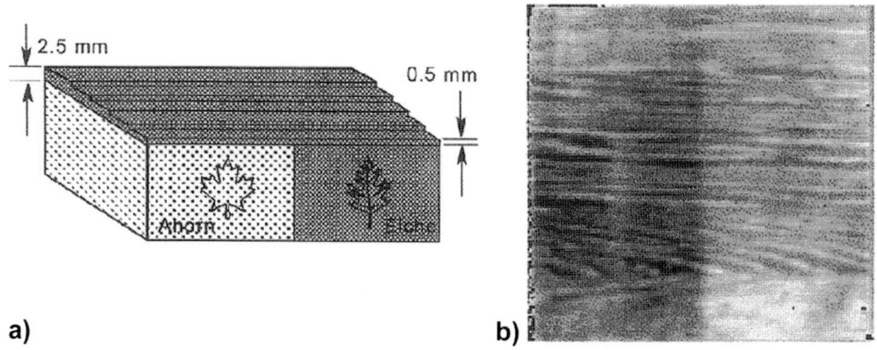

Fig. 3.40a,b. Detection of blocks of different species under the veneer sheets. **a** Geometry of the specimen. **b** Thermal image. (Wu and Busse 1995, with permission)

Table 3.8. Particleboard production measuring task and corresponding methods used for quality control. (Data from Schulte et al. 1995, courtesy of Tappi J)

Operation	Measuring task	NDT methods for evaluation and control	Thermal methods
Storage of raw material	Species mixture biodegradation	No method	
Chipping	Chip dimensions	Image analysis	
Drying	Moisture content of wet and dry chips	Dielectric method	Infrared reflectance spectroscopy
Gluing	Resin content and distribution	Image analysis Glue color	Near-infrared spectrum at various frequencies
Mat forming	Chip size distribution mat weight	Gamma ray densitometry Belt scale	
Panel quality control	Thickness	Roller head	Phase imaging
	Blow detection	Ultrasonic attenuation	Lock-in thermography
	Weight	Belts scale	
	Density profile	X ray, nuclear radiography	
	Mechanical properties		
	Surface properties	Ultrasonic velocity	Thermography

- blow detection could be performed with lock-in thermography
- mechanical properties such as the modulus of elasticity of the plate of real size can be measured using a resonance frequency method. The plate is tapped with an electronic hammer and the resonance frequency is detected by a laser probe by means of intensity modulated laser radiation. The width of the resonance curve at 3 dB is a measure of the vibration damping and of the corresponding viscoelastic properties of the board.

3.4.4 Imaging of Defects in Different Types of Lumber Joints

From an engineering point of view, it has been recognized that there is a permanent need for the development of different types of lumber joints for frame systems of large spans and for members of different widths and shapes for architectural applications. For structural members it is necessary to insure the quality control of each element, to optimize the design values and to establish the strength.

Metal plate-connected wood trusses are largely used in agricultural, commercial and residential constructions. Loading transfer from the metal connectors to the wood member is nonlinear and depends also on the grain orientation of the lumber. When this complex and heterogeneous structural member is loaded, it deforms as a whole element. Stress concentration occurs, and plastic deformations, micro cracking, slippage at component interface and fracture can be observed in wood or in metallic connectors. A better understanding of the behavior of the joints is necessary for upgrading the design procedure. Failure modes of the joint include failure of the wood member within the plated zone, tension or shear in the metallic connector, teeth pulling out of the wood, buckling of the plate in gaps between wood members, etc.

Luong (1996) used infrared thermography with an infrared camera to detect the damage in splice joints under tension and shear loading before failure, as can be seen in Fig. 3.41. The infrared detector has a response time shorter than 1 μs. The temperature patterns are represented by distinct color hues. Each of them corresponds to 0.2 °C. The energy emitted as infrared radiation is a function of thermal conduction, thermoelastic coupling effects and intrinsic dissipations generated by viscosity and /or plasticity. Intrinsic dissipation (Luong 1995, 1998) seems to be the most accurate indicator of damage to a mechanical, loaded specimen.

The development of glued laminated timber during the 20th century had a significant effect on wood use and opened new opportunities for creative products. The capability to make engineered structural products of a variety of forms became possible with finger joints and with relatively small timber of high quality.

The thermographic detection of defects in finger-joint lumber was studied by Masuda and Takahashi (2000) with Douglas-fir lumber specimens of 35×75-mm cross section and 600-mm length under bending cyclic loading. The maximum temperature rise was 3 °C for a 1.5 J energy loss. In a defective sample, local variations of temperature are different in tension and compression zones.

The defective area has high stress concentration in which heat is generated by the hysteresis loop of the stress-strain curve. The neutral axis zone has a rather constant temperature, while in the tension area the temperature rises from 16.75 to 18.5 °C at 1000 N repeated load. In specimens with better

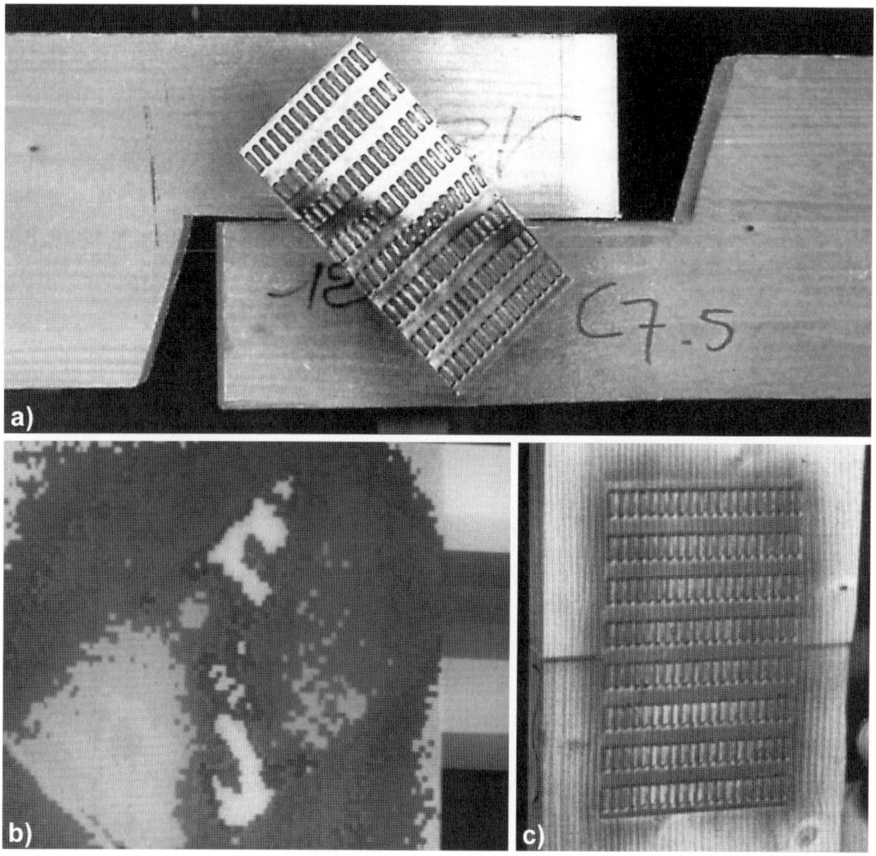

Fig. 3.41a,b. Infrared images of splice joints under shear loading. **a** Before loading. **b** Shear stress before failure. Color hue corresponds to 0.2 °C. **c** Splice joints under tension. (Luong 1996, with permission)

adhesion, the dispersion of experimental values around the neutral axis is less important.

3.5 Summary

Thermal wave imaging that provides a map of surface temperature distribution is a nondestructive testing technique that uses heat flux to probe the physical properties of wood with active or passive heating procedures. The main thermographic methods for solids are: pulse thermography, heating-up thermography and lock-in thermography. Pulse thermography records the transient thermal response of an object to a thermal pulse, provided by a flash tube.

Heating-up thermography utilizes heat sources such as commercial photographic lighting devices, hot liquids, laser beams, etc. The equipment for thermal wave imaging uses an amplitude-modulated focused laser beam directed onto the surface of the sample. At the point of light absorption, some energy is absorbed by the sample, giving rise to a modulated heat source. This heat flux propagates as a thermal wave through the material in all directions and interacts with the structure. The propagation phenomena are influenced by the thermal impedance of the material, which varies as a function of the interaction of the thermal wave with the structure.

As an alternative to heating, cooling can also be used with aerosol freezer spray. The transient character of the temperature field demands a dynamic tuning of the thermo-vision system. The development of infrared video cameras has extended the wavelength range of visible light video cameras to a thermal infrared range between 3 and 12 μm. Lock-in thermography can provide three types of images: thermographic, phase and magnitude images. The development of lock-in thermography allows the utilization of thermographic equipment to image the average temperature distribution of oscillating components.

In wood science, thermography is a relatively new field and thermal imaging of subsurface temperature distribution in wood-based composites to ascertain the integrity of subsurface structures were first produced less than 10 years ago.

The active heating procedure, or stress-generated thermal field under cycling loading, was used in early work on wood thermographic imaging. The advantage of this technique is that the influence of defects on mechanical properties of wood can be easily examined. With lock-in thermography, a large sample depth range can be observed. This method has been used for detection of delaminations in wood-based composites because of its sensitivity to subsurface defects and imaging capability within a short time span.

The passive heating method has a large field of applications for knot detection, for the detection of the slope of grain, for imaging of moisture content distribution in lumber, for imaging of wood rupture phenomena, for imaging of cavities in trees, etc. The thermal image is the result of a very complex interaction between the heating source, the material and the defects. The rate of heat application and the mode of heating by contact or by radiation are factors of major importance for obtaining good images with correct resolution.

An advantage of passive heating over the active heating procedure is its ability to produce temperature distributions without resorting to mechanical loading of the material. The thermal stress or the thermal shock is relatively low and does not damage the material.

The disadvantage of the passive method is that the thermal images are transient and require a fast recording system to capture the most interesting images during the test. Electronic systems (with an infrared camera) or chemical systems (with liquid crystals) can be used. The most widely used method is scanning with an infrared camera, which produces an image of an object through electronic detection of infrared radiation emitted from the object.

The scanning infrared imaging technique seems today to be the most convenient remote sensing method for thermographic inspection of solid wood and wood-based composites in situ.

4 Microwave Imaging

4.1 Introduction

The microwave scanning technique is a relatively new tool for the characterization of wood materials. During the last 15 years, considerable progress has been achieved in the design of scanning probes and in image interpretation (Martin et al. 1987; Filippini et al. 1990; Chazelas 1991; Yokoyama and Norimoto 1996; Golosovsky et al. 2000; Kaestner and Baath 2000).

For the correct interpretation of microwave imaging, it is necessary to know the response of the material to electrical magnetic fields. It is also necessary to understand the mechanism of contrast and the interaction between the sample and the probe and to model and to measure the distribution of the electric and magnetic fields around the probe. Consequently, progress achieved in the basic knowledge of the dielectric properties of wood and of wood-based composites will allow new industrial applications of the microwave imaging technique. Systematic studies on the dielectric properties of different species were reported from 1948 by Skaar and later by Hearmon and Burcham (1954) and Brown et al. (1963). The following basic points were discussed:

- the behavior of water and the principal constituents of wood, cellulose and lignin;
- the determination of the fiber saturation point of wood–water interactions and the hygroscopic characteristics of wood;
- the measurement and prediction of wood density and moisture content.

The dielectric properties of wood were related to microwave processing. The advantages of microwave heat processing are numerous, such as: fast processing, small space occupied by the equipment, less temperature degradation of the products compared with other conventional heating techniques, dimensional stability of the products, selective heating of materials and, lower heat loss to the surroundings (Dench 1973).

The industrial applications were originally oriented primarily toward improving of drying and gluing technology. Today, the industrial applications of the microwave imaging technique are related to the detection of internal defects such as knots, spiral grain, slope of grain, structural discontinuities, etc. of logs, lumber, and wood-based composites. Another interesting field of application of microwave imaging is the dielectric behavior of vegetative materials, including leaves, stalks and trunks under various moisture and temperature conditions over a wide range of frequencies (Tan 1981; Ulaby and Jedlicka 1984; Sieber 1985; El-Rayes and Ulaby 1987).

In the first part of this chapter, the dielectric properties of wood are discussed, including the effects of experimental factors and of physical parameters on these constants.

The equipment for laboratory measurements and imaging is presented in the second part of this chapter. The final part is devoted to the applications for inspection of forests, logs, lumber and wood-based composites.

4.2 Basic Aspects

The basic aspects of dielectric behavior of wood are treated in several reference books and articles (James 1989; Skaar 1988; Torgonikov 1990). The dielectric behavior of a solid (Bahr 1982; Zoughi 1990, 1996; Ida 1992; Zoughi et al. 1996) is determined by the polarization of atoms and molecules under the influence of an applied electrical voltage. The polarization can occur at electronic, atomic, dipole and interfacial levels, and can have different causes that are strongly dependent on the frequency (Fig. 4.1). Electronic polarization is determined by the displacement of electrons with respect to the nucleus within the atoms and occurs in the ultraviolet frequency region. Atomic polarization is induced by the bending of bonds between the atoms and occurs in the infrared frequency range. Dipole polarization is caused in wood by the hydroxyl groups of bound or free water molecules (that are permanent dipoles), which rotate in the electric field at a relatively low frequency. Interfacial polarization is induced by the accumulation of charged ions at interfaces between different regions within the cell wall of the wood (i.e., between the middle lamella and the secondary wall, or between crystalline and amorphous regions).

The mechanism of electrical conduction of wood has been explained by several theories (Skaar 1988) and it is generally accepted that the electrical charges are carried by the ions present in the wood cell walls and in cellulose. In a direct current electric field, cellulose exhibits typical ionic conduction polarization phenomena. Temperature increase in dielectric solids is produced by an increase in the energy of the continuous random vibrations of molecules. When an electric field is applied to a dielectric material, the randomly oriented dipoles align themselves in a direction opposite to that of the applied external field. In this configuration, the energy supplied by the field is stored in the molecules as potential energy. By the removal of the electric field, this potential energy of ordered dipoles can be converted back into the kinetic energy of disordered dipoles. The rise in vibrational motions of molecules in a microwave field is determined by the mechanism of ionic conduction and by dipole rotation, that is, a rotation of the polar molecules under the influence of the external electric field. The electric field applied to wood at microwave frequencies

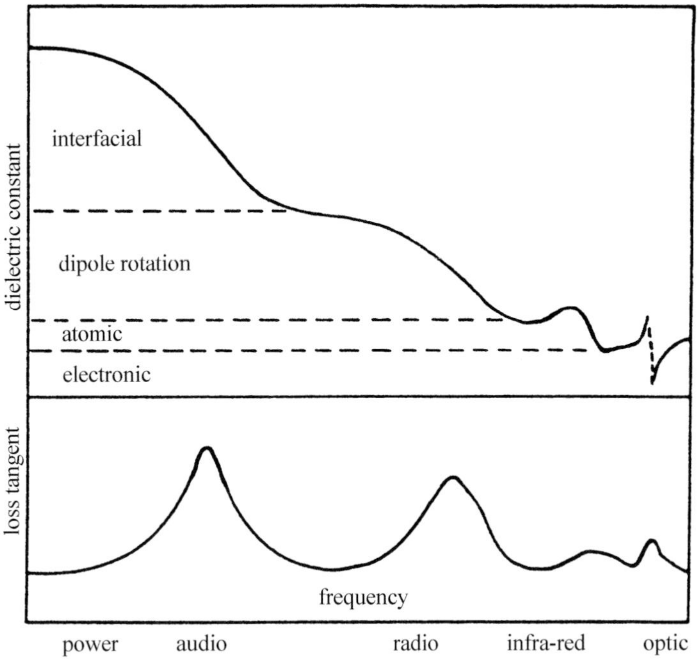

Fig. 4.1. Relationship between the frequency and the dielectric constants of solids. (Skaar 1988, with permission)

induces a dissociation of molecules and a migration of ions, by delivering ordered kinetic energy. The collisions of the migrating ions with un-ionized molecules produce heat. The ordered energy is then converted into heat that is a disordered kinetic energy produced by collisions of the migrating ions with un-ionized molecules. A continuous dipole rotation can be obtained by an external alternating electric field. Each time the field is reversed the polar molecules reverse their position. This transfer of energy leads to an increase in the temperature of the solid. Dipole rotation is dependent on temperature and frequency.

Dipole polarization also called the Debye effect depends strongly on several wood physical parameters, such as species, density, moisture content, temperature and structural orientation versus the anisotropic axes. The relationship between the dielectric constants in a dielectric material containing permanent dipoles subjected to a sinusoidally varying electric fields of strength E, are shown in Fig. 4.2. The dielectric displacement D has two components, the first noted D′ in phase with ωt, where the angular frequency $\omega = 2\pi f$ and t is the time and the second noted D″ lagging ωt by $\pi/2$ radians.

The dielectric constants ε′ and ε″ are the components of the complex dielectric constant ε*. They are defined as:

$$\varepsilon' = D'/E^* \tag{4.1}$$

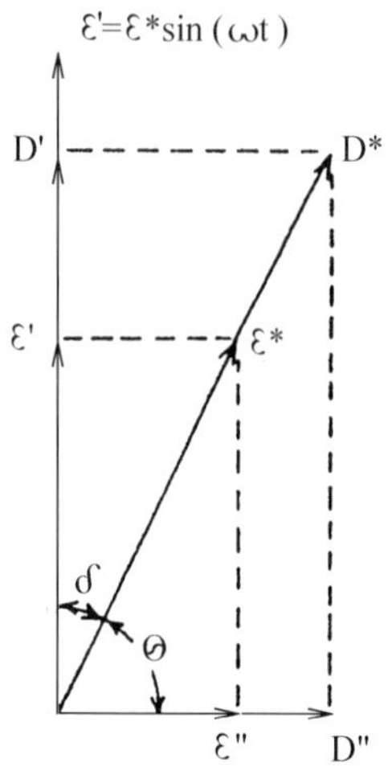

Fig. 4.2. Relationship between the complex dielectric displacement, dielectric constants phase angle θ and loss tanδ angle. (Skaar 1988, with permission)

and

$$\varepsilon'' = D''/E^{*} \tag{4.2}$$

where the strength field E = E* sin ωt and E* is the peak amplitude of the sinusoidally varying electric field. In complex notation, we have:

$$[\varepsilon^{*}] = [\varepsilon'] - j[\varepsilon''] \tag{4.3}$$

In other words, in an alternating electric field the amount of energy that can be stored by wood is expressed by its dielectric constants [ε*], which are tensors and complex constants composed of a real constant [ε′] and an imaginary constant [ε″] and are related to Eq. (4.3). The imaginary constant [ε″] expresses the loss of the kinetic energy that can be dissipated as heat and is also called the constant of dielectric absorption. The physical meaning of the imaginary part of the dielectric constant [ε″] is related to the attenuation of the wave and can be correlated with the moisture content (Tiuri et al.1980). The ratio of real and imaginary constants ε″/ε′ is called tan δ and is currently used in laboratory measurements.

It is generally accepted that wood is an orthotropic material and the dielectric tensor $[\varepsilon_{ij}]$ corresponding to this symmetry (Lin 1973) can be written as:

$$[\varepsilon_{ij}] = \begin{bmatrix} \varepsilon_{11} & \varepsilon_{12} & \varepsilon_{13} \\ \varepsilon_{21} & \varepsilon_{22} & \varepsilon_{33} \\ \varepsilon_{31} & \varepsilon_{32} & \varepsilon_{33} \end{bmatrix} \tag{4.4}$$

Each term of the tensor is a complex number that can be written with a real and an imaginary part.

If the principal anisotropic directions (1, 2, 3 or L, R, T) coincide with the direction of the electrical field, then the tensor $[\varepsilon_{ij}]$ is written as:

$$[\varepsilon]=\begin{bmatrix}\varepsilon_{11} & 0 & 0\\ 0 & \varepsilon_{22} & 0\\ 0 & 0 & \varepsilon_{33}\end{bmatrix} \tag{4.5}$$

and

$$[\tan\delta]=\begin{bmatrix}\tan\delta_{11} & 0 & 0\\ 0 & \tan\delta_{22} & 0\\ 0 & 0 & \tan\delta_{33}\end{bmatrix} \tag{4.6}$$

Out of the principal directions of elastic symmetry, the dielectric constants are expressed by Eqs. (4.5) and (4.6):

$$\varepsilon'_{\alpha}=\varepsilon'_{L}\cos^2\alpha_{L}+\varepsilon'_{R}\cos^2\alpha_{R}+\varepsilon'_{T}\cos^2\alpha_{T} \tag{4.7}$$

and

$$\tan\delta_{\alpha}=\tan\delta_{L}\cos^2\alpha_{L}+\tan\delta_{R}\cos^2\alpha_{R}+\tan\delta_{T}\cos^2\alpha_{T} \tag{4.8}$$

where α_L, α_R, α_T are the projection angles on the axes L, R and T.

It was demonstrated by James and Hamil (1965) that the coefficients in R and T directions are very similar (Table 4.1). Consistent results were obtained by Chazelas (1991) for spruce at a constant moisture content at 12% (Table 4.2).

When the transverse elastic symmetry of wood is considered, the values of the complex dielectric constants (diagonal and off diagonal terms of dielectric tensor) are given in Table 4.3. The off diagonal terms of dielectric tensor express the spatial distribution of the depolarization in wood.

The linearly depolarized field E tilted to 45° versus the direction of the fibers is expressed as:

$$[E]=\begin{bmatrix}E_{//}\\ E\perp\\ 0\end{bmatrix} \tag{4.9}$$

Table 4.1. Dielectric properties of Douglas fir at 8.53 GHz, as a function of structural orientation and anisotropic directions for different moisture content. (James and Hamill 1965; courtesy of Forest Products J)

Moisture content (%)	Direction ε	L Tanδ	Direction ε	R Tanδ	Direction ε	T Tanδ
7	1.9	0.14	1.7	0.07	1.8	0.09
10	2.1	0.17	1.9	0.10	1.9	0.11
12	2.6	0.22	1.9	0.11	2.1	0.13
16	2.9	0.26	2.1	0.18	2.3	0.21
22	4.2	0.45	2.6	0.26	3.0	0.25

Table 4.2. Dielectric constants at 9.6 GHz as a function of wood anisotropy for spruce at 12% moisture content. (Chazelas 1991, with permission)

Specimen	Thickness (cm)	Density (kg/m³)	ε'_L	ε'_R	ε'_T	ε''_L	ε''_R	ε''_T
LR 6 cm	6.13	471	2.51	1.88		0.52	0.22	
	6.13	466	2.46	1.87		0.51	0.21	
	6.12	462	2.44	1.87		0.50	0.21	
LT 6 cm	4.16	462	2.36		1.94	0.46		0.23
	6.07	451	2.22		1.85	0.42		0.19
	6.07	455	2.20		1.85	0.39		0.17
LR 4 cm	3.98	468	2.23	1.78		0.44	0.18	
	4.47	465	2.36	1.83		0.49	0.18	
	3.98	458	2.31	1.77		0.42	0.18	
LT 4 cm	3.98	462	2.46		1.97	0.46		0.25
	4.47	494	2.36		1.95	0.51		0.23
	4.48	491	2.38		1.95	0.50		0.21
LR 3 cm	2.51	475	2.30	1.86		0.47	0.21	
	3.02	467	2.43	1.84		0.47	0.22	
	2.00	460	2.34	1.90		0.41	0.19	
LT 3 cm	2.01	461	2.37		2.01	0.36		0.20
	3.01	495	2.47		1.94	0.48		0.25
	2.51	495	2.32		1.96	0.48		0.25

Table 4.3. Dielectric complex constants of Douglas fir at 4.8 GHz frequency and 6 and 12% moisture content, calculated with the hypothesis of transverse anisotropy with $\varepsilon_{22} = \varepsilon_{33}$. (James et al. 1985; courtesy of Forest Products Society)

Moisture content (%)	ε_{11}	ε_{22}	ε_{12}	ε_{21}
6	2.35 – j 0.15	1.9 – j 0.09	0.0018 + j 0.0075	0.0027 – j 0.001
12	2.73 – j 0.07	2.16 – j 0.031	0.0052 + j 0.0078	0.0123 + j 0.018

where $E_{//}$ is parallel to the fibers and $E_\perp$ is perpendicular to the fibers. The depolarization is expressed by the ratio $D = E_{///}E_\perp$ called also "polarization ratio" and can be calculated using the Stokes parameters defined as:

$$D = (Q^2 + U^2 + V^2)^{1/2} I^{-1} \quad (4.10)$$

where Q, U, V and I are given by the following equation

$$Q = E_{//}E^*_{//} - E_\perp E^*_\perp \quad (4.11)$$

$$I = E_{//}E^*_{//} + E_\perp E^*_\perp \quad (4.12)$$

$$U = E_{//}E^*_\perp + E_\perp E^*_{//} \quad (4.13)$$

$$V = -j(E_{//}E^*_\perp + E_\perp E^*_{//}) \quad (4.14)$$

where * denotes the complex conjugate of the quantities. The values of D range between 0 and 1 $[0 \le D \le 1]$. When D = 1, the wave is completely polarized.

To express the anisotropic dielectric behavior Togovnikov (1990, 1993) defined the coefficients expressed by Eqs. (4.15), (4.16), (4.17), for the case of the transverse anisotropy:

$$k_1 = \varepsilon'_{//}\varepsilon'_{\perp} \tag{4.15}$$

$$k_2 = \tan\,\delta_{//}\tan\,\delta_{\perp} \tag{4.16}$$

$$k_3 = \varepsilon''_{//}\varepsilon''_{\perp} \tag{4.17}$$

The coefficients k_1, k_2 and k_3 are given in Table 4.4 as a function of the frequency. The increase in frequency determined the decrease in all coefficients for both softwood and hardwood species. The modification of dielectric properties with frequency in polar dipole materials is caused by the interfacial polarization also called the Maxwell-Wagner effect. We shall see later that the dielectric properties of wood depend in a very complex way on several factors such as the frequency and power of the electric field, the moisture content of the specimen, the temperature, the density of the species and the structural orientation on which the constants are measured.

According to reference books in physics (Hippel 1961; Kittel 1968) for dielectric materials, the parameters that can be measured are the permittivity (the complex dielectric constants), the conductivity, the capacitance, and the susceptibility of the materials.

It is beyond of the purview of this book to refer in detail to all these parameters.

The dielectric response of wood material in terms of complex capacitance is expressed by Eq. (4.18)

$$C^*(\omega) = C'(\omega) - j\,C''(\omega) \tag{4.18}$$

where C' (ω) is the real part of the capacitance being out of phase with the applied voltage and C'' (ω) is the imaginary or loss component being in phase with the voltage ω is the angular frequency. The loss component can be written as:

Table 4.4. The coefficients of dielectric anisotropy k_1, k_2, k_3 as a function of frequency, at 20 °C, where $k_1 = \varepsilon'_{//}\varepsilon'_{\perp}$, $k_2 = \tan\delta_{//}\tan\delta_{\perp}$ and $k_3 = \varepsilon''_{//}\varepsilon''_{\perp}$, for wood density ranging between 200 and 1000 kg/m^3. (Torgovnikov 1990)

Species		Frequency kHz									
		10^2	10^3	10^4	10^5	10^6	10^7	10^8	10^9	10^{10}	10^{11}
Softwood	k_1	1.60	1.54	1.50	1.46	1.42	1.38	1.32	1.20	1.16	1.14
	k_2	1.60	1.70	1.75	1.80	1.90	1.80	1.70	1.50	1.30	1.25
	k_3	2.6	2.6	2.6	2.6	2.7	2.5	2.2	1.8	1.5	1.4
Hardwood	k_1	1.38	1.36	1.32	1.28	1.26	1.22	1.16	1.14	1.12	1.11
	k_2	1.35	1.50	1.50	1.55	1.75	1.65	1.55	1.40	1.30	1.25
	k_3	1.90	2.00	2.00	2.00	2.20	2.00	1.80	1.60	1.50	1.40

$$C''(\omega) = \varepsilon_o \varepsilon''(\omega)(A/d) + G/\omega \tag{4.19}$$

where ε_o is the free space permittivity, A is the area of the sample, d is the thickness and G is the conductance of the sample.

The relationships between the dielectric constants, the permittivity, the capacitance and the susceptibility of the material are given by Eqs. (4.20) and (4.21):

$$C^*(\omega) = \varepsilon_o (A/d)\,[\varepsilon'(\omega) - i\,\varepsilon''(\omega)] - i\,G/\omega \tag{4.20}$$

and

$$C^*(\omega) = \varepsilon_o (A/d)\,[\chi'(\omega) + \varepsilon(\infty) - i\,\chi''(\omega)] - i\,G/\omega \tag{4.21}$$

where $\chi'(\omega)$ and $\chi''(\omega)$ are the real and imaginary parts of the susceptibility and $\varepsilon(\infty)$ is the permittivity at infinite frequency.

In recent years, modeling microwave nondestructive testing procedures of dielectric materials have received new attention because of their suitability for nonmetallic composites (Poliszko and Hoffmann 1985; Ida 1992; Lebowitz, Eoughi 1996; Thompson 1996) and ceramics (Schneider et al. 1996). The models must take into account the wave propagation effects and a combination of induction and propagation phenomena. For accurate modeling of the existing environment, Ida and Wang (1996) proposed two models that take into account the anisotropy of dielectrics, assuming that the media are linear. A thorough discussion of these models is beyond the purview of this book, but we can note that the proposed methods can be used for monitoring production processes.

Another parameter that can be measured is the power of the electric field dissipated in the specimen as heat. It can be calculated with Eq. (4.22):

$$P = 2\pi\,\omega\,E^2\,\varepsilon'\,\tan\delta \tag{4.22}$$

where: ω is the angular frequency and E is the electric field strength.

Before discussing the principles of defect detection using microwave techniques it is important to consider first some factors related to the experimental variables and also to analyze the influence of the physical properties of wood on dielectric constants.

4.2.1 Effect of Experimental Factors

During the propagation of electromagnetic waves in wood, the initial parameters of transmitted waves are modified by interaction with the wood structure. As noted by King (1978), the main wave parameters that interact strongly with wood density and moisture content are the amplitude and the phase. The complex propagation parameter (γ) of a steady-state plane wave can be written as:

$$\gamma = \alpha + j\beta \quad (4.23)$$

where α is the attenuation constant (Nepers/m) and β is the phase constant (radians/s). After corresponding calculation, for $\tan\delta \langle\langle 1$, it was conclude that:

$$\alpha = 0.5\beta_o\sqrt{\varepsilon'}\tan\delta \quad (4.24)$$

and

$$\beta = \beta_o\sqrt{\varepsilon'}.\,[1 + 0.125\tan^2\delta] \quad (4.25)$$

where βo is the phase constant of air, defined by

$$\beta_o = 2\pi/\lambda_o \quad (4.26)$$

where λo is the wave length of the electromagnetic wave in air.

In the case of wood material, it has been demonstrated (James 1975; Wert et al. 1984; James et al. 1985) that ε' depends strongly on wood density and $\tan\delta$ on moisture content.

The loss of wave intensity is reflected by the reduction of signal amplitude and by the phase retardation that are caused by the absorption of wave energy and by scattering. In addition, the anisotropic structure of wood induces depolarization of the incident wave. The electric vector of the incident wave is decomposed into two components in the directions of the maximum and minimum dielectric constants. When the wave is elliptically polarized, the minor axis of the ellipse corresponds to the maximum dielectric constant, which coincides with the directions of the fibers. The polarization is a function of the thickness of the specimen. Perfect elliptical polarization is obtained only when the specimen thickness is optimum. For lesser thicknesses the polarization angle is smaller than the true slope of the grain by a factor corresponding to the depolarization index, as can be seen in Fig. 4.3.

The effect of density on the phase and on the attenuation of waves was demonstrated by Kharadly (1985) for hemlock (Fig. 4.4) at 11–12% moisture content. The corresponding regression equations are given in Table 4.5. For both parallel and perpendicular polarization, an increase in the density increased both the attenuation and phase shift. These experimental data were used for the abacus presented in Fig. 4.5, from which it is possible to determine both wood density and the moisture content of a board by measuring the attenuation and the phase shift of a microwave regression equation ($r = 0.87$) of the phase shift for different species as a function of the density, ranging from 400 to 800 kg/m^3 (Fig. 4.6) at 9% moisture content as was reported also by Martin et al. (1987). The effect of moisture content on the attenuation, phase angle and depolarization index was studied by James et al. (1985) with Douglas fir as can be seen in Figs. 4.7, 4.8 and 4.9. Increasing moisture content caused an increase in all these parameters.

Values of attenuation coefficients for Indian wood species with nearly 9% moisture content, as reported by Jain and Sanyal (1996), are given in Table 4.6. The measurements were performed at 8.2 GHz in the X band microwave region

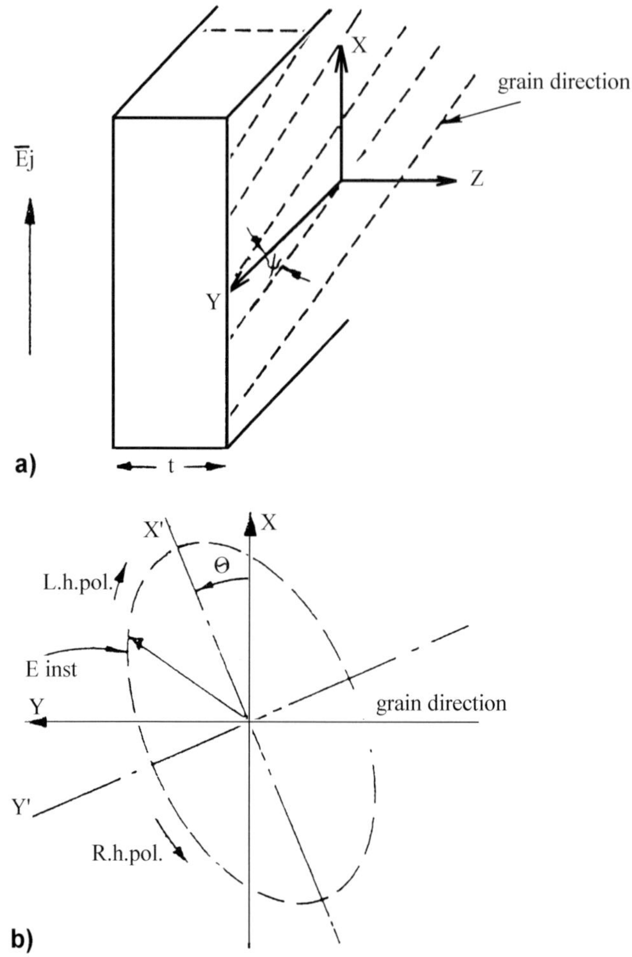

Fig. 4.3. Slope of grain and the elliptic locus of the instantaneous electric field. **a** Grain direction at angle Ø in the plane *Y–Z* in a specimen of thickness t. **b** Elliptical polarization of transmitted instantaneous field. The ellipse is inclined at angle Θ; the sense of rotation depends upon whether the *X*-field component leads or lags the *Y*-field component in the time phase. The locus becomes a circle when the two components coincide grain direction and instantaneous E inst. (King 1978; courtesy of Forest Products Society)

on an E_{10} mode of vibration. Figure 4.10 shows the linear increase of attenuation with density in the range of 400 to 500 kg/m^3. At a density greater than 510 kg/m^3, the attenuation appears to be independent of the density.

Purslow (1971) reported attenuation measurements at 10 GHz for three species: redwood, hemlock and whitewood, as a function of moisture content in the range of 6 to 30%. He stated that the relationship between the moisture content and microwave attenuation is sigmoid, with three main zones, namely, zone I between 0 and 8%, zone II between 8 and 30% and zone III over 30%.

In the first zone, the attenuation increases very slowly and is probably due to the macromolecular water being very strongly bonded to the wood structure. In the second zone, corresponding to the presence of "bound water", the attenuation dramatically increases with moisture content. Over 30% "free water" is

Table 4.5. Relationships between the variation of amplitude (ΔA) and of phase (ΔΦ) of the microwave signals and density (ρ) of hemlock wood samples at 20 °C and with the moisture content M, constant. (Kharadly 1985; courtesy of Forest Products Society)

Polarizatioin	Regression equation	Correlation coefficient	Standard deviation
Parallel	$\Delta\Phi = 38 + 540.73\rho$	0.95	10.92
	$\Delta A = 2.1 + 21.16\rho$	0.81	1.02
	$\Delta\Phi = 44.51\Delta A$	0.87	
Perpendicular	$\Delta\Phi = -25.52 + 497.81\rho$	0.93	12.36
	$\Delta A = -1.07 + 15.67\rho$	0.84	0.67
	$\Delta\Phi = 52.19\rho$	0.90	

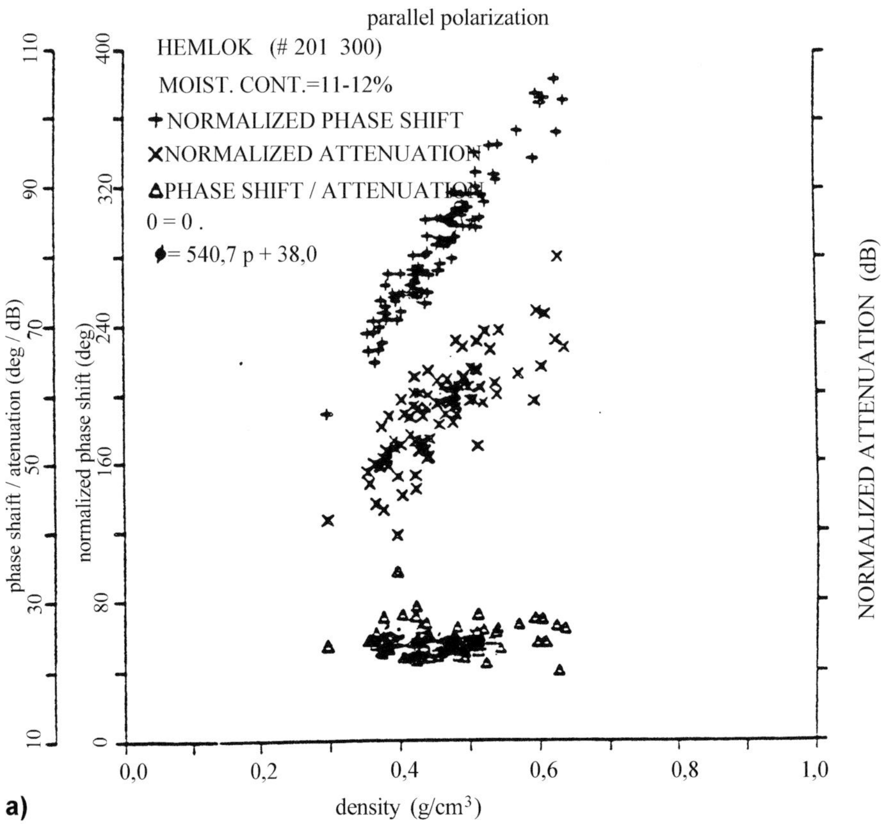

a)

Fig. 4.4a,b. Attenuation and phase shift as a function of density. (Kharadly 1985; courtesy of Forest Products Society)

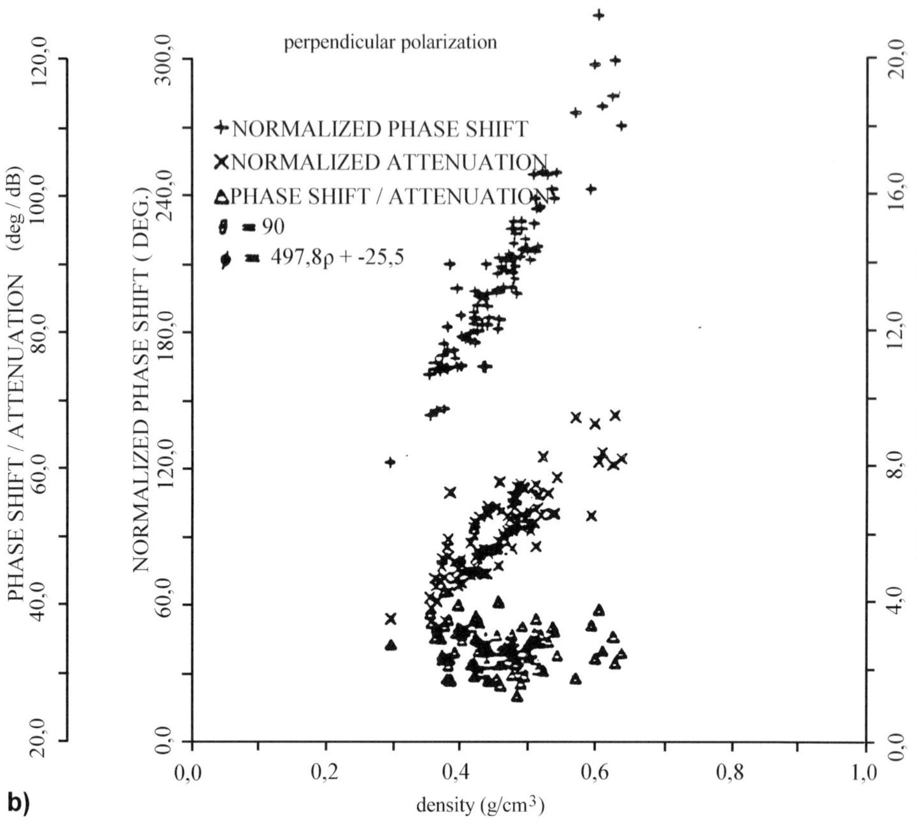

b)

Fig. 4.4a,b. *Continued.*

Table 4.6. Attenuation of microwaves at 8.2 GHz frequency measured on several Indian species. (Jain and Sanyal 1996, with permission)

Species	Density (kg/m^3)	Moisture content (%)	Attenuation constant[a]
Abies pindrow	390	9.6	0.321
Jacaeanda acutifolia	440	9.5	0.516
Toona ciliata	480	9.2	0.708
Alibizia lebbek	600	9.4	0.798
Dalbergia sissoo	630	8.3	0.804
Acacia catechu	850	8.9	0.819

[a] The attenuation constant was calculated with the relationship; $\alpha = 2.303\ X_2^{-1}.\ (\log_{10} F_2 - \log_{10} F_1)$ (4.27), where F_2 and F_1 are the corresponding intensities of microwaves for wood samples. The measurements were performed in the X-band region of microwaves on the E_{10} mode of vibration.

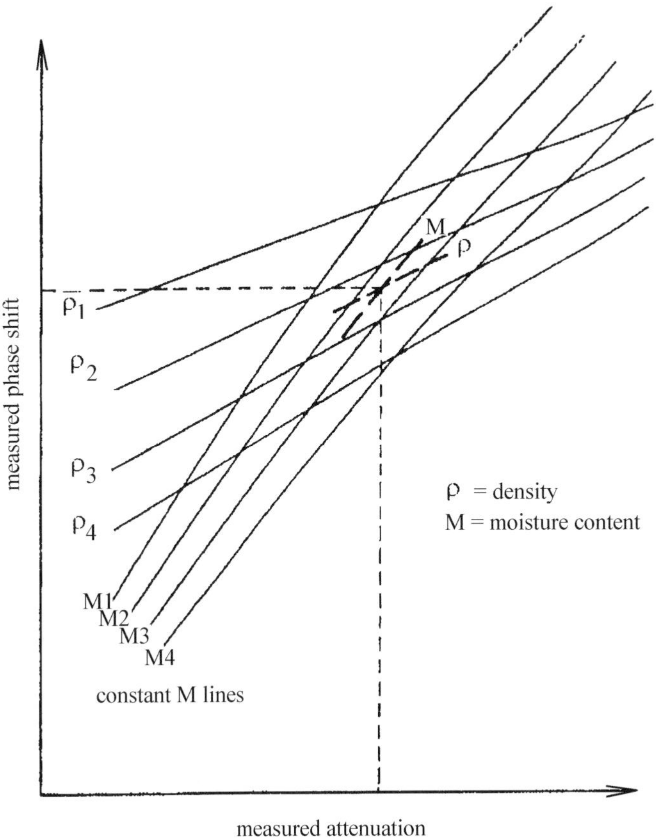

Fig. 4.5. Abacus for the determination of unknown density and moisture content as a function of variation of phase and amplitude of the microwave signal for hemlock wood samples at 20°. (Kharadly 1985; courtesy of Forest Products Society)

present, and the slope of the regression relating the attenuation and moisture content is less steep than in the second zone. The fiber saturation point was found to be between 26.9 and 28.2%, as the intersection between the regression lines in the second and in the third zones.

4.2.2 Effect of the Physical Properties of Wood

The basic physical parameters of wood related to the detection of defects with a microwave technique are: the moisture content, the density and the grain angle (the slope of grain). These three parameters can be determined from the dielectric tensor. The relationships between the diagonal terms of the dielectric tensor and the moisture content, density, grain angle and chemical constituents

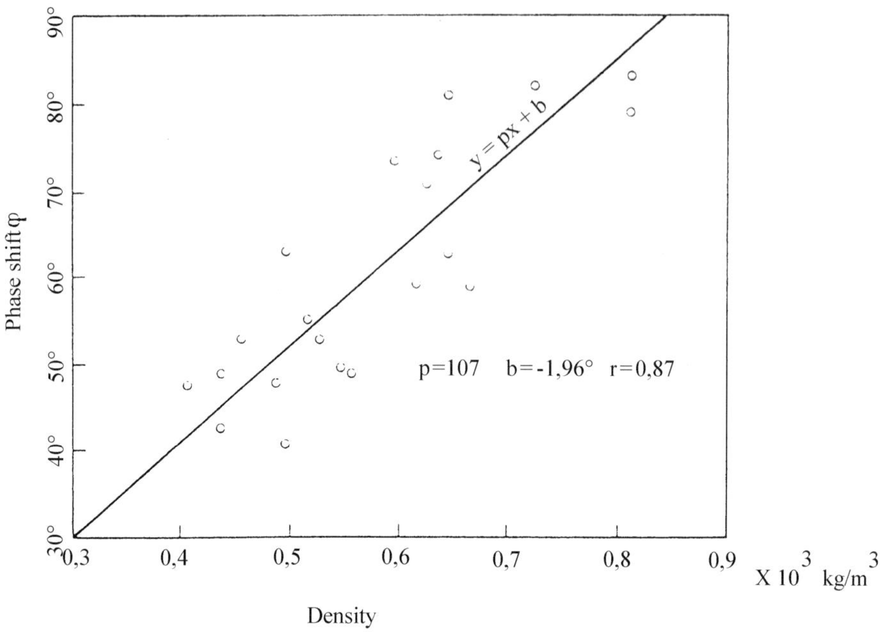

Fig. 4.6. Relationship between the phase shift and the density of different wood species, with density ranging from 400 to 800 kg/m^3. (Martin et al. 1987; courtesy of Wood Science and Technology)

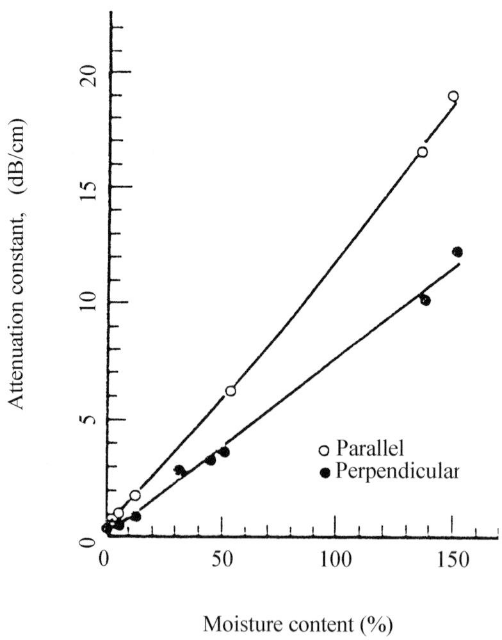

Fig. 4.7. Attenuation of the signal as a function of moisture content in two anisotropic directions, parallel and perpendicular to fibers. (James et al. 1985; courtesy of Forest Research Society)

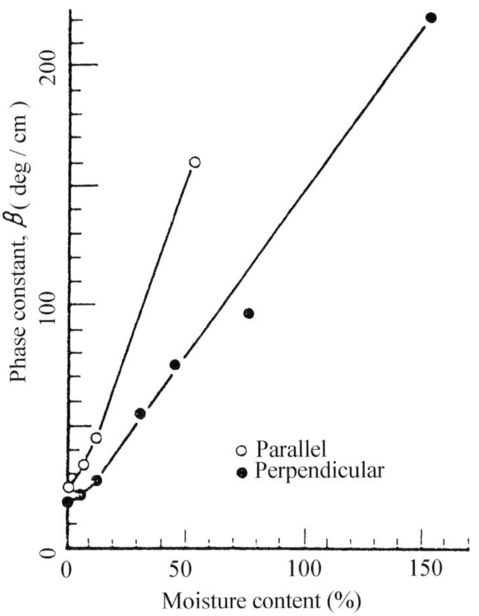

Fig. 4.8. Phase shift as a function of moisture content in two anisotropic directions, parallel and perpendicular to the fibers. (James et al. 1985; courtesy of Forest Products Society)

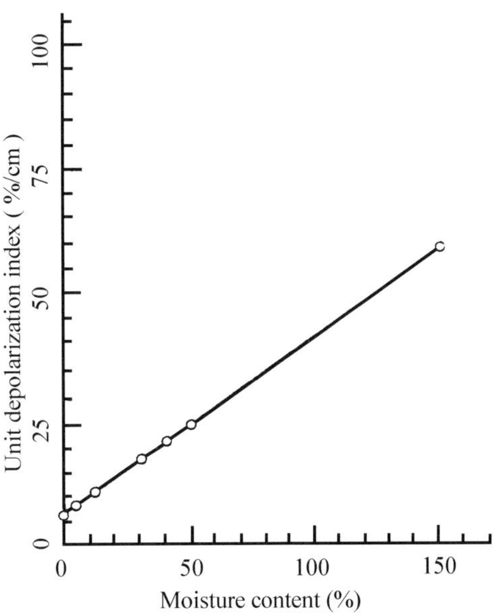

Fig. 4.9. Depolarization index as a function of moisture content in two anisotropic directions, parallel and perpendicular to the fibers. (James et al. 1985; courtesy of Forest Products Society)

have been studied by many investigators. Values of dielectric constant measurements for various species and at various frequencies (ranging from 1 to more than 20 GHz, and also in the low frequency range from 10^{-2} to 10^{6} Hz) at different temperatures, moisture contents and anisotropic directions, were reported for different wood species, originating in the temperate zone in

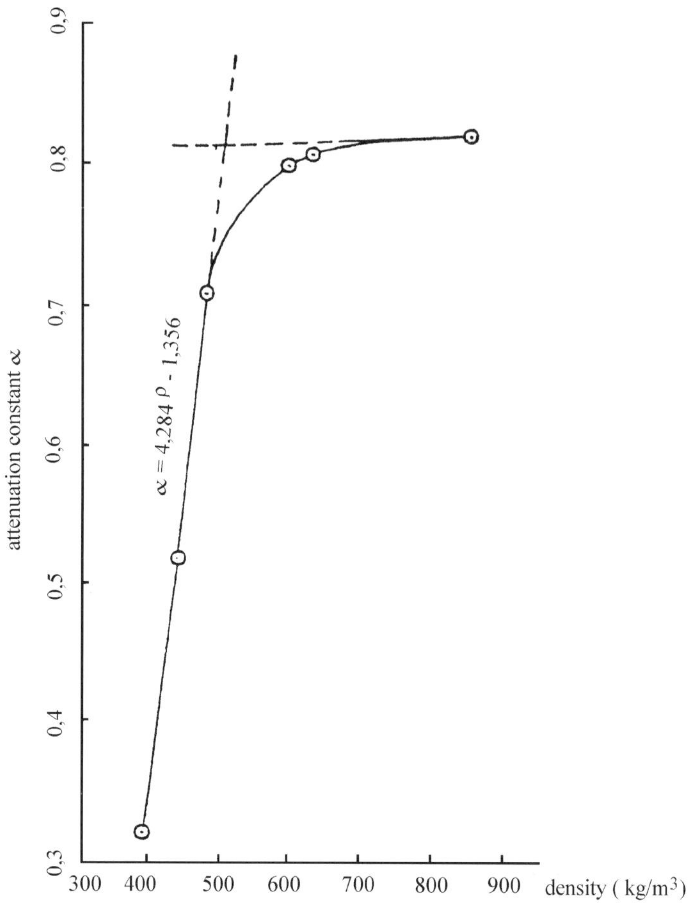

Fig. 4.10. Attenuation as a function of density for different Indian tree species with density ranging from 400 to 900 kg/m^3. (Jain and Sanyal 1996; with permission)

Europe, USA and Japan, and also from the tropical zones in India and Malaysia (Hearmon and Burcham 1954; Uyemura 1960; James and Hamill 1965; Lin 1967, 1973; Norimoto 1976, James 1975; Norimoto et al. 1978; Peyskens et al. 1984, Chazelas 1991; Choffel et al. 1992; Goy et al. 1992; Jain and Sanyal 1996; Obataya et al. 1996; Dubey and Deorani 1997; Kabir et al. 1998; Greaves 1998; Zielonka and Dolowy 1998).

For brevity and simplicity, from this important literature related to the dielectric constants of wood, the discussion will be limited to considering the relationship of the dielectric constant to the moisture content, to the structural orientation that induces anisotropy and to wood density.

4.2.2.1 Effect of Moisture Content

The effect of moisture content on the dielectric constant of Douglas fir at 3 GHz was reported by James and Hamill (1965). The dielectric constant ε′, both parallel and perpendicular to the grain, increased continuously with moisture content in the range from 8 to 75% (Fig. 4.11). This relationship was considered linear below the fiber saturation point. The constant tan δ behaved differently, increasing rapidly at low moisture levels to a maximum value at fiber saturation. A concave downward trend of tanδ was observed with increasing moisture content in the range of 25 to 75%.

For oven-dry wood Norimoto and Yamada (1972) reported that ε′ of *Chamaecyparis obtusa* varies with the grain angle, as can be seen from Fig. 4.12. In the longitudinal direction the ε″ value is approximately twice its value in the radial direction. Very little difference was observed between the ε′ in the radial and tangential directions.

Dielectric properties of three softwood species, pine, hemlock and spruce, were studied by Peyskens et al. (1984) below the fiber saturation point. It was observed that ε′ and ε″ increased with moisture content from 5 to 35%, for all anisotropic directions (Fig. 4.13). The curve corresponding to tan δ reached a

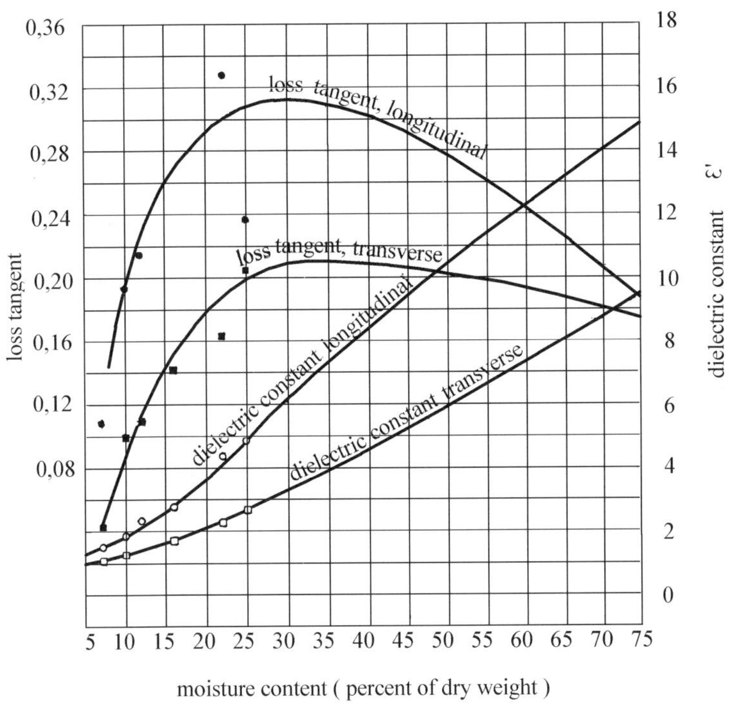

Fig. 4.11. Dielectric constant ε′, and tanδ of Douglas fir as a function of the moisture content. (James and Hamill 1965; courtesy of Forest Products J)

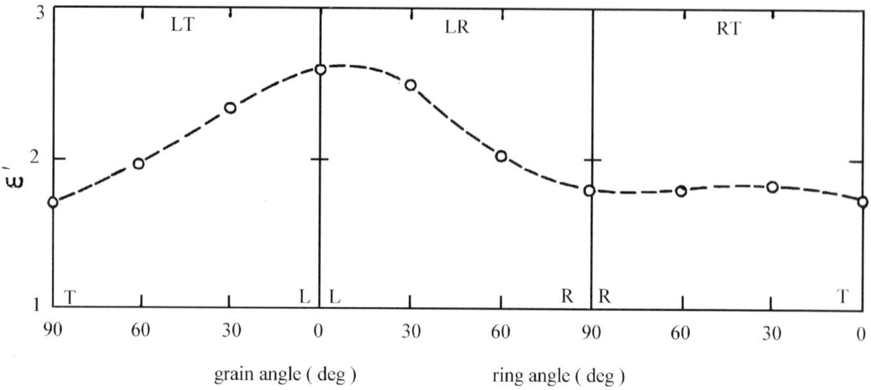

Fig. 4.12. Variation of the dielectric constant in three anisotropic planes. (Norimoto and Yamada 1972, with permission)

maximum between 20 and 25% moisture content. The increasing values of tan δ with a moisture content <25% are probably due to two factors:

- the increase in water within the wood matrix, since water is characterized by high dielectric values
- the polar components of the cell wall and the cellulose itself acquire more freedom of rotation at higher moisture contents and contribute to a more pronounced dielectric behavior.

When the moisture content of the specimen increases in the range >25%, in the vicinity of the fiber saturation point, the polar groups in the cell wall do not increase any more because their freedom of rotation is at a maximum and, consequently, the curve of tan δ that is indeed $\varepsilon''/\varepsilon'$, decreases.

The strong effect of the moisture content on the dielectric constant suggests that a specific drying pattern can be obtained with microwave energy. High moisture content zones with high values of ε'' above the fiber saturation point will dry first, compared with the surrounding zones containing less moisture. This is in contrast with conventional drying methods in which the surface dries more rapidly than does the moist interior.

4.2.2.2 Effect of Anisotropy

The effect of anisotropy can be expressed as a ratio between the dielectric constants, as demonstrated by Makoviny (1988) for two frequencies and four species: spruce, pine, beech and oak (Table 4.7). For all species it was noted that $\varepsilon'_L > \varepsilon'_R > \varepsilon'_T$. In the transverse plane, the ratio $\varepsilon'_R/\varepsilon'_T$ ranges from 1.02 and 1.08 for softwoods and between 1.13 and 1.16 for hardwoods. This means that from a dielectric point of view the softwood species are less anisotropic than

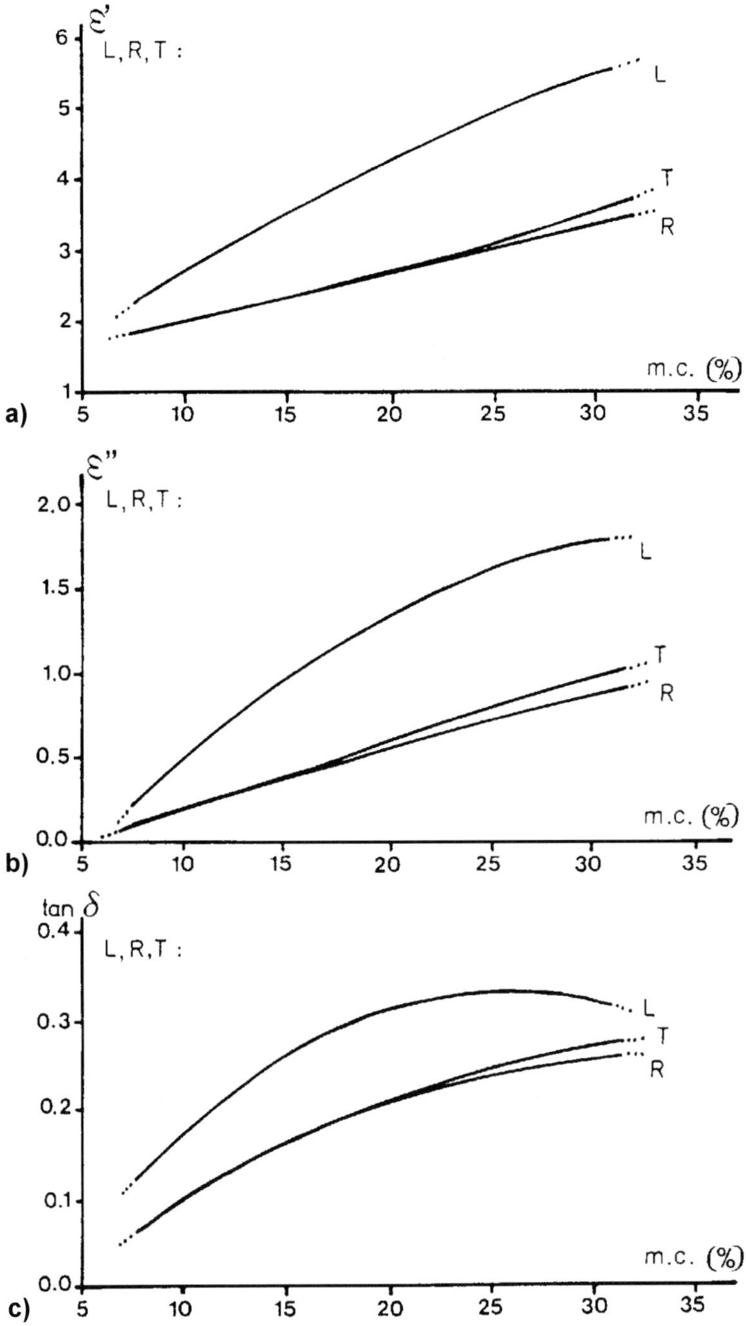

Fig. 4.13a–c. Dielectric constants as a function of moisture content and anisotropic direction in pine. **a** ε′ as a function of moisture content; **b** ε″ as a function of moisture content; **c** tanδ as a function of moisture content. (Peyskens et al. 1984; courtesy of Wood Science and Technology)

Table 4.7. The anisotropy of wood expressed by the dielectric constants of four species (spruce, pine, beech and oak) with densities ranging from 416 to 677 kg/m^3. (Makoviny 1988)

Frequency	Anisotropy	Spruce	Pine	Beech	Oak
1 MHz	$\varepsilon'_L/\varepsilon'_T$	1.72	1.54	1.51	1.42
	$\varepsilon'_R/\varepsilon'_T$	1.08	1.08	1.18	1.14
20 MHz	$\varepsilon'_L/\varepsilon'_T$	1.54	1.56	1.33	1.35
	$\varepsilon'_R/\varepsilon'_T$	1.02	1.06	1.16	1.13

Table 4.8. Attenuation coefficients of microwaves at 10 GHz as a function of sample thickness and of the anisotropic direction. (Chazelas 1991, with permission)

Range of thickness	Thickness[a] (mm)	Attenuation in L direction (Neper/m)	Attenuation in T direction (Neper/m)
From 28 to 20 mm	27.9	48.21	26.07
	25.15	47.72	25.97
	23.11	51.59	29.55
	19.57	51.78	28.32
From 30 to 22 mm	30.41	60.33	34.02
	27.43	60.41	34.65
	24.79	64.78	34.69
	22.10	65.35	37.97

[a] To avoid the effect of wood structure on the measurements, the thickness of the sample was progressively reduced.

hardwood species in the transverse plane. The influence of the frequency is more important for the ratio $\varepsilon'_L/\varepsilon'_T$ than for $\varepsilon'_R/\varepsilon'_T$. The highest anisotropy was observed at 1 MHz in spruce, for which $\varepsilon'_L/\varepsilon'_T = 1.72$ and the lowest anisotropy was found for beech, for which $\varepsilon'_L/\varepsilon'_T = 1.33$ at 20 MHz. This behavior was explained by an electrical model of the annual ring.

The effect of anisotropy on the attenuation coefficients as a function of specimen thickness for spruce is shown in Table 4.8. The samples with a thickness of 30 to 22 mm have a greater attenuation than the 28- to 20-mm samples. Increasing the thickness tended to increase the attenuation coefficient.

The effect of wood anisotropy on the dielectric constants of a large number of softwood and hardwood species was extensively studied by Norimoto and his coworkers at Kyoto University, Japan (Tanaka et al. 1975a,b; Norimoto et al. 1978; Nishino and Norimoto 1990). This research indicates that the dielectric constants in the longitudinal direction are higher than those in the transverse plane. Norimoto et al. (1978) noted that this behavior results from the arrangement of the cell wall and lumen, which is in agreement with Lin's (1967) statement that the hydroxyl groups of the cellulose should have more freedom of rotation in the longitudinal direction than in radial and tangential directions. The study of the effect of anisotropy on ε″ clearly demonstrated that during drying the absorbed power can be increased by an optimum orientation of the specimens versus the microwave antenna.

4.2.2.3 Effect of Density

The effect of density on the dielectric properties of wood was extensively studied by Skaar (1948), Rafalski (1966), Lin (1973) and Vermaas (1973). Skaar (1948) showed that the density has a consistent effect on the dielectric constants of different species. He noted that the low variability of density within a single species minimizes the variability in dielectric constants among species. Rafalski (1966) observed a linear effect of density, ranging between 700 and 1400 kg/m^3 on ε″ and tanδ.

Lin (1973) studied only one species, western hemlock, using a stepwise analysis of the logarithmic values of dielectric permittivity, and resistivity as they varied with moisture content and density. He found that the moisture content alone accounted for more than 94% of the variability in dielectric constants. Incorporation of density as an additional variable improved the regression coefficient only by several percent. Vermaas (1973) established regression equations of the dielectric constants of *Pinus pinaster* versus density, moisture content (0; 5; 10; 15 and 20%), frequency (10, 30 and 50 MHz), grain direction and extractives (resin content). The proposed models of regression equations contained linear, quadratic, and third power terms as well as first order interaction terms. He concluded that the dielectric constant ε′ for all directions increased with increasing density and moisture content and decreased with increasing frequency. The loss tangent for all anisotropic directions increased with increasing density, frequency and moisture content from 0 to 8%. From 8 to 20% an important decrease was observed only in the longitudinal direction. The extractives had no influence on either ε′ or the loss tangent. As can be seen, the effect of density was not clearly demonstrated for all species. The sorptive capacity and the permeability of each species seems to be more important than the density for the dielectric properties of wood.

From all dielectric measurements, the parameter able to express in a better way the molecular structure of wood is the dielectric loss (Norimoto and Zhao 1993; Yokoyama and Norimoto 1996, 1997; Yokoyama et al. 1999, 2000a). The dielectric relaxation process induced by the motions of methylol groups in the amorphous zones of the cell wall was demonstrated by Yokoyama and Norimoto (1996) with absolutely dry spruce wood, within a wide range of frequencies and temperatures (−160 to +20 °C). It was found that the maximum dielectric loss in the relaxation process was around 10 MHz. The iso-dielectric loss parameter was calculated for hinoki, from the Cole–Cole equation, as a function of frequency and temperature. The effect of relative humidity on relaxation phenomena was demonstrated by varying the relative humidity, in

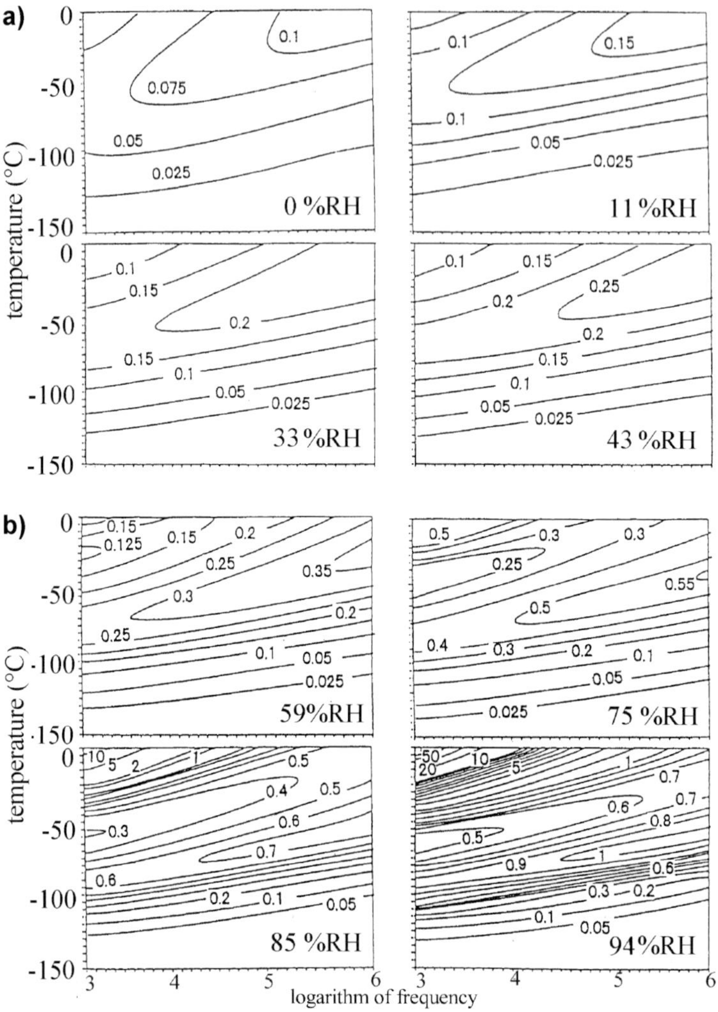

Fig. 4.14a,b. Iso-curves of the dielectric loss as a function of temperature and frequency for the relative humidity, ranging from 0 to 94%, calculated for hinoki. (Yokoyama et al. 2000a; courtesy of the Japan Wood Research Society)

which the wood samples were tested at constant temperature (Fig. 4.14). The distribution of relaxation times narrowed with increasing relative humidity.

4.2.2.4 Effect of Chemical Constituents

The effect of the chemical constituents of wood such as crystalline cellulose, hemicelluloses and lignin, on the anisotropy of hardwood dielectric constants, was studied by Nishino and Norimoto (1990). The theoretical development was based on models in which the proportion of chemical constituents of wood was deduced from the rule of mixtures. The dielectric constants of the principal constituents are given in Table 4.9. These values were used for estimating the dielectric constants of the cell wall shown in Table 4.10. Closer examination of these data and of a simplified model of the anatomic structure of hardwood leads to the theoretical values of the dielectric constant ε' as a function of moisture content in the range 0–15%, for three anisotropic directions. The estimated values in the longitudinal direction were higher than those in the radial and tangential directions and were in good agreement with data measured with mizume (*Betula grossa*).

Table 4.9. Theoretical values of dielectric constants of chemical constituents of wood. (Nishino and Norimoto 1990, with permission)

Moisture content (%)	ε_1 Noncrystalline cellulose in direction 1	ε_2 Noncrystalline cellulose in direction 2	Lignin	$\varepsilon_{P\ for}$ pentosan	$\varepsilon_{M\ for}$ mannan matrix
0	8.21	8.00	4.04	4.01	4.02
10	10.73	10.51	4.45	6.52	5.46
20	12.87	12.63	4.70	8.64	5.56
30	14.99	14.73	4.95	10.78	7.63
40	17.48	17.24	5.25	13.28	8.86
50	20.64	20.37	5.80	16.49	10.52
60	24.02	23.74	6.30	19.88	12.19
70	28.32	28.00	7.00	24.15	14.37
75	30.58	30.33	7.30	26.48	15.49

Table 4.10. Calculated dielectric constants of cell wall of hardwoods in longitudinal (ε_{Lcw}) and transverse direction (ε_{Tcw}), calculated for a microfibril angle of 13°. (Nishino and Norimoto 1990, with permission)

Moisture content (%)	ε_{Lcw} In longitudinal direction	ε_{Tcw} In transverse direction
0	5.60	4.68
10	6.42	5.84
20	7.54	6.72
30	8.62	7.54
40	9.90	8.49
50	11.58	9.72
60	13.34	10.97
70	15.65	12.60
75	16.87	13.64

4.3 Equipment for Dielectric Measurements and for Microwave Imaging Technique

The technique of microwave imaging of materials was initiated in 1980 with the development of microwave holography and of radar and remote sensing applications (Bolomey and Pichot 1990). During the last 10 years, microwave imaging of penetrable objects in complex environments has been shown to be a complementary technique to X-ray and ultrasonic tomography.

The recent development of microwave imaging was made possible because of three factors:

- the development of probing devices of greater flexibility and accuracy at a reasonable cost
- advanced wave front processing related to the enhancement of image quality, considering simultaneously the spatial and contrast resolution and the solution of theoretical problems of inverse scattering.
- the relevant technological progress in microwave and computer technologies.

The algorithms for microwave imaging are similar to those for ultrasonic tomography. After the amplitude, phase and polarization probing a of wave front or of a scattered field have been obtained in the transmission or in the reflection modes, different procedures can be applied in order to obtain information about the structure of an object. The image can be obtained by the direct use of projection. (Fig. 4.15a). The expansion of the scattered field during propagation is determined by the precise localization of the target in the transverse plane. The domain of validity for the direct use of projections is restricted

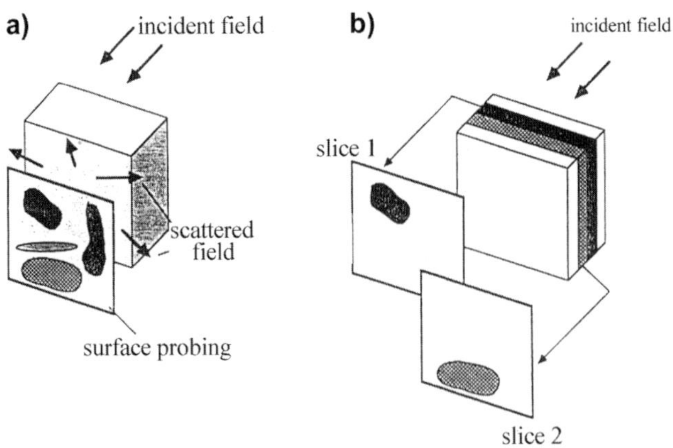

Fig. 4.15a,b. Microwave imaging of an object. **a** With projection produced by direct measurements; **b** with tomography obtained by numerical processing of the wave front. (Bolomey and Pichot 1990, with permission)

to planar objects, for which the distance of observation must be within one wavelength.

In contrast to projection imaging, tomographic imaging is obtained with a more elaborate processing of a microwave wave front and gives a cross-sectional representation of the object (Fig. 4.15b). Either a space domain iterative approach or a spectral formulation for diffraction tomography can be used. In the latter case, using the Born approximation, the image is obtained after taking the inverse Fourier transform. For inhomogeneous solids, the procedure of reconstruction is subjected to various artifacts due to diffraction effects. The most promising approach for microwave imaging is the space domain iterative formulation, which is very sensitive to the accuracy of local measurements. The profile of the object is obtained from the complex permittivity values. The domain of validity of this method depends directly on the signal-to-noise ratio. For industrial applications, cost considerations are of great importance and a compromise between the cost of the equipment and the rapidity of the performance must be found.

In the following, we will analyze first the equipment for laboratory measurements of dielectric constants of wood and, second, the equipment for imaging the structure of wood during processing under industrial conditions.

4.3.1 Equipment for Laboratory Measurements of Dielectric Constants

There are two basic microwave techniques for dielectric measurements: the transmission and the reflection methods. In the first configuration, the dielectric material is located between the transmitting and receiving horn antennas, while in the second case both antennas are located in front of the sample.

The instrumentation for laboratory measurement of dielectric constants and the loss tangent of wood is a resonant method, as described in ASTM D 150-74 (1986). The main components of an experimental device are:

- the source of continuous microwaves (Klystron microwave generator) that produce the electromagnetic wave at relatively low power and high frequency.
- the waveguide.
- two pyramidal horn antennas, one acting as a transmitter and the other as a receiver The distance between the two horn antennas must be kept greater than $2D2/\lambda$, where D is the horn aperture and λ is the operating wave length.
- the specimen, located between the transmitter and the receiver horns.
- the detectors of the signal amplitude, phase and degree of depolarization.

The equipment for measurements of the variation of amplitude and phase of the electromagnetic wave as described by Kharadly (1985) is shown in Fig. 4.16a. The system works as a bridge circuit balanced in two steps, without

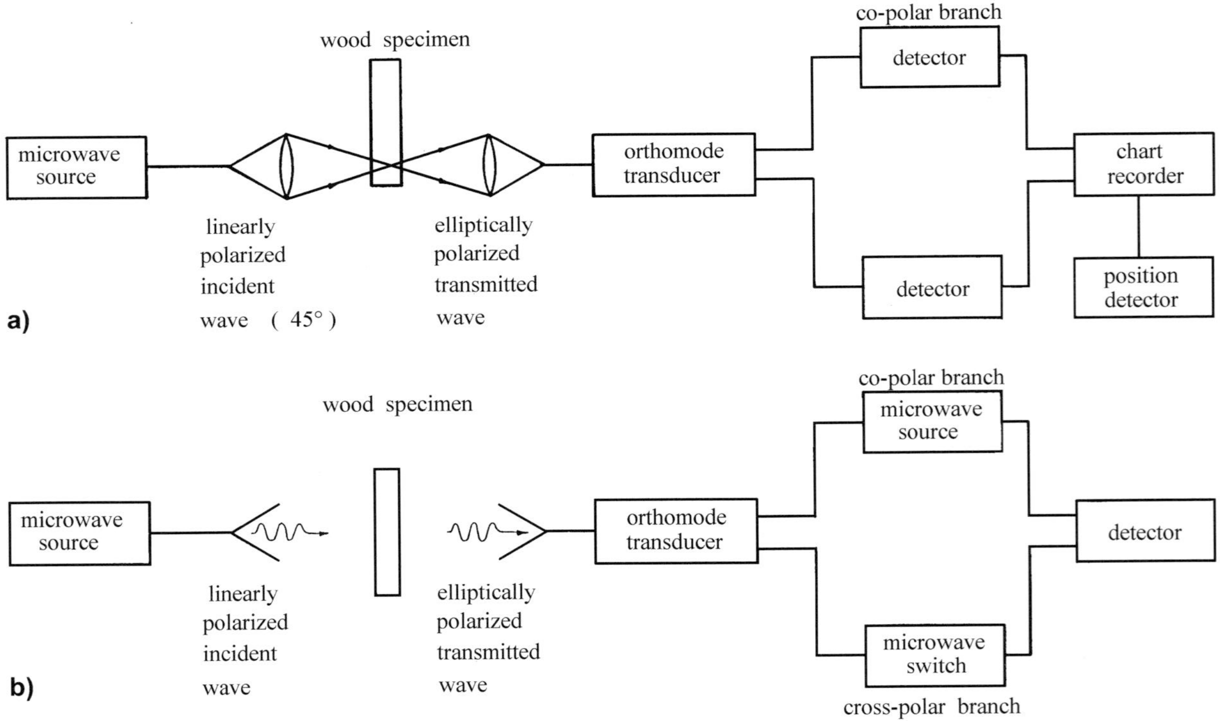

Fig. 4.16a,b. Microwave systems for the location and detection of different parameters of wood (Kharadly 1985). **a** System for the measurement of attenuation and phase shift for detection of wood density and moisture. **b** System for measurements of polarization of transmitted waves for the detection of the slope of grain.

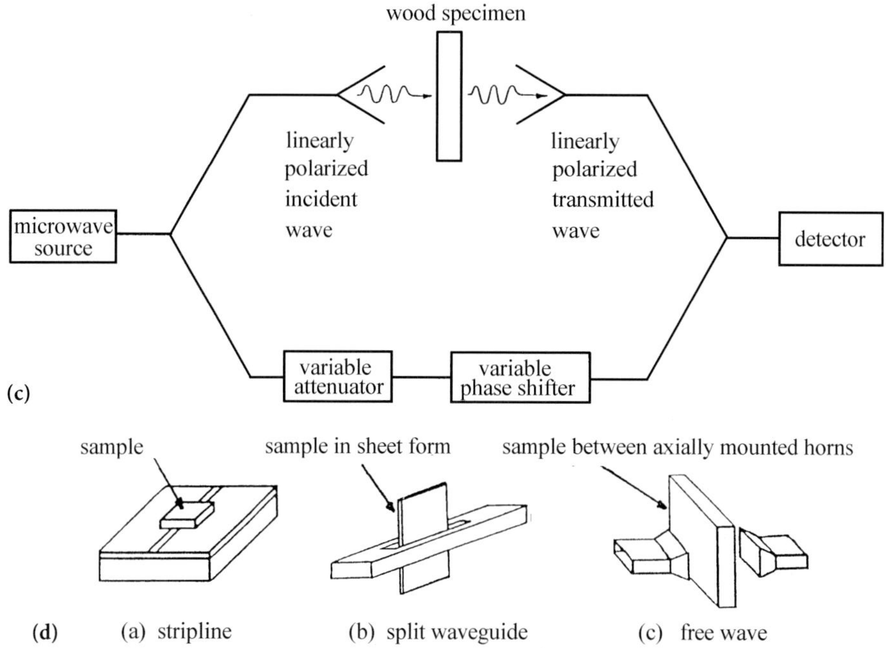

Fig. 4.16c,d. **c** System for the detection of the discontinuities introduced by the presence of knots (courtesy of Forest Research Society). **d** Sensors for transmission measurements (Thompson 1996, with permission)

and with a sample, by adjusting the variable attenuator and phase shifter. This system was designed to measure the moisture content and density of wood.

The slope of grain measurements are based on the measurements of the variation of the polarization of a transmitted wave that is elliptically polarized by the wood structure (Fig. 4.16b). The orthomode transducer separates the components of the transmitted wave into a co-polar component and a cross-polar component, which are alternately sampled with a high-speed microwave switch before processing of the corresponding signal by the detector.

For the detection of defects, such as knots, both the variation of density and the slope of grain must be measured. For this purpose, a combination of the previous systems can be used, as shown in Fig. 4.16c. The incident wave is polarized at 45° versus the direction of the fibers, which generates an elliptically polarized wave with two components (the co-polar and the cross-polar), which are detected separately. In this way, a typical response signature is obtained, indicating the presence of the defect.

Based on a similar approach, Forrer and Funck (1998) used a dielectric spectrometer operating in a frequency domain at 10 MHz, for detection and 3-D imaging of knots. The electrode of the spectrometer was 19.5 mm in diameter and was mounted on a translation table. The scanning line was chosen to

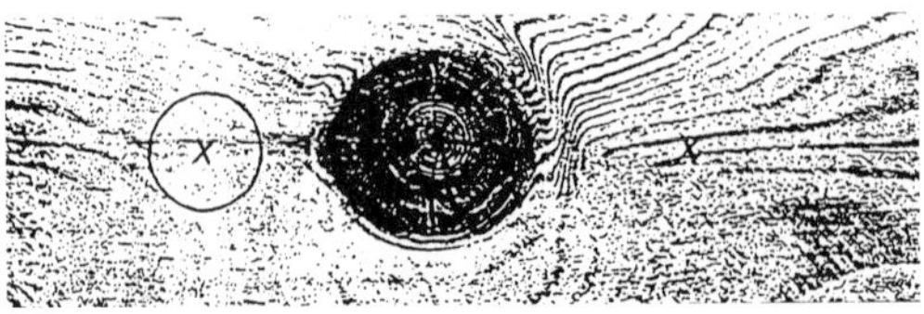

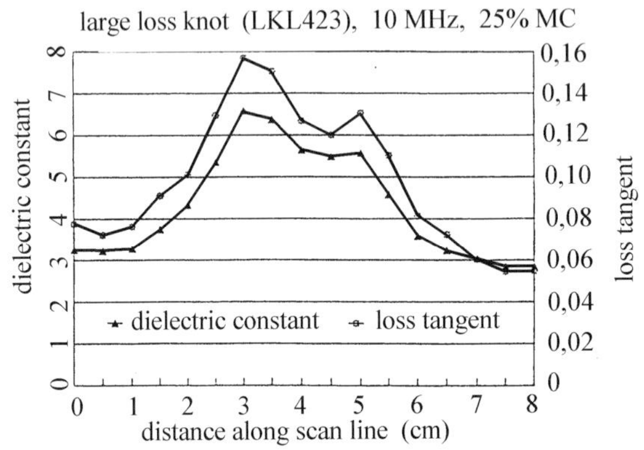

Fig. 4.17. Signature of a loose knot expressed by the dielectric constants and by the loss tangent. *Above* The location of the scanning line on a loose knot. *Below* Dielectric constants as a function of the position on the sample. (Forrer and Funck 1998; courtesy of Holz als Roh und Werkst.)

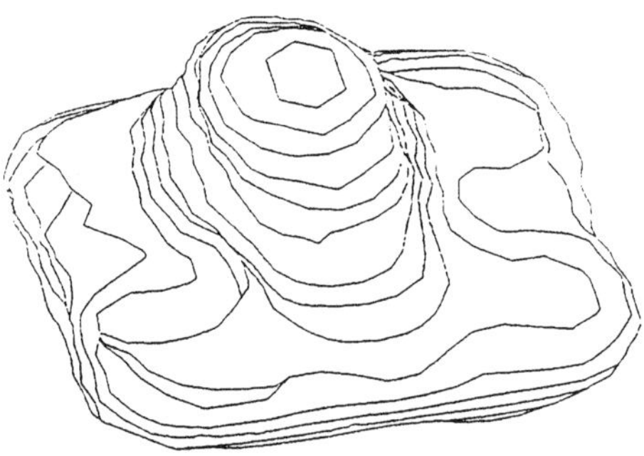

Fig. 4.18. Three-dimensional images of a tight knot 14 mm in diameter. (Forrer and Funck 1998; courtesy of Holz als Roh- und Werkst.)

cover a zone of approximately 5 cm^2 in the grain direction with the electric field applied through the specimen in the radial and tangential directions as can be seen in Fig. 4.17. The location of the loose knot is precisely determined by both the dielectric constant and the loss tangent. A 3-D image of a tight knot of 14 mm diameter is shown with good resolution in Fig. 4.18.

The limitation of microwave through-transmission or reflection methods is caused by the low sensitivity of this technique to differentiate between different types of defects, as can be seen in Fig. 4.19. Forrer and Funck (1998) studied the dielectric constants of knots, pitch pockets, clear wood, blue stain and open holes as functions of moisture content and frequencies (Fig. 4.20). From this

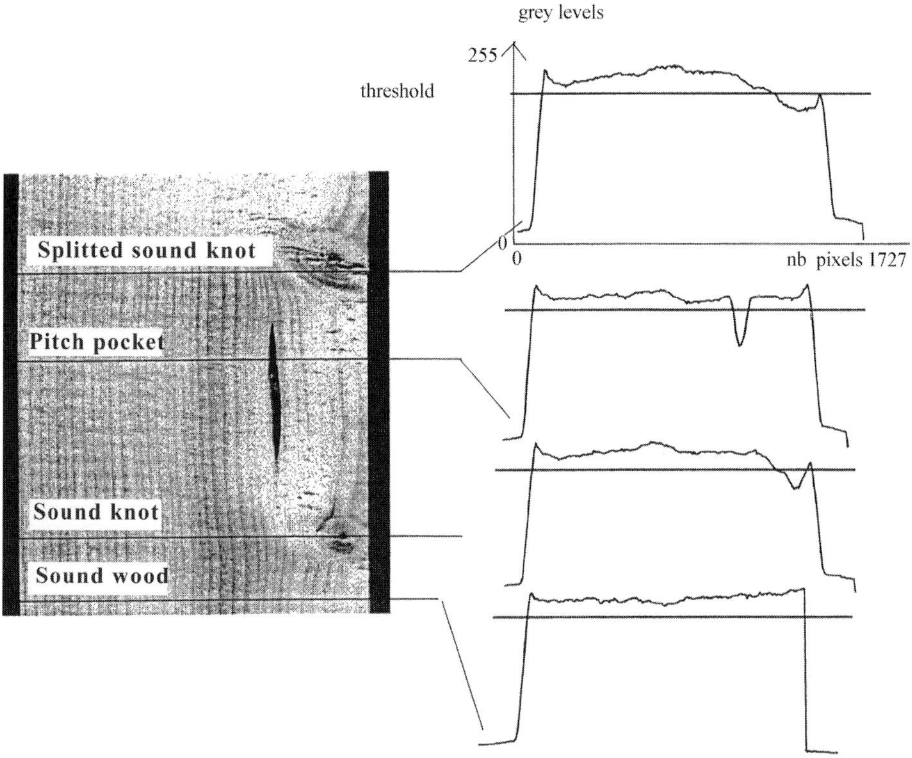

Fig. 4.19. Microwave signals for different types of defects. (Choffel 1999, with permission)

graph, it was impossible to define precisely a typical defect, but a general separation in three main groups (air gaps, pitch pockets and tight knots) occurred with increasing moisture content and frequency. In a large cloud cluster of data, open holes and loose knots are in opposite position to tight knots. Between them, clear wood, blue stain, and pitch pockets can be observed. Forrer and Funck (1998) noted that under laboratory conditions, the main sources of errors associated with dielectric measurements of wood containing defects is induced by the moisture content variation in the defect zone compared with the area of surrounding sound wood and by the variability of the signal introduced by the wood sensor interface.

4.3.2 Equipment for Online Imaging of Wood Structure

Microwave imaging is a quantitative nondestructive evaluation method for characterizing surface and subsurface defects of a wood sample. The defects can be imaged by scanning in the proximity of the samples with a resolution that depends on the size of the aperture. The images can be obtained in the far

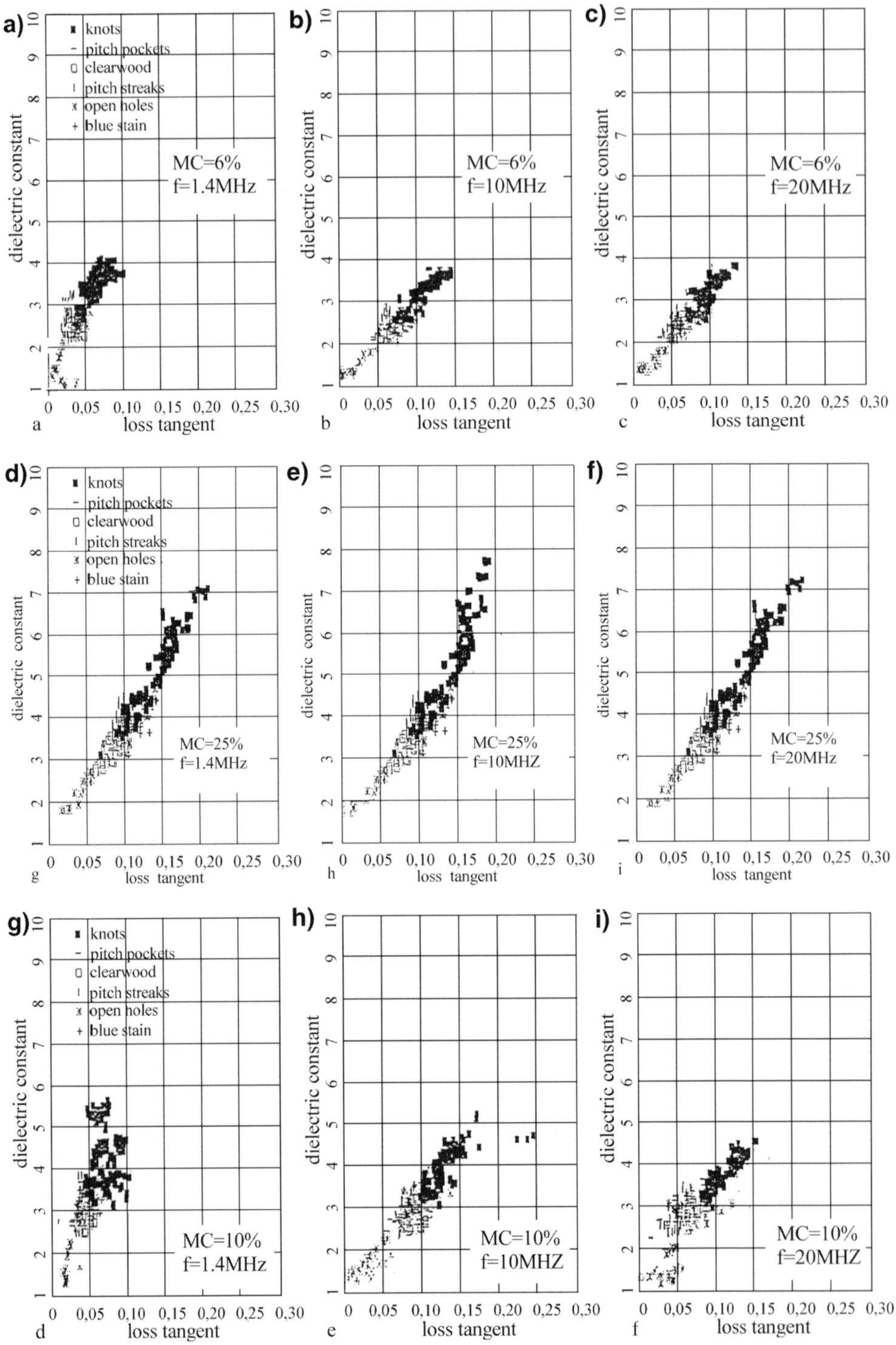
a)
knots
pitch pockets
clearwood
pitch streaks
open holes
blue stain
MC=6%
f=1.4MHz
dielectric constant
loss tangent
a
b)
MC=6%
f=10MHz
b
c)
MC=6%
f=20MHz
c
d)
MC=25%
f=1.4MHz
g
e)
MC=25%
f=10MHZ
h
f)
MC=25%
f=20MHz
i
g)
MC=10%
f=1.4MHz
d
h)
MC=10%
f=10MHZ
e
i)
MC=10%
f=20MHZ
f
0 0,05 0,10 0,15 0,20 0,25 0,30
1 2 3 4 5 6 7 8 9 10

or in the near field. For far-field imaging the smallest detectable defect is determined by the ratio $d/\lambda > 1$, where d is the length of the defect in the plane normal to the microwave propagating vector and λ is the microwave wavelength. This condition is known as the Rayleigh criterion, which is applicable to all electromagnetic imaging in the far field. For near-field imaging, the resolution can be as high as $\lambda/1000$, because in this case Rayleigh's criterion does not apply.

The probes used to evaluate the properties of the samples can be scatterometers or radars that measure the scattering properties of the medium, or can be reflectometers that measure the reflectivity of the sample due to its inhomogeneities and defects. The probes can operate in the proximity of the medium as open-ended coaxial lines, cavity resonators, antennas, etc. By moving the position of the transmitter across the sample, an image of the structure can be obtained. The development of microwave systems for online measurements of the moisture content in wood and wood-based composites is directly related to the economic benefit in the sawmill and particle board industries for front-end processing, in-process control and output quality control.

For online wood processing, the microwave equipment must satisfy several criteria, including:

- the recognition, location and mapping of defects, such as knots, reaction wood, slope of grain and other major discontinuities of wood structure;
- the capability of operating over a wide range of temperature, wood moisture content and density of different species, with sharp resolution;
- the normal speed of operation required by the processing;
- a noncontact and nonhazardous technique, with no susceptibility to vibrations and with simplicity of operation;
- inexpensive training for the operators.

The actual microwave equipment for the inspection of logs, lumber and composite boards is discussed below.

The inspection of logs is performed with the experimental device (Kaestner and Baath 2000) shown in Fig. 4.21. In front of a wide-band antenna, the log is rotated and translated. The frequency band used for measurements was between 4 and 8 GHz. More details about the experimental results obtained

◄

Fig. 4.20a–i. Relationships between the dielectric constant and loss tangent for knots, pitch pockets, clearwood, pitch streaks, open holes and blue stain at moisture contents ranging from 6.6 to 25% and for frequencies ranging from 1.4 to 20 MHz. **a** Moisture content 6% and frequency 1.4 MHz. **b** Moisture content 6% and frequency 10 MHz. **c** Moisture content 6% and frequency 20 MHz. **d** Moisture content 10% and frequency 1.4 MHz. **e** Moisture content 10% and frequency 10 MHz. **f** Moisture content 10% and frequency 20 MHz. **g** Moisture content 25% and frequency 1.4 MHz. **h** Moisture content 25% and frequency 10 MHz. **i** Moisture content 25 and frequency 20 MHz. (Forrer and Funck 1998; courtesy of Holz als Roh- und Werkst.)

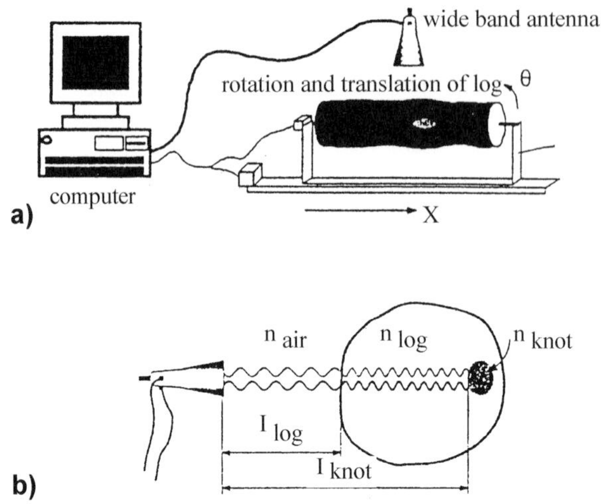

Fig. 4.21a,b. Scanning of logs with a microwave device. **a** Components of the device. **b** Refraction indices at the interface between log and knot. I_{log} and I_{knot} are the distances from the antenna to the log and the knot; n_{air} and n_{knot} are the refraction indices at the interface. (Kaestner and Baath 2000; courtesy of Forest Research Society)

with this equipment are given in Section 4.4.2. Technical details concerning the equipment used for the measurement of phase and amplitude for the inspection of lumber are given by King (1978), King and Yen (1981) and Martin et al. (1987).

While microwave techniques have shown a great potential for inspection of lumber, their applicability seems to be limited by their inability to recognize the nature of the defects. To overcome this drawback (Choffel and Martin 1966; Choffel 1999), it was proposed to supplement the equipment with a CCD camera, thus providing a coupling of a vision and a microwave device (Fig. 4.22). In this system, wood structural data are correlated with lumber strength and compared with classification grades. The assignment of a grade to a piece of lumber was based on the detection and location of defects and on three physical and mechanical parameters: the density, the strength and the stiffness. The camera provides the profile to calculate the dimensions of the board and to detect the presence of defects, while the microwave sensor shows the internal structure of the wood. The network is described by Beatty et al. (1999) and is based on the identification of the characteristic microwave signature of lumber that distinguishes one piece from another. The accuracy of the inspection of the internal structure depends on lumber thickness, operating frequency and the input power level.

The system described here has some advantages, such as: relatively low cost, satisfactory design complexity, good spatial and measurement resolution, no extensive need for signal processing and no radiation hazard. The system requires no contact and has a rapid response to automatic process control.

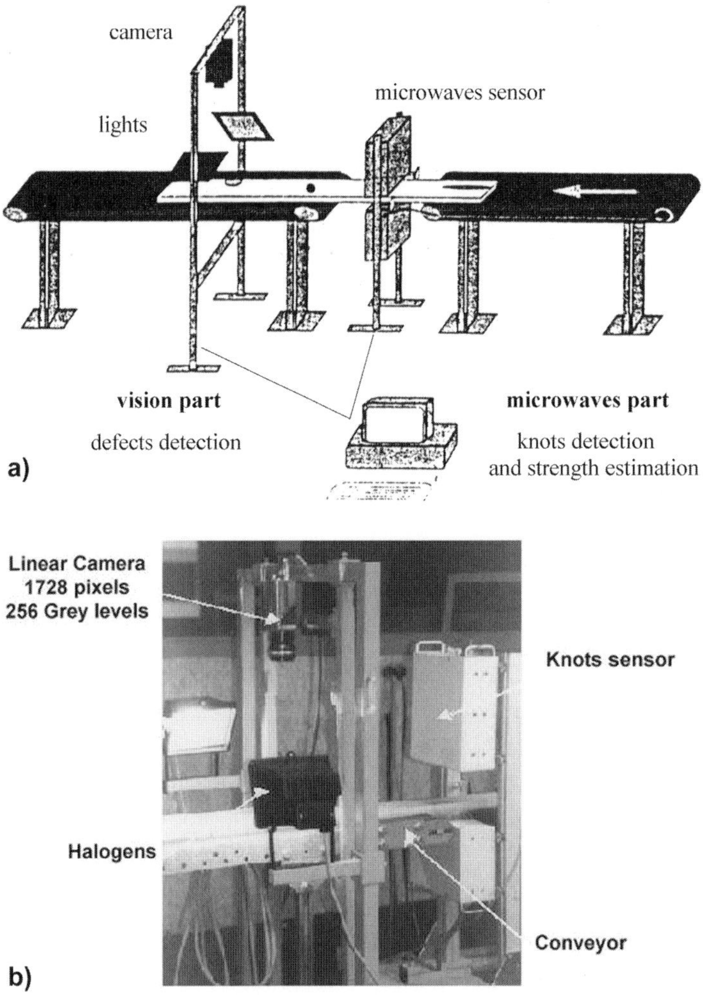

Fig. 4.22a,b. A vision and microwave-coupled device for detection of defects in lumber. **a** The device used for lumber inspection. **b** Details of the system of microwave probe and vision camera. (Choffel 1999, with permission)

For the inspection of composite boards (fiberboard, flakeboard and particleboard), King and Bausel (1993) developed a homodyne instrumentation with a continuous-wave source of 10 GHz. (Fig. 4. 23). The microwave signal received is amplitude modulated at 455 kHz and then mixed in the homodyne receiver with a portion of a continuous wave carrier signal. The wave amplitude and phase are converted to the 455 kHz base band signal for detection. Based on all of the collected experimental data, the computer processes the moisture content, density and dielectric constants. More recently, Völgyi (2000) has proposed a nondestructive method based on a microwave free-

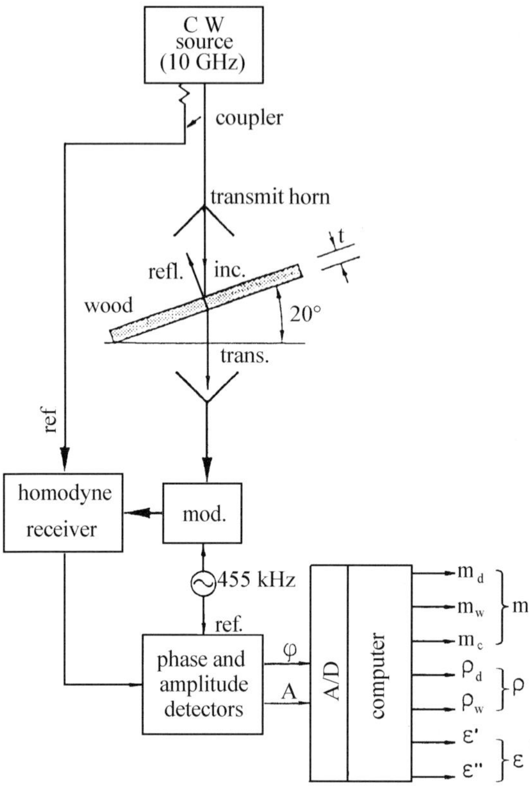

Fig. 4.23. System for inspection of composite boards with a homodyne instrumentation. (King and Bausel 1993; courtesy of Forest Products J)

space/double transmission reflection for the measurement of the dielectric complex tensor for continuous quality prediction of particleboard using monostatic transmission sensors in reflection-aperiodic-open configuration. Images of moisture distribution in real-size particleboard were obtained yielding an average value of 4% with a standard deviation of 0.48%.

4.4 Applications

This section is intended to cover a wide range of topics related to the application of microwave imaging techniques for the inspection of forests, logs, lumber and wood based composites. The microwave parameters used for image reconstruction are: the amplitude, the phase and the polarization. The system involves transmission of microwave energy through the samples or objects and measurements of reflected, refracted or diffracted fields. Two- or three-dimensional images of the objects can be obtained without physical sectioning.

As with other imaging techniques, inspection of large areas are possible. At the same time, wood structural information can be obtained on a scale ranging from meter to centimeters. The breadth of the discussion is intended to aid in deciding which approach would be best to make a specific series of measurements for the purpose of a well-defined imaging purpose.

4.4.1 Microwaves for Inspection of Forests

Microwave dielectric properties of plant materials were extensively studied at Kansas University in Professor Ulaby's laboratory. As noted by Ulaby and Jedlicka (1984), "from the standpoint of wave propagation, a vegetation canopy is a dielectric mixture of discrete dielectric inclusions (leaves, fruits, stalks, etc.) distributed within a host or background material such as air." On the other hand, the canopy is an inhomogeneous and anisotropic medium. The size of the canopy inclusions compared with the wavelength in the microwave region determines the absorption and scattering of microwaves.

The absorption and scattering microwave coefficients depend on the both wave polarization and the angle of incidence of the microwaves to the canopy and are related to the dielectric constant, volume fraction and geometry of the canopy inclusions (Ulaby et al. 1990; Richards 1990; Baker et al. 1994). It was demonstrated by Ulaby et al. (1983) that the canopy absorption coefficient can be calculated from the canopy equivalent dielectric constant ε_c where ε_c is a tensor with two components associated with the horizontal and vertical polarization directions. The waveguide transmission method can be used for the measurement of this dielectric tensor.

The two main factors influencing the dielectric tensor of plant materials are the dielectric constants of the water component of the plant material and the temperature of the media. El-Rayes and Ulaby (1987) used a broadband measurement technique to determine the dielectric properties of vegetation (leaves and trunks) over a wide range of frequencies, from 0.2 to 20 GHz. As can be seen from Fig. 4.24a, the dielectric spectrum of a balsam fir tree shows an important effect of frequency on ε' between 0.5 and 6 GHz. A smaller frequency effect was found for ε''.

To evaluate the radial variation of dielectric constants, measurements were performed from the pith to the bark on a freshly cut trunk of balsam fir, as can be seen in Fig. 4.25. The variations of the complex dielectric constant ε^* values are represented by concentric circles as a function of two parameters, the position on the transverse section of the tree and the corresponding moisture content. The values reported for each circle represent the average of 16 sample measurements. From the above figures, we observe a decrease of the ε^* values with increasing moisture content from the pith to the bark. The complex

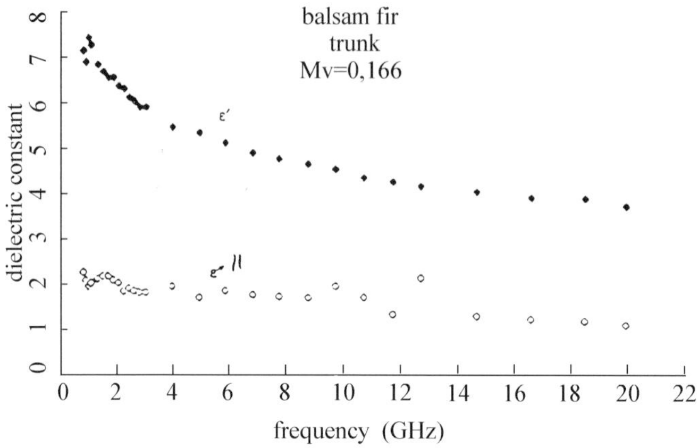

Fig. 4.24. Dielectric spectrum for ε′ and ε″ of a balsam tree trunk in a frequency range from 1 to 22 GHz. (El-Rayes and Ulaby 1987; copyright 2001 IEEE)

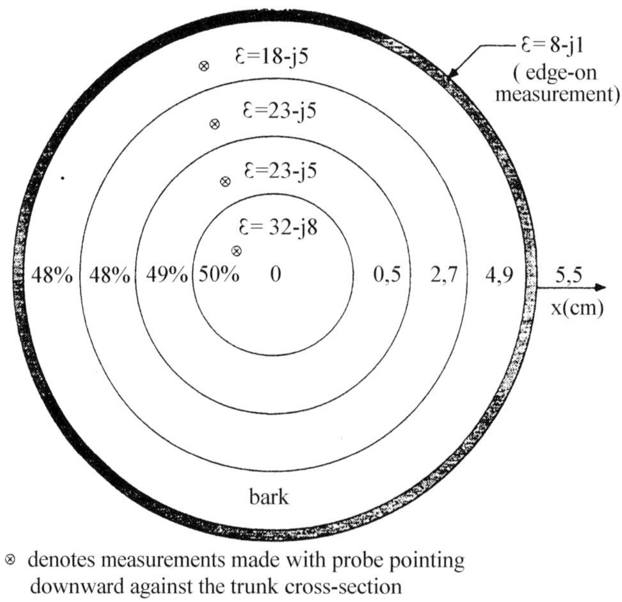

Fig. 4.25. Radial variation from pith to bark of the complex dielectric constant ε*, of a fresh cut balsam fir tree. (El-Rayes and Ulaby 1987; copyright 2001 IEEE)

dielectric constant in the pith region is $\varepsilon^* = 32 - j\,8$ and in the bark region is $\varepsilon^* = 8 - j\,1$.

The applicability of microwave techniques for detection and classification of forest vegetation was possible by using radar imaging in the microwave frequency range (Sieber 1985; Kessler 1987; Kasischke and Melack 1997). It was demonstrated that the estimation of the woody plant biomass and the monitoring of different temporally dynamic processes are possible by using spaceborne synthetic aperture radars (SAR). The best performance for biomass estimation was achieved with the lower frequency (P- and L-bands) radar systems.

The penetration measurements at the L-band for coniferous and deciduous trees are given in Fig. 4.26, for horizontal (HH) and vertical (VV) polarization directions using the experimental arrangement shown in Fig. 4.27. The data were collected at 1-m distance along blocks of ten trees. The mean backscattering cross section was calculated as a function of the height, ranging from the ground to 6 m. For coniferous trees (spruce) the backscattering between 1.5 and 4 m is relatively constant at −20 dB for both HH and VV channels and increases dramatically between 4 and 6 m. The attenuation is higher for horizontally (HH) polarized L-band waves than for vertically polarized waves (VV). The behavior of hardwood trees is different, and the hypothesis that the radar acts as a shape filter was advanced by the authors. They proposed that, for hardwood trees, the volume of branches is large enough to act as prime scatterers. For coniferous trees, however, the tree area projected in the direction of the incident radar wave seems to be the most important parameter. The backscattering characteristics of spruce trees were investigated as a function of polarization angle, ranging from 0° to 100°. At 0° the electrical field vector was oriented parallel to the trunk. The signal obtained was different, depending on whether or not the trunk was illuminated (Fig. 4.28). The stem strongly scatters the incident wave, and the increasing polarization can be related to the decrease of its cross section with height. The slope of the curve between 0° and 35° can be related to the diameter of the trunk. At a 43° polarization angle, a maximum of the curve was observed. This angle was presumed to correspond to the preferred orientation of the major branches. The peak at 43° was also observed for a tree whose trunk was not illuminated. For polarization angles larger than 43°, both curves (HH and VV) have the same appearance. This aspect was related to the high probability that other branches were oriented at an angle of 60°.

Hardwood trees have a different backscattering behavior because of the different size and distribution of their branches (with or without leaves) in the canopy. The signature of a deciduous tree is very different from that of a conifer tree, as can be seen for a walnut tree (Fig. 4.29), a very asymmetric tree. In this case, a minimum of −13 dB was observed at a polarization angle of 60°, followed by an increase to −9 dB at 85°. The decreasing part of the curve, between 0° and 60°, is due to the thick branches down to 45° from nadir. The second part of the curve for polarization, above 60°, was related to the connection angles between the trunk and branches. The modification of illumination geometry indicated a different orientation and branch structure (Fig. 4.29). Sieber (1985) concluded that the selection of an appropriate polarization with an X-band and an L-band radar permitted the detection and identification of deciduous and coniferous trees. The experimental data reported by Sieber

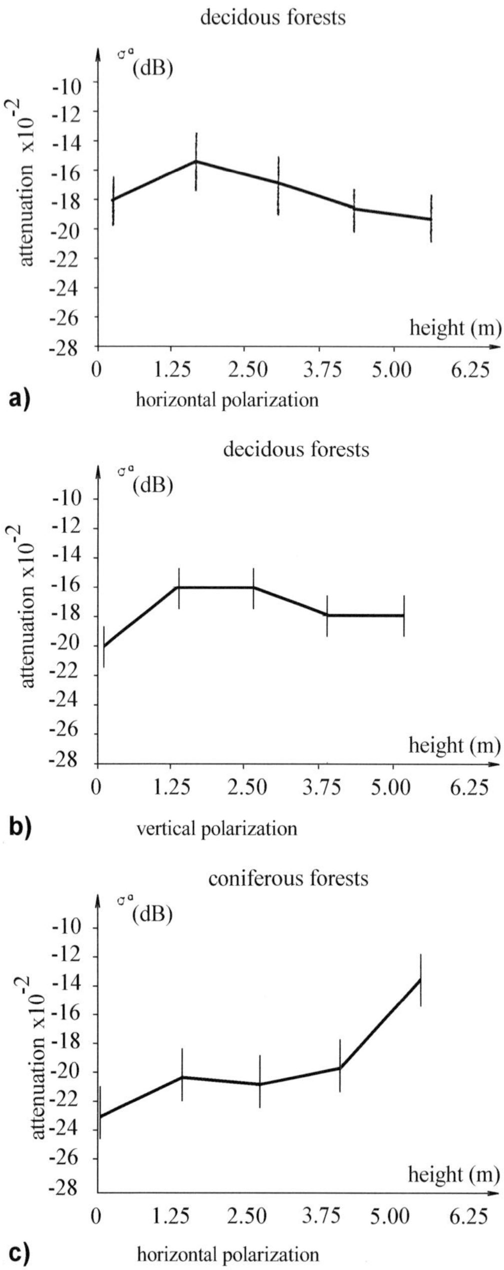

Fig. 4.26a–d. Penetration measurements at L-band radar for coniferous and deciduous forests. **a, c** Horizontal HH polarization. **b, d** Vertical VV polarization. (Sieber 1985, with permission)

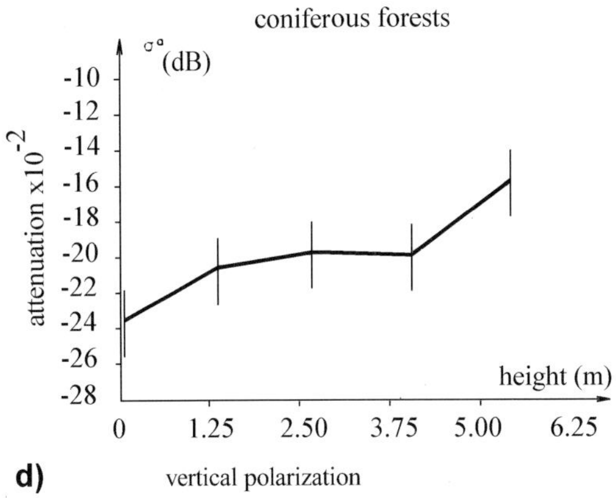

Fig. 4.26a–d. *Continued.*

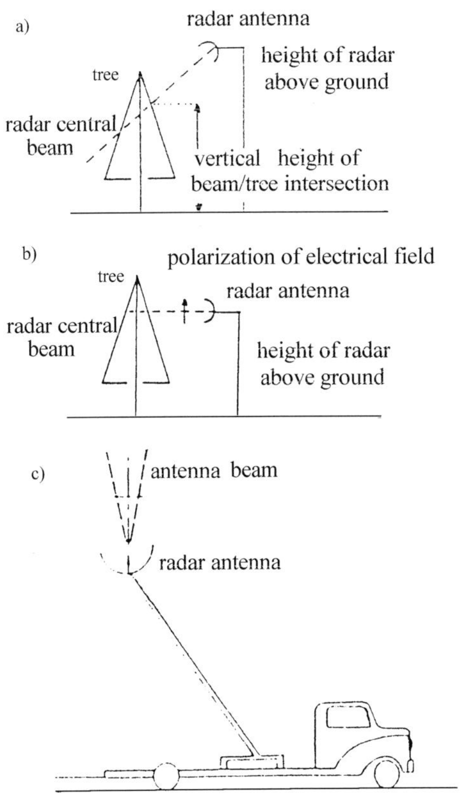

Fig. 4.27a–c. Experimental arrangement for penetration measurements (Sieber 1985; courtesy of ESA Journal). **a** The position of radar antenna and of the tree. **b** Position of the radar central beam and of the polarization direction versus the tree. **c** The INSCAT (Institute for Navigation's Scatterometer) sensor mounted on a truck. (Ulaby and Jedlicka 1984; copyright 2001 IEEE)

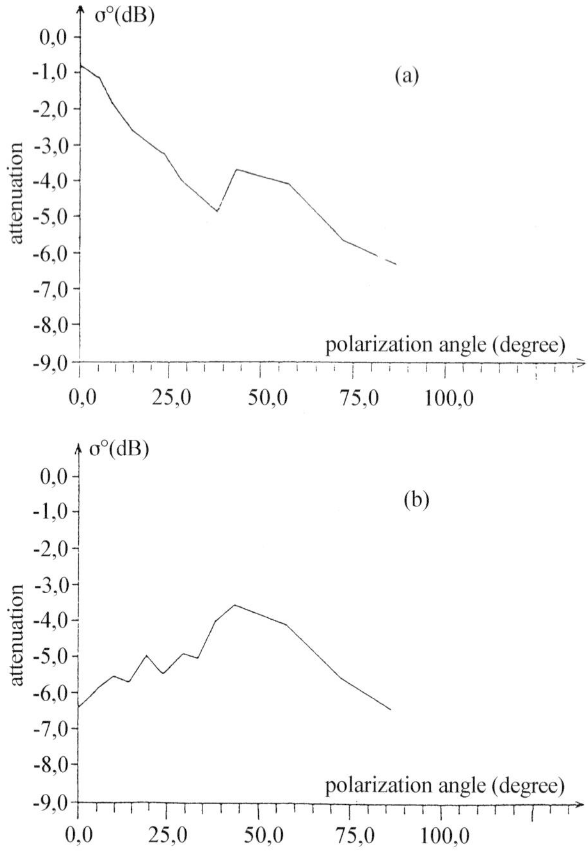

Fig. 4.28a,b. Backscattering characteristics for a spruce tree as a function of polarization angle. **a** Tree illuminated by radar beam. **b** A tree not illuminated by radar beam. (Sieber 1985; courtesy of ESA Journal)

(1985) were collected in the Rhine Valley, Germany, with the INSCAT (Institute for Navigation's Scatterometer) sensor mounted on a truck (Fig. 4.27c).

4.4.2 Microwaves for Internal Inspection of Logs

The inspection of the internal structure of logs by the microwave technique has been reported by Kaestner and Baath (2000) The detection of knots in logs

Fig. 4.29a–d. Backscattering characteristics for a walnut tree as a function of polarization angle. **a** Tree illuminated by radar beam. **b** Tree not illuminated by radar beam. **c** Idealized profile of the tree. **d** Detail of branches. (Sieber 1985; copyright 2001 IEEE)

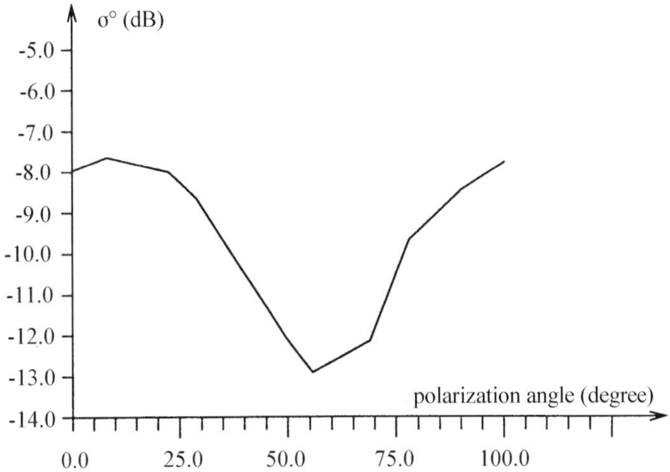
σ° (dB)
-5.0
-6.0
-7.0
-8.0
-9.0
-10.0
-11.0
-12.0
-13.0
-14.0
polarization angle (degree)
0.0
25.0
50.0
75.0
100.0

(a)

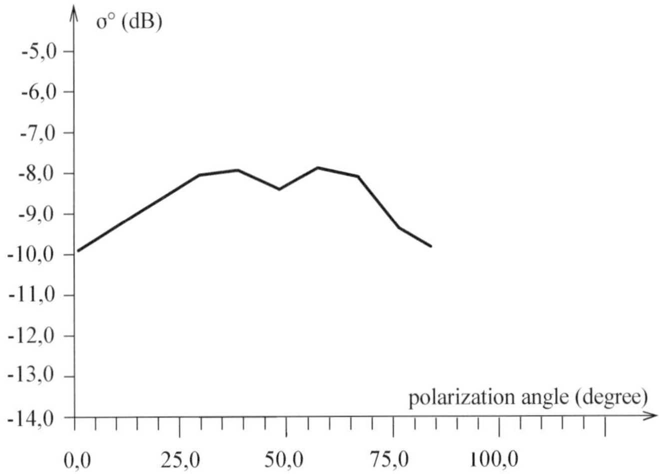
σ° (dB)
-5,0
-6,0
-7,0
-8,0
-9,0
-10,0
-11,0
-12,0
-13,0
-14,0
polarization angle (degree)
0,0
25,0
50,0
75,0
100,0

(b)

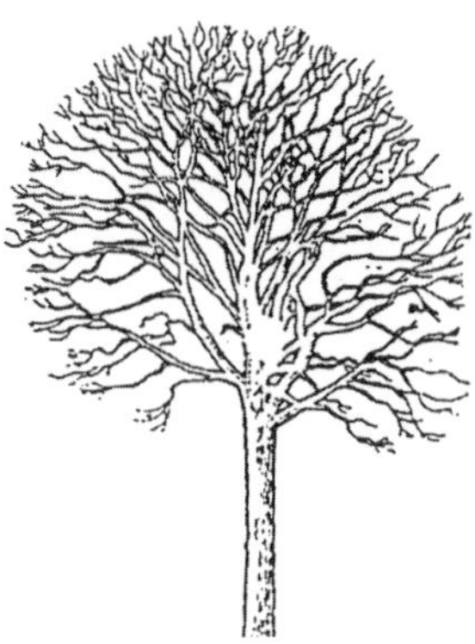

c)

d)

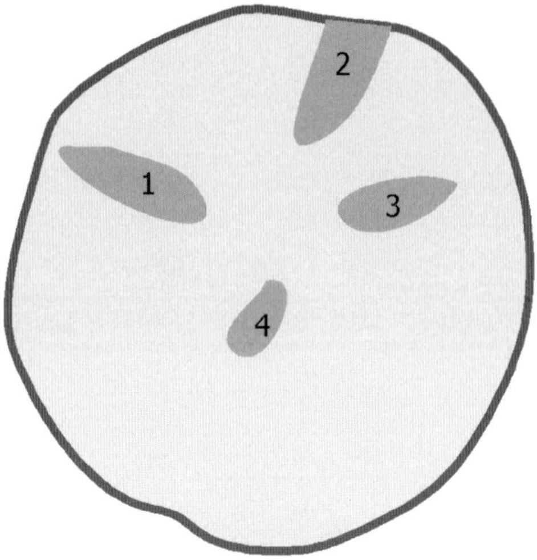

Fig. 4.30. Distribution of knots on the transverse section of the bolt. (Kaestner and Baath 2000, with permission)

of various species was performed in the frequency band between 4 and 8 GHz. In this section, we discuss the imaging technique used for the detection of knots in a spruce bolt of 17.5 cm diameter and 18 cm length. The actual distribution of knots is sketched in Fig. 4.30 where we observe three large knots designated as a, b, and c. They are disposed in a radial position at about 90° between knots a and b and at 60° between knots b and c. Knot b is externally visible.

Figure 4.21 shows the experimental device for knot detection. It is composed of three main parts: the microwave device with a wide band antenna, the conveyer for bolt rotation and translation, and the computer for equipment control and for storage of experimental data. The image reconstruction is based on the following dielectric parameters of the wood: the polarization ratio and the phase difference of two components of the electrical field, one parallel to the fibers and the other perpendicular to the fibers. The wave propagation phenomena that take place in the sample are very complex and a simplified hypothesis must be introduced to facilitate the understanding of these phenomena. In a first approximation it is assumed that the received signal at the antenna is the sum of two signals, the first emitted from the log and the second emitted from the knots. Secondly, it is assumed that for the continuous case, the sum can be replaced by an integral and the dielectric constants by a tensor function.

Continuous wave radar measurements of the depolarization of reflected waves were used to reconstruct the image. In the frequency range from 4 to 8 GHz, it was possible to obtain a discrete frequency spectrum. Using the inverse Fourier transform, the discrete spectrum measurements were transformed into a time delay spectrum. This spectrum shows the spatial distribution of the electrical field and of the polarization-related parameters that are used for image reconstruction.

The images obtained with the polarization ratio (Fig. 4.31a) have less resolution than those images obtained with phase difference (Fig. 4.31b), on which the knots are better defined. An improvement of the image resolution was obtained using iso-surface topographic slices calculated for the polarization ratio and for the polarization angle between the two components of the electric field (Fig. 4.32a,b). Knots a and b are readily visible in both cases, while knot c is readily visible only in Fig. 4.32b, where data based on the polarization angle were used.

Further research is needed for improvement of the image resolution and for a better understanding of possible correlations between the dielectric tensor

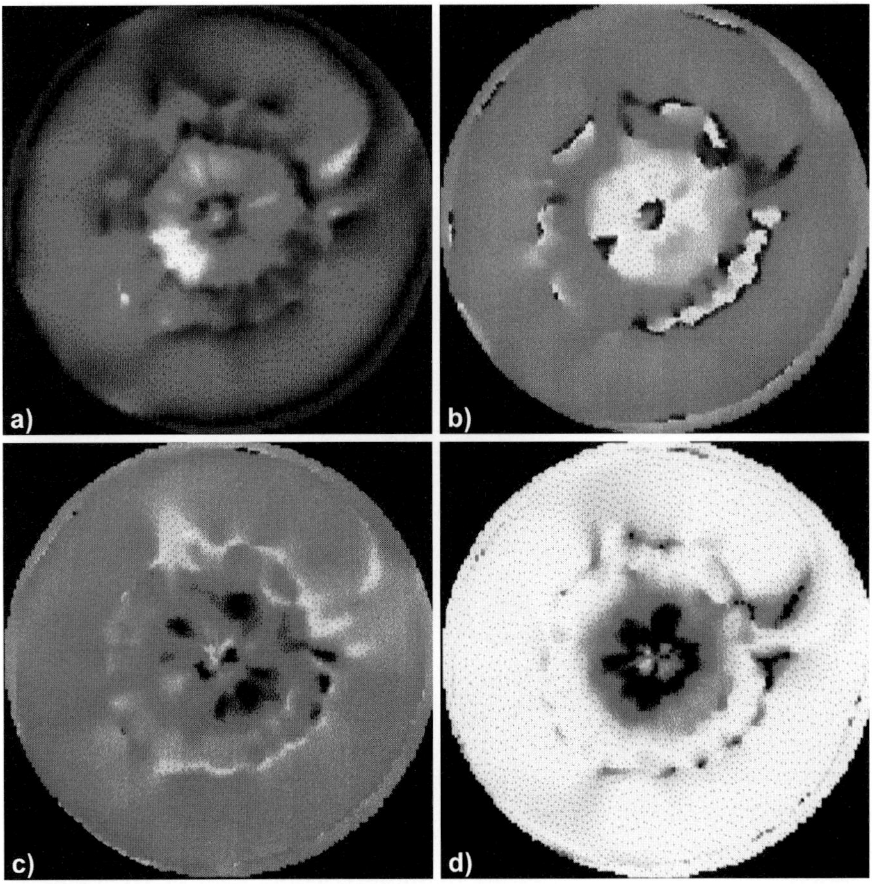

Fig. 4.31a–d. Iso-surface on a topographic slice. **a** $|\rho|$ The amplitude; **b** γ the argument (γ, arg ρ). **c** ε The ellipticity angle which gives the phase difference between two wave components and describes the degree of elliptical polarization. **d** τ The angle between the two components of the wave, corresponding to the tilt of the linear polarization. (Kaestner and Baath 2000, with permission)

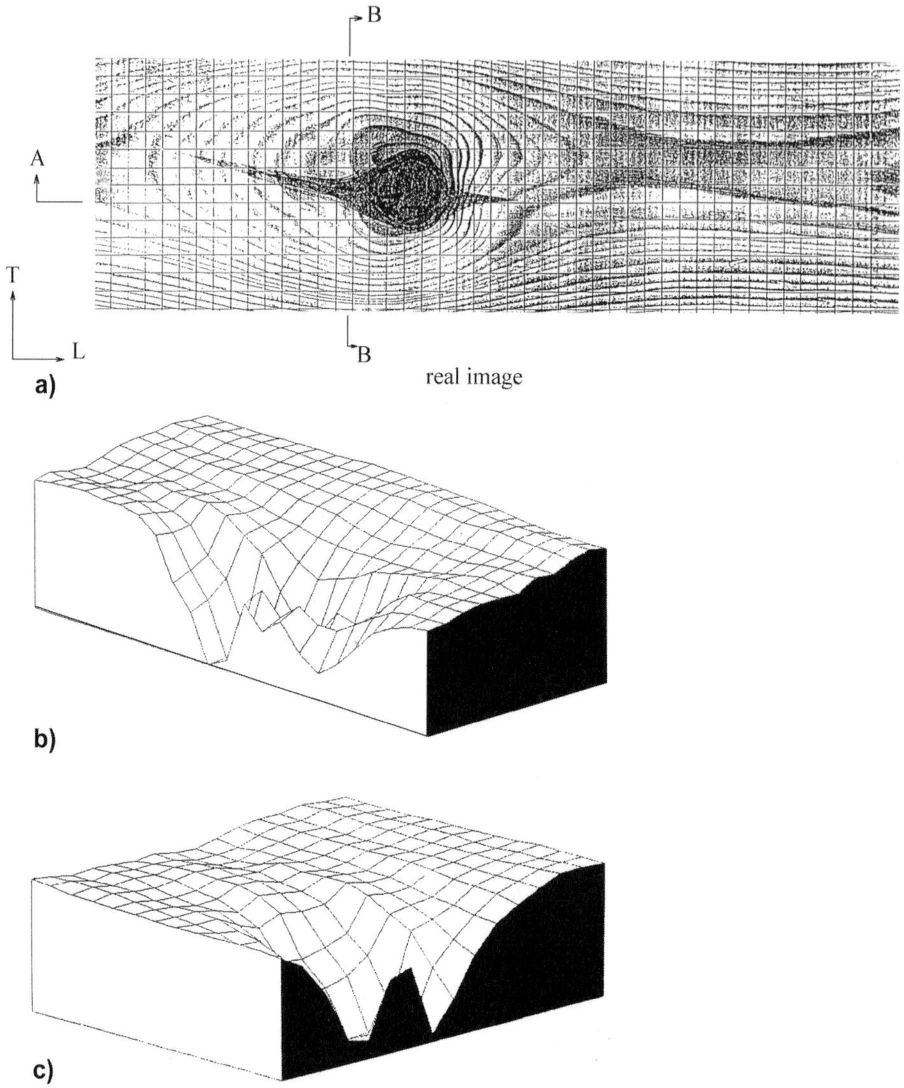

Fig. 4.32a–c. Microwave imaging of a knot in softwood lumber. **a** Real image of the board. **b** Longitudinal-tangential microwave imaging of the section through AA. **c** Radial–tangential microwave imaging of the section through BB. (Chazelas 1991, with permission)

and the physical parameters of the wood, such as the slope of grain and the variation of moisture content. Today the technology is available to build a microwave scanner that can operate at industrial rates. The only important step to be fulfilled is the improvement of image resolution.

4.4.3 Microwaves for Mechanical Grading of Lumber

Microwave nondestructive techniques have increasingly attracted the interest of wood industry practitioners for many years because of the ability of microwaves to penetrate the wood and to inspect the internal structure without contact. The microwave measurement systems are relatively inexpensive and can operate in industrial environments. The major applications were originally oriented toward the development of continuous systems for drying of lumber with microwave energy (Barnes et al.1976). However, microwave techniques are also being considered today as an "emerging technique" for quality control of lumber. The technical approach when conducting microwave inspection of lumber consists of using the wave properties in the far-field of the transmitter. In this case, access to both sides of the lumber is not required. Coherent phase information and the ability to polarize the waves as a function of wood structure or wood defects have significantly improved the capabilities of the microwave method for the evaluation of lumber quality.

King and Yen (1981) proposed a microwave analog and digital system for nondestructive testing of dimensional lumber. The versatility of the system was obtained by using a combination of two elliptically polarized waves, that take into account the reflection, refraction and diffraction of waves by wood, which is an anisotropic medium. The elliptically polarized waves were measured following the major and minor axes of the field. The ratio of the amplitudes at the major and minor axes ranged between 15 and 40 dB. Lumber thickness, slope of grain and moisture content are critical parameters for lumber flexural strength. In a microwave field wood becomes very attenuating material when it has a high moisture content or is too thick.

The system proposed by King and Yen (1981) has several advantages:

- the low cost of the components;
- the signal processing is done at audio frequency ;
- a single probe antenna is used to measure polarization and there is no need for a probe calibration;
- amplitude, phase and polarization angles are available in real time, simultaneously and independently.

The main disadvantage of the system is that at high speed inspection rates of lumber, the mechanical vibrations disturb the measurements of the polarization ratio and increase the noise in the electrical field.

The implementation of microwave scanning techniques for defect detection in the wood industry is only 20 years old (Szymani and McDonald 1981; Glos 1982). The scanner produced in Finland, for an inspection speed ranging from 60 to 300 m/s, used the polarization of microwaves for detection of the slope of grain, and the phase velocity of microwaves for detection of knots in softwood lumber.

Automatic radio frequency scanning of lumber with knots, disordered slope of grain, clear wood, and wood with different defects has been demonstrated by many scientists (McLauchlan et al. 1973; Samson 1988; Niemz 1989; Choffel et al.1992; Rice et al. 1992; Steele et al. 1992, 2000; Tremblay 1995). The slope of grain indicator produced by Metriguard (USA) has the ability to evaluate the defects on unplaned and planed hardwood lumber. All defect types can be evaluated with the exception of rot. The capacitance head of the device has an 18-mm diameter and can be positioned at 3 mm from the wood surface. The evaluation of dielectric properties of wood by the device is made with 3600 rpm and provides 60 grain-angle measurements per second.

Since in a sawmill the profit crucially depends upon choosing the right sawing patterns, it is important to be able to determine the size of the lumber accurately. Microwave measurement systems are capable of operating in industrial environments without requiring access to both sides of the lumber, the microwave attenuation being exponentially related to the thickness.

Numerical values of attenuation of microwaves at 10 GHz in the longitudinal and tangential directions of wood, for spruce samples of different thicknesses are given in Table 4.8. It is informative to note the important variability of these values and to mention that interface reflection phenomena were not considered. The 3-D imaging of a knot in a spruce sample with a laboratory device is shown in Fig. 4.33. The position of the knots is clearly located and the deviation of the slope of grain around the knot is well represented. A microwave imaging of a spruce board containing different defects (a sound knot, a pitch pocket, cracks in a knot) and sound wood is presented in Fig. 4.34. The defects are well identified but their nature is not quite clear.

4.4.4 Microwaves for Inspection of Wood-Based Composites

In this section we will analyze the microwave nondestructive technique for testing of wood-based composites, such as fiberboard, oriented flakeboard and particleboard. Relevant contributions have been reported by Musial (1988), King and Basuel (1993), Wissing and Welling (1995) and Völgyi (2000). The dielectric parameters selected for the inspection of wood-based composites were the dielectric constants of the material and the amplitude and phase of the wave. These parameters were correlated with the moisture content, density, temperature and mechanical properties of the boards.

The moisture content measurement in wood-based composites using microwaves is based on their large water permittivity compared to that of other constituents. In Fig. 4.35 the variation of the relative permittivity of pure water at 25 °C versus frequency and wavelength are presented. We note the inverse variation of both ε' and ε'' versus frequency in the range from 1 to 10 GHz. At a frequency greater than 20 GHz , both ε' and ε'' have the same decreasing ten-

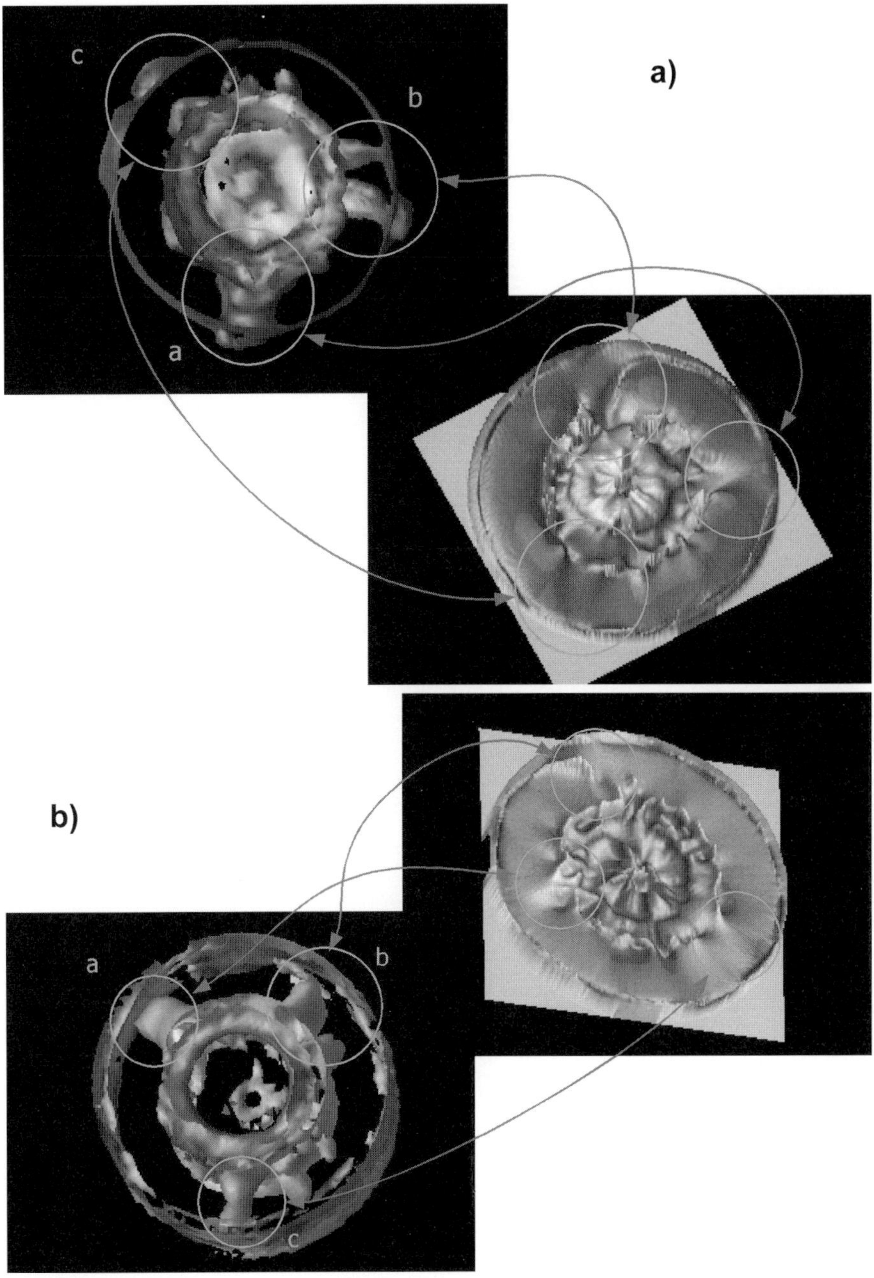

Fig. 4.33a,b. Images of knots reconstructed with **a** the polarization ratio; **b** the phase difference between the two components of the electric field, parallel and perpendicular to the fibers. (Kaestner and Baath 2000, with permission)

dency. It is well known that the moisture in wood is composed of both water and dissolved ions. For a hypothetical liquid containing both water and ions, the variation of ε″ versus frequency is different than that of pure water as can be seen from Fig. 4.36. The bound water does not have the same behavior as free water. The bound water reached a maximum value between 10^8 and

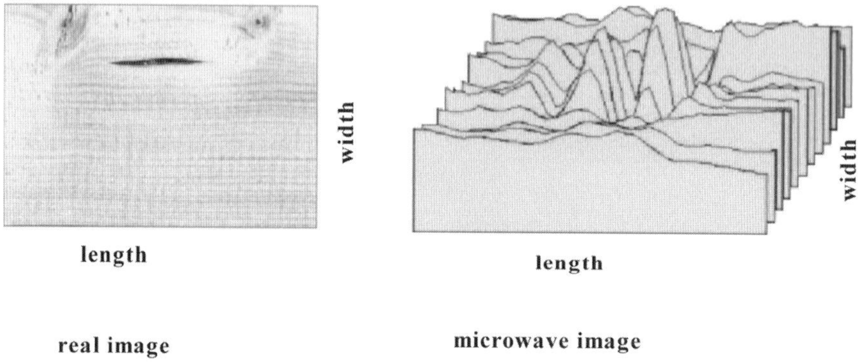

Fig. 4.34. Microwave imaging of a softwood board (100 × 40 × 1000 mm) (Choffel 1999, with permission)

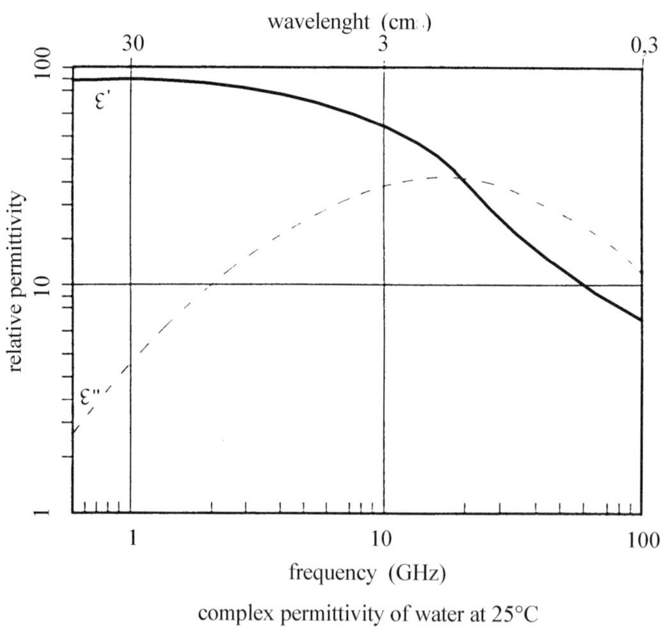

Fig. 4.35. Relationship between the permittivity and frequency for pure water at 25 °C. (Thompson 1989, with permission)

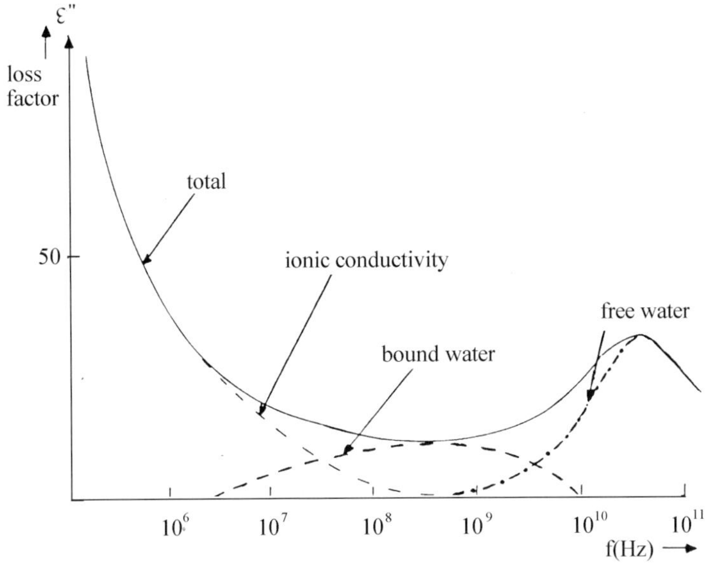

Fig. 4.36. ε″ as a function of frequency for a hypothetical dielectric material containing water and dissolved ions. (Thompson 1989, with permission)

10^9 Hz, in the same domain in which free water reached a minimum. The influence of free water is more important than that of bound water at a frequency higher than 1 GHz. This graph (Fig. 4.36) can help in selecting the appropriate frequency for wood drying because of the capability of the electromagnetic radiation, directed through the bulk material to quantify the water content in real time (Thompson 1996).

From a large number of references we can see that the dielectric constants were extensively used for measurements of the moisture content in wood and wood-based-composites (Nanassy 1970, 1972; Norimoto and Yamada 1972; Tiuri et al. 1980; Skaar 1988; Kuroda and Suzuki 1996; Yokoyama et al. 1999, 2000a,b).

Thompson (1989) found that the volume of water present in the dielectric material can be deduced from the complex permittivity of the sample ε^*, and can be expressed as follows:

$$\varepsilon^* = V_m \varepsilon_W{}^* + (1 - V_m)\varepsilon_d{}^* \tag{4.27}$$

where: V_m is the fraction volume of water; $\varepsilon_w{}^*$ is the complex permittivity of water, and $\varepsilon_d{}^*$ is the complex permittivity of the dielectric material such as dry wood substance. In this equation, all parameters are known except the volume fraction of water, V_m.

For in-process quality control, King and Basuel (1993) developed a non-contact microwave transmission technique for measurement of the moisture content of fiberboard, based on the assumption that the variation of the attenuation and phase of the microwave signal are linearly dependent on the dry masses of solid components and of water.

The absolute moisture content, $MC_{\mu w}$, deduced by the microwave technique, is calculated as the ratio of the partial basis weights of the water present to the partially dry wood weight.

$$MC_{\mu w} = m_w + m_d \tag{4.28}$$

where $MC_{\mu w}$ is the absolute fractional moisture content of the sample, m_w is the weight of water, and m_d is the weight of wood on the basis of the absolutely dry fraction.

The total weight m_{total} can be deduced as the sum of the components:

$$m_{total} = m_w + m_d = m_d(1 + MC) \tag{4.29}$$

The variation of the amplitude and of the phase as functions of m_d and m_w are given by:

$$\Delta A = a_1 m_w + a_2 m_d \tag{4.30}$$

$$\Delta\phi = a_3 m_w + a_4 m_d \tag{4.31}$$

where a_1 and a_2 are experimental calibration constants for amplitude, and, a_3 and a_4 are experimental calibration constants for the phase.

After calculation, the expression of the moisture content as the ratio between m_w , the weight of water, and m_d, the weight of wood, on the basis of an absolutely dry fraction, the moisture content is expressed by:

$$MC_{\mu w} = m_w / m_d = (a_3\Delta A - a_1\Delta\phi)/(a_2\Delta\phi - a_4\Delta A) \tag{4.32}$$

The values of the coefficients a_1 a_2 a_3 and a_4 are given in Table 4.11.

Figure 4.37 gives the variation of $MC_{\mu w}$ as a function of the gravimetric moisture content measurements in the range of 0 to 20%. A breakpoint in the variation of amplitude and phase were observed at 6% moisture content. Under

Table 4.11. Values of the coefficients a_1, a_2, a_3, a_4 calculated by a least square regression for fiberboard at constant temperature of 24 °C. (King and Basuel 1993; courtesy of Forest Products J)

Moisture content	m_d (g/cm^2)[a]	a_1 (dB cm^2/g)	a_2 (dB cm^2/g)	a_3 (deg cm^2/g)	A_4 (deg cm^2/g)
>6%	0.614	–0.936	53.15	76.94	469.62
	0.777	–0.181	48.90	77.36	513.68
	0.948	0.031	43.59	78.28	515.64
	Mean value	–0.268	47.39	76.75	515.39
<6%	0.614	0.594	27.63	81.16	377.22
	0.777	0.905	30.04	83.42	408.05
	0.948	1.204	29.04	85.17	454.66
	Mean value	0.962	27.81	83.65	409.34

[a] m_d (g/cm^2) is the weight of wood per unit area of board, on the basis of absolute dry fraction. $m_d = \rho_d t$; where ρ_d (g/cm^3) is the dry bulk density of the board of thickness t (cm).

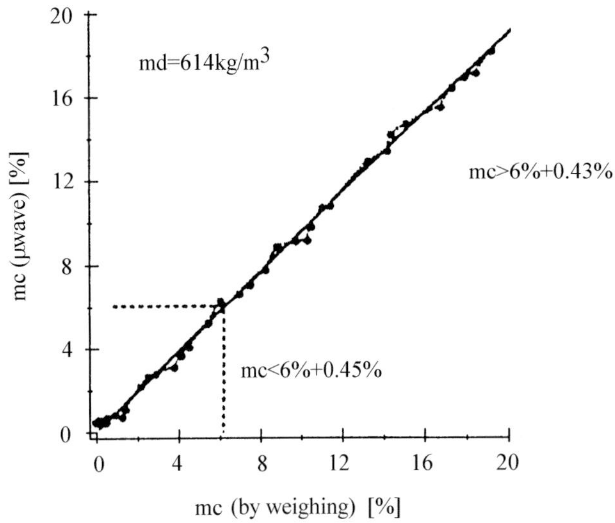

Fig. 4.37. Relationship between the moisture content determined with a microwave technique and the routine moisture content measurement for a fiberboard of 614 kg/m^3 density, at 24 °C. (King and Basuel 1993; courtesy of Forest Products J)

6% moisture content, the water molecules are strongly bound to the chemical constituents of wood and the attenuation of the microwaves is reduced and exhibits a nonlinear behavior. Increase in the moisture content from 6% to the fiber saturation point activates weaker bonds of hydroxyl groups of the hemicelluloses. At the fiber saturation point, free water is present in the cell lumina. From the electromagnetic point of view, there appears to be little or no difference between the polarizabilities of weakly bonded and free water, so that ΔA and $\Delta\phi$ are linearly dependent on the moisture content above 6%.

The influence of moisture content on the complex dielectric constants of the fiberboard is shown in Fig. 4.38. The real dielectric constant ε' is linearly dependent on the moisture content, whereas the dielectric constant ε'' is nonlinearly dependent on the moisture content. During the processing, the temperature of a fiberboard varies over a very large range (from 200 °C after emerging from the press, to 20 °C in the final stage). The influence of temperature on the amplitude and phase variations at different levels of moisture content, ranging from 0.3 to 10.4%, is shown in Fig. 4.39. Both the amplitude and phase variations increase linearly with the temperature. The phase variations seem to be more sensitive to an increase of temperature when the moisture content increases from 5 to 10%.

Microwave attenuation was used by Musial (1988) to study the degree of flake orientation of OSB under the assumption that the physical characteristics of flakes are similar to those of solid wood from which the flakes had been produced. On the other hand, Musial (1988) found that the influence of resin

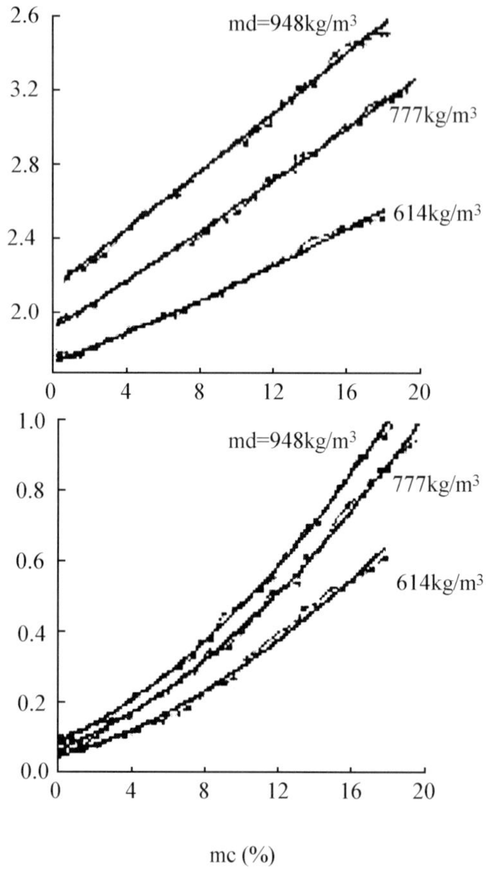

Fig. 4.38. Relationship between the components of the complex dielectric constants and the moisture content of fiberboard at 10 GHz and a constant density and temperature of 24 °C. (King and Basuel 1993; courtesy of Forest Products J)

and water repellent on the dielectric properties of the board can be neglected. The third point is that there is no electrical interaction between the board and the individual flakes. The dielectric properties of the flakes are not significantly modified during board production. The anisotropy of the boards is expressed by the ratio between the attenuation parallel and perpendicular to the direction of board formation. Considering the coordinate system x, y and z for the board and x′, y′ and z′ for the flakes as shown in Fig. 4.40a, the attenuation anisotropy ratio is a linear function of flake orientation (Fig. 4.40b). This linear relationship can be used in further development of a continuous nondestructive monitoring technique in the processing of boards.

The results previously described are based on the microwave transmission technique. More recently, the development of monostatic transmission sensors in reflection-aperiodic-open configuration (King 2000) permitted the continuous control of large dielectric boards. Völgyi (2000) proposed the utilization of a multifrequency permittivity monitoring system, in which the attenuation and the phase are measured and correlated with the moisture

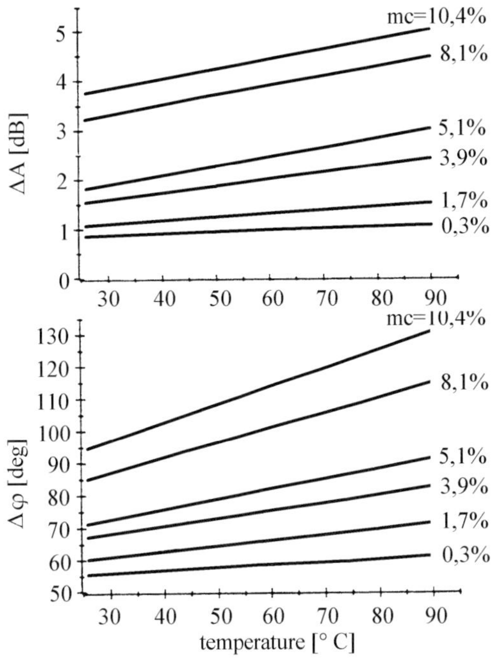

Fig. 4.39. Influence of temperature at a constant moisture content on amplitude and phase variations, measured on a fiberboard. (King and Basuel 1993; courtesy of Forest Products J)

content and density of the board and with the corresponding mechanical properties. Imaging of the distribution of the moisture content, resin proportion and other parameters can be obtained for each board. Numerical analysis of microwave measurements was based on the implementation of Kasa's circle fit procedure (Corral and Lindquist 1998) that minimizes the errors of antenna pattern measurements.

In the future, the holographic technique of a multifrequency microwave method as generally described by Drobakhin (2000), for dielectric material, could be an important advancement in the nondestructive control of large wood-based composite boards.

4.5 Summary

The imaging of wood structure using microwave techniques is based on the determination of its dielectric properties. This chapter has presented an overview of microwave imaging applications for wood, describing theoretical and experimental results. Systematic studies of the dielectric properties of various wood species have been reported by many authors who have discussed wood–water interactions and the measurement and prediction of wood density and moisture content.

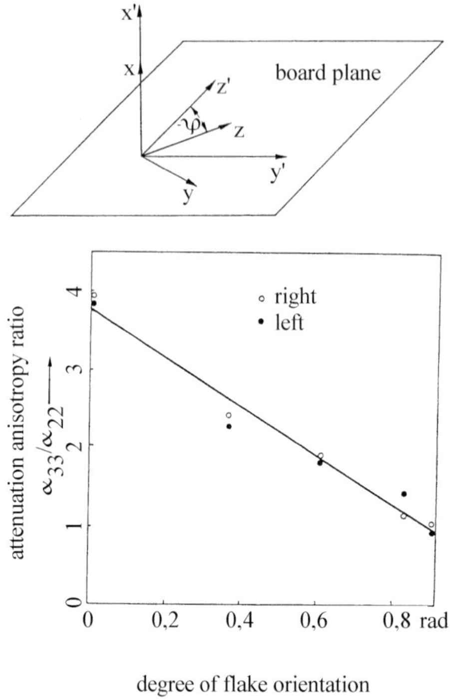

Fig. 4.40. Microwave attenuation anisotropy ratio and the degree of flake orientation in OSB at 13.6 GHz (Musial 1988). *Above* Orientation angle of flakes in the board. *Below* Linear negative relationship between the degree of flake orientation and the attenuation anisotropy ratio. (Courtesy of Wood Science and Technology)

There are two basic microwave techniques, the transmission and the reflection technique. Signal analysis of probes is relatively simple and is related to the measurement of amplitude, phase and the polarization of the waves. These parameters are used for image reconstruction of wood structure. Today, the areas that may benefit from microwave imaging techniques are related to the internal defect detection of logs, lumber and wood-based composites, and, also to the imaging of vegetation including leaves, stalks and stems under various moisture and temperature conditions over a wide range of frequencies.

The imaging of wood structure can be obtained by scanning in the proximity of the sample with a resolution given by the aperture size. The images can be obtained in the far field or in the near field. In the far field, the smallest detectable defect size is determined by the ratio $d/\lambda > 1$, where d is the length of the defect in the plane normal to the microwave vector and λ is the wavelength. The probes used to evaluate the properties of the medium can be either scatterometers or radars, which measure the scattering properties, or reflectometers, which measure the reflectivity of the medium due to its inhomogeneities and defects. The probes can operate in the proximity of the medium as open-ended coaxial lines, cavity resonators or antennas. By moving the position of the transmitter, an image of the medium can be obtained. While microwave techniques have shown great potential for lumber inspection, their applicability seems to be limited by the impossibility of identifying the nature of the defects. To overcome this difficulty a vision system has been coupled to microwave antennas.

The development of the microwave imaging technique is based on several advantages such as:

- a noncontact operating system and relatively small size of antennas, determined by the wavelength at microwave frequencies;
- fine resolution of measurements that can increase the ability of this technique to detect defects;
- wave parameters, such as amplitude, phase and polarization can be measured in real time.

The difficulties of applying microwave imaging to wood material arise from inherent material properties, such as the anisotropy, heterogeneity, and the presence of natural defects in wood. The main disadvantage of the microwave system is that at a high rate of inspection the mechanical vibrations of logs or lumber disturb the measurements of polarization. Technical limitations, associated with the advantages of microwave imaging techniques, illustrate the potential and capability of this method for new practical applications.

5 Ultrasonic Imaging

5.1 Introduction

Ultrasonic imaging is a technique widely used to reconstruct the properties of the materials under inspection from global wave propagation data. Acoustic tomograms provide an excellent noninvasive means of obtaining information about inhomogeneous media as noted by many investigators (Kak 1979; Herman 1980; Greenleaf 1981; Crawford and Kak 1982; Devaney 1982, 1986; Dean 1983; Denis et al. 1986; Tomikawa et al. 1986; Dumoulin and de Belleval 1987; Falls et al. 1989; Frick et al. 1997; Douglas Mast 1999).

As for X-ray computed tomography, ultrasonic tomography refers to the cross-sectional imaging of an object from either transmission or reflection data collected by illuminating the sample from different directions.

Ultrasonic tomographic reconstruction techniques can be classified as:

- techniques based on the projection-slice theorem (filtered back-projection and direct Fourier transform), which are fast, but restricted to projection data that are sets of straight rays.
- techniques based on iteration procedures (algebraic reconstruction techniques and simultaneous iterative reconstruction techniques) that are relatively slow, but may be used with complex sampling geometries and a bending ray path.

Different types of waves can be used for ultrasonic imaging of solids (Schomberg 1982; Schechter et al. 1994, 1996; Sulivan et al. 1996). The most common are the longitudinal waves but shear waves and surface waves like Rayleigh or Lamb waves can also be used (Hutchins et al. 1993; Pei et al. 1995; McKeon and Hinders 1999). First, a precise analysis of the wave type launched into the sample must be made to understand the propagation phenomena in an inhomogeneous solid wood specimen, with structural defects. Second, the possible mode conversion must be studied and avoided, because of the strongly induced degradation of ultrasonic images by such phenomena.

The resolution of the ultrasonic imaging techniques is very much limited by the wave length and by the size of the transducers. However, unlike X-rays, ultrasonic rays do not travel in a straight line through heterogeneous and anisotropic materials. For this reason, the computational requirements of this technique are much more important than for X-ray computed tomography and an increase in speed and accuracy of the reconstruction algorithms is necessary.

Ultrasonic tomography applied to wood is an important challenge for wood scientists because of the natural variability, the anisotropy and the inhomogeneity of this material. Ultrasonic imaging techniques applied to wood must be able to distinguish between the natural structure of the material and its pathological features. Ultrasonic velocities and attenuation in different anisotropic directions, the reflective properties of wood surfaces and the back scatter of ultrasonic waves from the inhomogeneities must all be considered.

Proper signal processing methods must be chosen according to the structural characteristics of wood at macroscopic and microscopic scales. The computationally most intensive part of the ultrasonic tomography technique, based on the ray theory, is the tracing of the acoustic ray paths through the medium. Commonly, ray tracing approaches are based on Snell's law (Lytle and Dines 1980). The ray bending is based on Fermat's principle of minimum propagation time (Kline and Wang 1992; Wang and Kline 1994). In anisotropic media like wood, Snell's law is difficult to formulate because it is associated with the wave normals while the wave paths are defined by the energy propagation. Figure 5.1 shows the energy flux deviation for oak in three anisotropic planes. Each wave has an energy flux that deviates from the wave normal. If an arbitrary oscillatory displacement is applied to the surface of a wood specimen, the displacement will be resolved into three components which are then propagated with individual phase velocities and energy fluxes. The variation of the polarization vector when the propagation direction is outside of the principal directions of the material symmetry is shown in Fig. 5.2. To display the propagation phenomena in three dimensions, the numerical modeling methods suggested by Lanceleur et al. (1998) were used . In Fig. 5.2a (with three color codes), the propagation vector is along r as defined by the local reference basis. At the same time, three polarization directions are represented, corresponding to three waves deduced from the Christoffel equation, such as one quasi-longitudinal, very rapid wave, having the propagation direction parallel to the polarization direction and two quasi-shear waves (the first one fast and the second relatively slow) having the polarization directions perpendicular to the propagation direction. From a theoretical point of view, the polarization direction of these three waves is mutually perpendicular. Figure 5.2b shows the variation in the acoustic properties of oak. The slowness surfaces corresponding to the three types of waves are represented. The inner surface corresponds to the quasi-longitudinal wave and the outer surface corresponds to the slow quasi-shear wave. The smaller slowness value corresponds to the velocity measured in the fiber direction (L). The inner slowness sheet is colored in red and exhibits a flat ellipsoidal shape. In the transverse plane, the color varies from red to yellow. This means that the polarization direction varies more and more when the angle approaches the third anisotropic axis of wood (axis labeled as T, tangential). The variation of the polarization direction of the waves as a function of the propagation direction is indicated by the color code. It should be noted that, for the quasi-longitudinal wave, the polarization vector is always

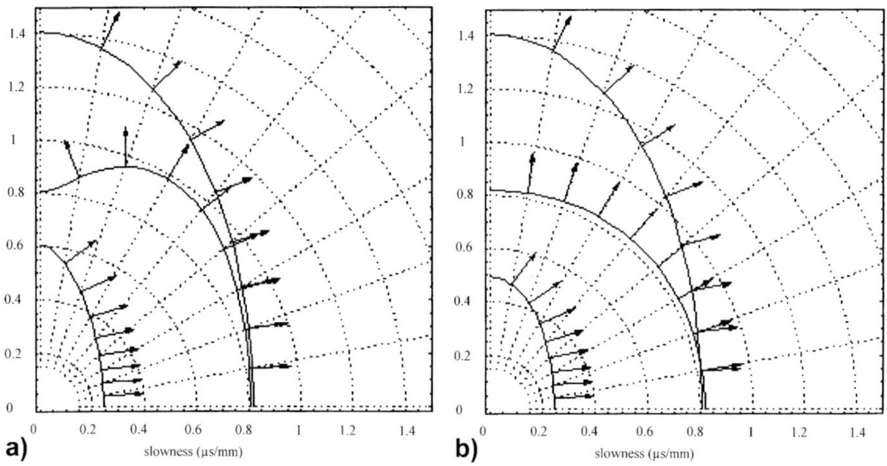

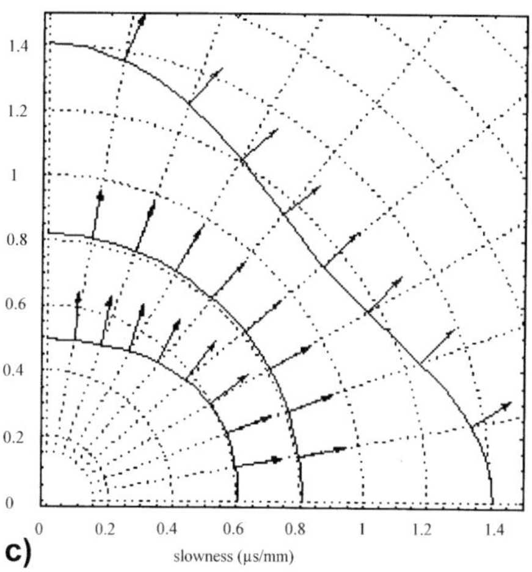

Fig. 5.1a–c. Energy flux deviation shown on slowness curves, as a function of direction of ultrasonic wave propagation in oak, in three anisotropic planes (Berndt, pers. comm.). **a** Plane LR; **b** plane LT; **c** plane TR. The slowness curves correspond to the quasi-longitudinal waves (the internal curve) and to the quasi-transverse curves (two external curves) deduced from Christoffel's equation. The numerical data were selected from Bucur and Archer (1984)

perpendicular to the propagation direction. This is not the case for quasi-shear waves, for which there is an important variation of color. This means that the polarization vectors of quasi-shear waves are more sensitive to the interaction with wood structural elements than is that of the quasi-longitudinal wave. The examination of the interaction between wood structure and the propagation phenomena has been discussed in detail by several investigators (Bucur 1980, 1995; Berndt and Johnson 1994; Berndt et al. 2000).

For media of interest to the geophysics community, the modeling of velocity propagation (Kline et al. 1994; Wang and Kline 1994) was performed using either finite difference or finite element formulations. In wood science, pioneering works on wood structure imaging reconstruction by scanning, from

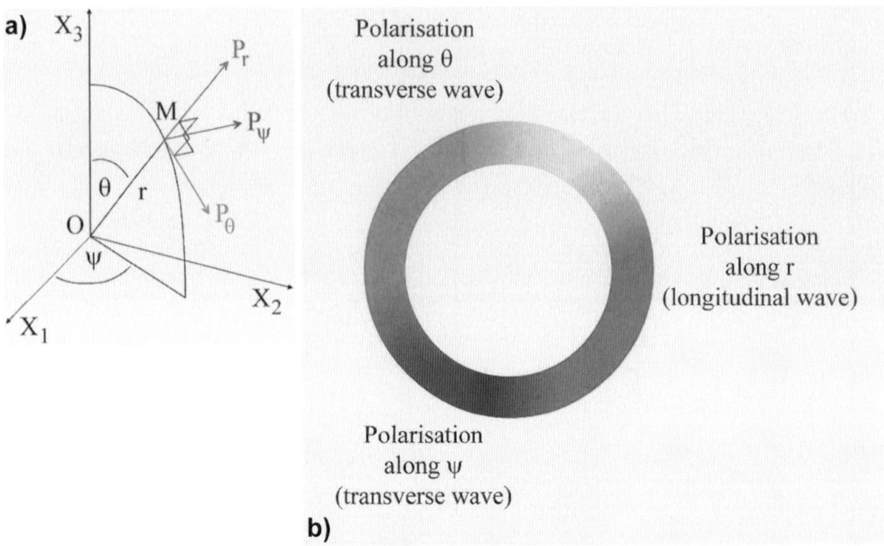

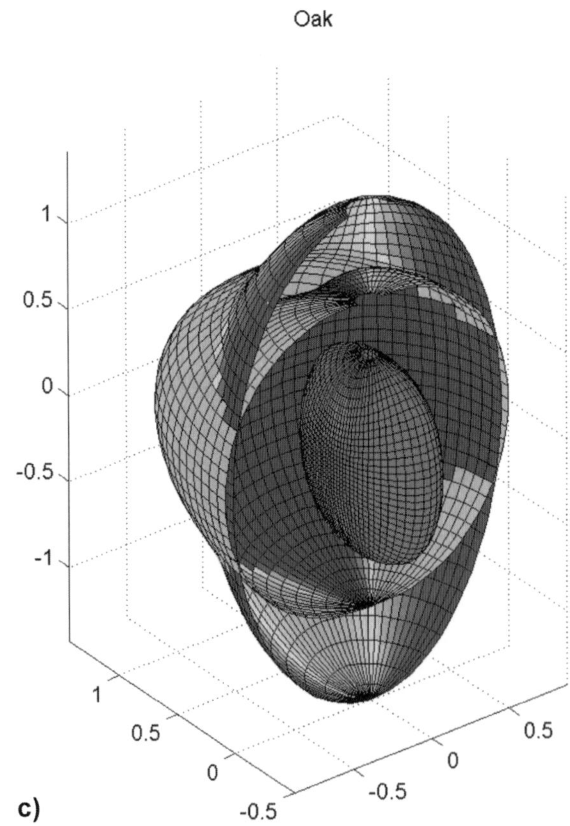

Fig. 5.2a–c. Three-dimensional representation of acoustic properties of oak (**c**). **a** Local basis and color code. **b** Variation of polarization angle on slowness surface for oak. (Bucur et al. 2001, with permission)

ultrasonic data as velocities and stiffnesses have been reported by McDonald (1978), Chazelas et al. (1988), and Biernacki and Beall (1993). More recently, high resolution ultrasonic imaging has been reported by Tomikawa et al. (1986), Biagi et al. (1994), Berndt et al. (1999), Comino et al. (2000), Rust (2000), Socco et al. (2002) and Martinis (2002). Stress wave tomography in standing trees was reported by Rust (2000).

This chapter is intended to cover the treatment of the subject of ultrasonic imaging of wood structure on a wide range of topics related to image processing and to the algorithms for the reconstruction of the internal structure of solid wood and wood-based composites. The breadth of the discussion is intended to aid the reader in deciding which technique would be the most appropriate for a specific purpose.

5.2 Basic Aspects

Ultrasonic tomography is a diffraction type tomography and is an important alternative to straight ray tomography with ionizing radiation. The main benefits of imaging techniques using acoustic and electromagnetic radiation are that they are noninvasive and safe at a low energy level. The accuracy of tomography with acoustic energy is affected by the diffraction or refraction of the energy field. In contrast to X-rays, acoustic waves do not travel along straight rays and the projections are not line integrals. The energy flow is described with the wave equation and can sometimes deviate from the straight-line trajectory.

There are three main types of algorithms that can be used to form tomographic images from ultrasonic data: transform techniques, iterative methods and direct inversion techniques. The parameters used in ultrasonic imaging are: time of flight, amplitude, frequency spectra of the waveforms, the integrated energy of the spectral peak making possible an estimation of attenuation and the "central value" of the spectrum, and the phase shift. The peak amplitude and centroid of each spectrum are sensitive parameters for the detection of significant structural differences. The evaluation of wood behavior in the frequency domain (Halabe et al. 1995) gives more information about its structure than it does in the time domain. Energy distribution and energy flow are good parameters for enhancing the contrast of the images.

In practice, approximate formalisms are employed to allow the use of the theory of homogeneous medium wave propagation for media with slight inhomogeneities. Different approaches can correct the way paths errors, such as:

1. Algebraic reconstruction algorithms of the image that use an initial approximation of the refractive index to estimate each ray path. For weakly refracting objects, the correction of the refractive index distribution is obtained

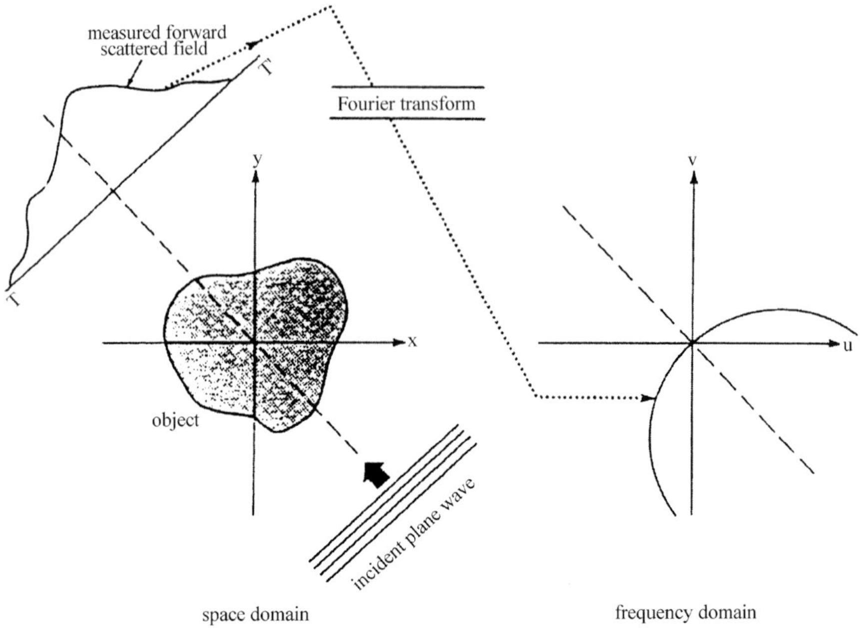

Fig. 5.3. Schematic representation of the Fourier diffraction theorem (Kak and Slaney 1988), which relates the Fourier transform of a diffracted projection to the Fourier transform of the object along a semicircular surface in the frequency domain. (Copyright 2001 IEEE)

after a few iterations. One can assume that the cross section of the object consists of an array of unknowns (Fig. 5.3). After setting up the corresponding algebraic equations for the unknowns in terms of the measured projection data, the image can be reconstructed.

The algebraic algorithms used most frequently are represented by the following acronyms:

- ART – algebraic reconstruction technique – on which each equation corresponds to a ray projection. The computed ray sums are a poor approximation of the measured ones and the image suffers from significant noise.
- SIRT – simultaneous iterative reconstructive technique – reduces the noise of ART by relaxation and produces better images than ART. The relaxation parameter becomes progressively smaller with increasing number of iterations. A comparison of ART, SIRT, LSQ (least-squares) and SVD (singular value decomposition) tomographic inversions for geophysical applications is given by McGaughey and Young (1990). ART and SIRT are iterative inversion techniques, while least squares and singular value decomposition are direct inversion techniques.
- SART – simultaneous algebraic reconstruction technique – combines ART and SIRT and yields image reconstructions in one iteration, using a

model of the forward projection process. This method is recommended in complex image reconstruction with curved rays with overlapping and nonoverlapping ray strips. The reader is referred to Herman (1980) for an exhaustive explanation of the algebraic reconstruction and to reference articles (Gilbert 1972; Herman et al. 1973; Dines and Lytle 1979; Andersen and Kak 1984; McGaughey and Young 1990).

2. Reconstruction of the image with an algorithm based on the Fourier diffraction theorem, when the sizes of inhomogeneities in the object are comparable to the wavelength. This theorem states that a projection yields the Fourier transform of the object over a semicircular arc in the frequency domain (Fig. 5.4). This is the fundamental algorithm of diffraction tomography. The wave equation and the Born and Rytov approximations are used to derive a mathematical expression that relates the scattered field to the object (Fink 1983). The mathematical approach to this subject is presented very clearly by Kak and Slaney (1988) and by Kline et al. (1996), and the algorithms in the frequency domain and the back projection method are discussed.

Diffraction and scattering in solids become important when the size of the object is comparable to the wavelength of the radiation (Delsanto et al. 1998). In this case, first-order Born or Rytov approximations for scattering are used. The Born approximation assumes that the scattered field is small compared to the incident field, whereas the Rytov approximation assumes that the phase change is small compared to the wavelength recommended or, in other words, that the phase gradient is small. The Rytov approximation is for large scatter, while the Born approximation works better with small scatters.

The factors that limit the accuracy of the images obtained with diffraction tomographic reconstruction are related to the theoretical approach of the approximations in the derivation of the reconstruction process and to the experimental limitations. The mathematical limitations imposed by the Born and Rytov approximations are severe, because the reconstruction process is limited (Norton and Linzer 1980) in the range of the objects that can be examined (cylinders, spheres, etc.). The experimental limitations are related to the finite amount of data that can be collected. Up to the limit introduced by the possible presence of the evanescent waves, it is possible to improve the reconstructed image by collecting more data. The effect of the Born or Rytov approximations can be estimated if the exact acoustic field is calculated or measured. The difference between the reconstructed image and the real object is a measure of the quality of the approximation. The Born reconstructions are appropriate at a large refractive index, as long as the phase shift of the incident field is less than π. The Rytov approximation is very sensitive to the refractive index and produces very good reconstructions for objects as large as 100λ (λ is the wavelength).

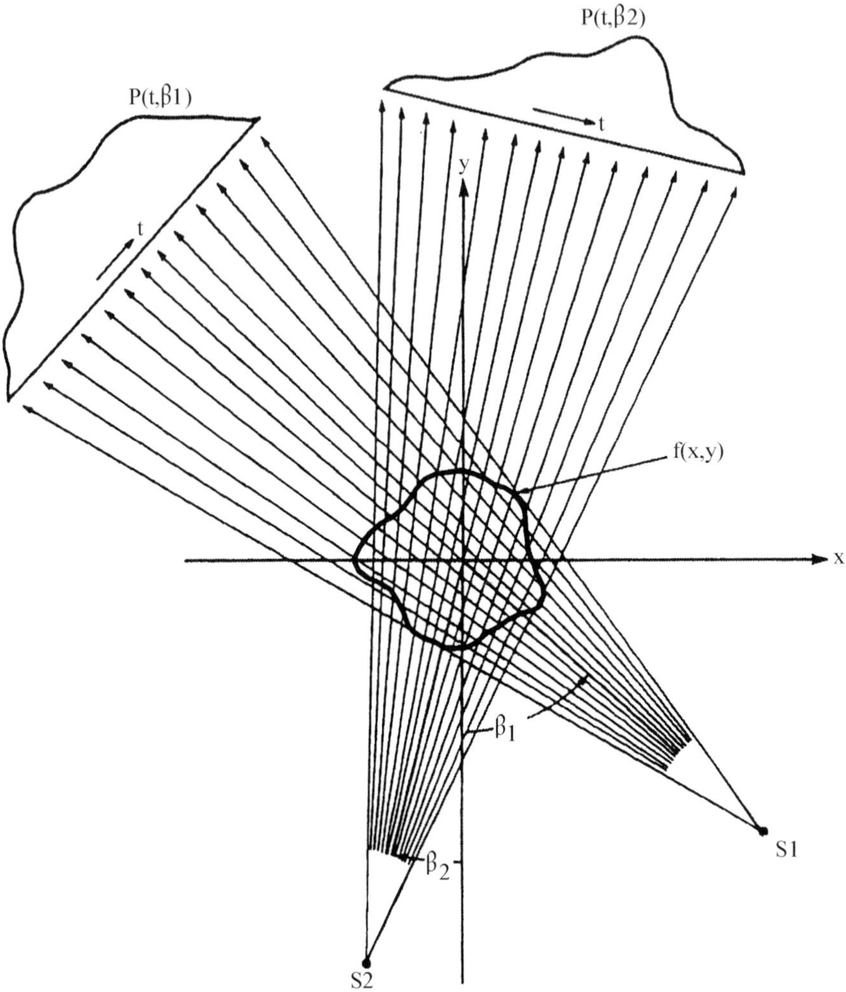

Fig. 5.4. Reconstruction of the image with fan beam projections collected in S_1, S_2, etc. (Kak and Slaney 1988); *x,y* coordinates related to the object in the space domain, β_1 and β_2 angles related to the position of S_1 and S_2; profiles P_1 (t, β_1) P_2 (t, β_2), etc. as a function of position and time t. (Copyright 2001 IEEE)

The experimental limitations are related to the equipment limitations and are caused by:

- ignoring the evanescent waves, that is, those that are attenuated over a distance of several wavelengths. The limitation is quantified by the highest received wave number $k_{maximum} = 2\pi/\lambda$;
- the finite received signal length, that is, a physical limitation on the amount of data collected;

- the number of sampling data along the receiver line, because the sampling is limited and a nonzero sampling interval must be chosen. With the Nyquist theorem, a low pass filtering operation can be introduced and the highest measured frequency can be calculated from $k_{measured} = \pi/T$, where T is the sampling interval.
- the limited views of the object under inspection, not estimated by the Fourier transform, can degrade the image in certain directions.

It can be noted that an enhancement of acoustic imaging of polymers can be obtained by colling (Yamanaka et al 1991).

High resolution images for wood of transmitted and reflected energy were presented by Berndt et al. (1999) and the group in the Geophysics Department at Politecnico di Torino, Italy, under the direction of Sambuelli and Socco (Martinis 2002) and the group from the University of Torino, under the direction of Nicolotti. Ultrasonic tomographic images were obtained with living trees, such as, for example, the tomography at 54 kHz by a direct transmission technique, shown in Fig. 5.5, which corresponds to the transverse section of a tree (*Platanus acerifolia*) of 40 cm diameter. The central zone of the section, about 10 cm in diameter, is degraded by a fungi. In this zone (Martinis 2002), the values of ultrasonic velocity are very low (600–1000 m/s) because of structural degradations induced by the fungi.

Berndt et al. (1999) reported images obtained with C scan by reflection and transmission with southern pine samples of $100 \times 70 \times (3\text{–}70)$mm, with 1 MHz unfocused and 5 MHz focused transducers immersed in a tank filled with water. On the ultrasonic images it was possible to recognize the earlywood (Fig. 5.6) and the latewood.

5.3 Equipment for Ultrasonic Imaging

There are three traditional ultrasonic experiments related to three visualization methods known as A-scan, B-scan and C-scan (Papadakis 1999). A-scan assumes a constant velocity and maps reflections where the depth of the reflection is proportional to the time of detection. Mapping of equal velocity zones permitted the detection of defects like knots in real size lumber pieces (McDonald 1978; Schmoldt et al 1996b; Niemz et al. 1999; Kabir et al. 2000a; Iancu et al 2000). B-scan attends the visualization with a set of waveforms along a line. This technique is preferred in medical applications. C-scan visualizes a set of line scans covering an area of the sample. This procedure is performed by an immersion technique or with air-coupled transducers and is largely used for nondestructive testing of materials for the aircraft industry. Work pieces of complex curved shape and very large size can be inspected automatically in correspondingly large tanks equipped with multi-axis manipulation devices. The experiments can be carried out on different sound paths, by varying the

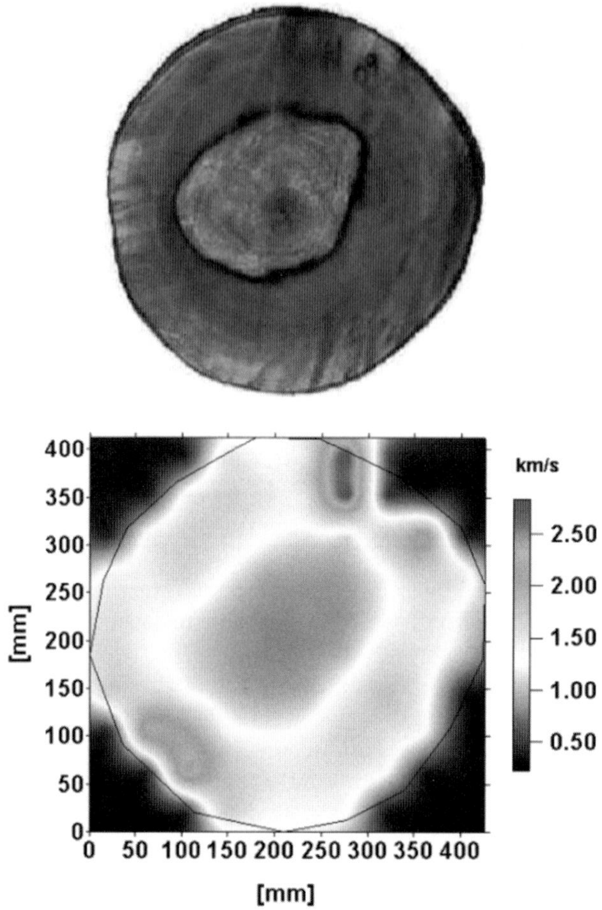

Fig. 5.5. High resolution images of the transverse section of a tree (*Platanus acerifolia*) (Martinis 2002, with permission)

location of the transmitter and the receiver. C-scan may be either of a reflection type or a through-transmission type. Modern digital computers can extract the signal corresponding to each point of the work piece and produce two dimensional images. Two-dimensional images in C-scan obtained with modern digital computers require advanced signal processing. Volumetric imaging can easily be obtained from a set of line scans covering an area of the sample for the benefit of understanding propagation phenomena.

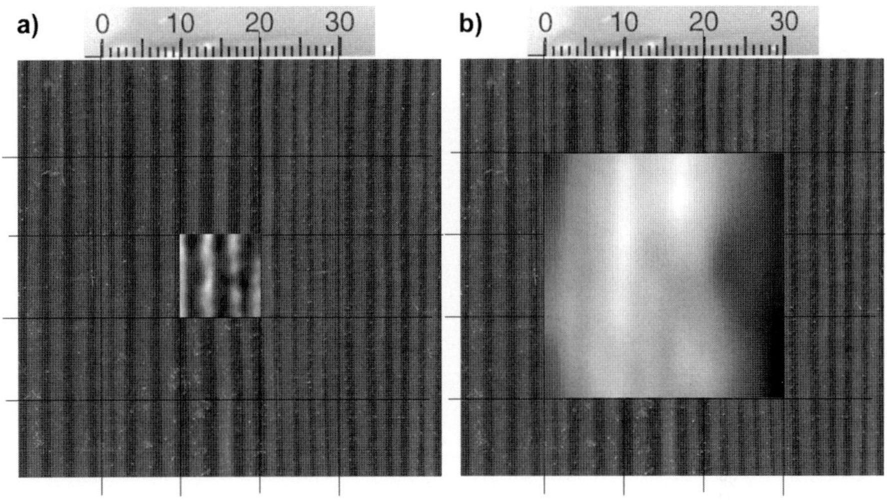

Fig. 5.6a,b. Earlywood and latewood in pine, imaged by ultrasonic reflection. **a** Ultrasonic and photographic image superimposition at a 5-MHz area scan. **b a** at 1-MHz area scan. (Berndt et al. 1999, with permission)

5.3.1 Equipment for Contact Scanning

In this section, we consider "contact" methods, immersion methods and non-contact methods such as procedures with air coupling transducers or similar. The equipment for direct contact scanning in A-mode was developed first (Harpole and McDonald 1981), As an example we have chosen the equipment designed for green and rough sawn lumber scanning in through transmission mode (Fig. 5.7). A pair of transducers of 1 MHz inspected the lumber piece immersed in water and the time of flight was measured. The scanning speed was about 50 m/min, which corresponds to 9000 board ft./h. The main disadvantage of this system was the relatively low sampling speed of lumber and the large quantity of water lost around the board at the input gate.

More recently, Schmoldt et al. (1996b) and Kabir et al. (2000a,b) at USDA Forest Service have proposed a scanning ultrasonic system equipped with rolling transducers, which operates in a range of frequency between 90 and 180 kHz. In view of the in-line detection of the specific differences between defects, several ultrasonic parameters (Fig. 5.8) were selected for the signal processing: time of flight centroid, time of flight energy, time of flight amplitude, pulse length energy value and energy/pulse value. These parameters were used for the detection of knots, cross-grain and bark pockets in yellow-poplar deck-boards. Energy value and energy/pulse value were the most sensitive parameters for defect detection. The scanning rate had little influence on the data

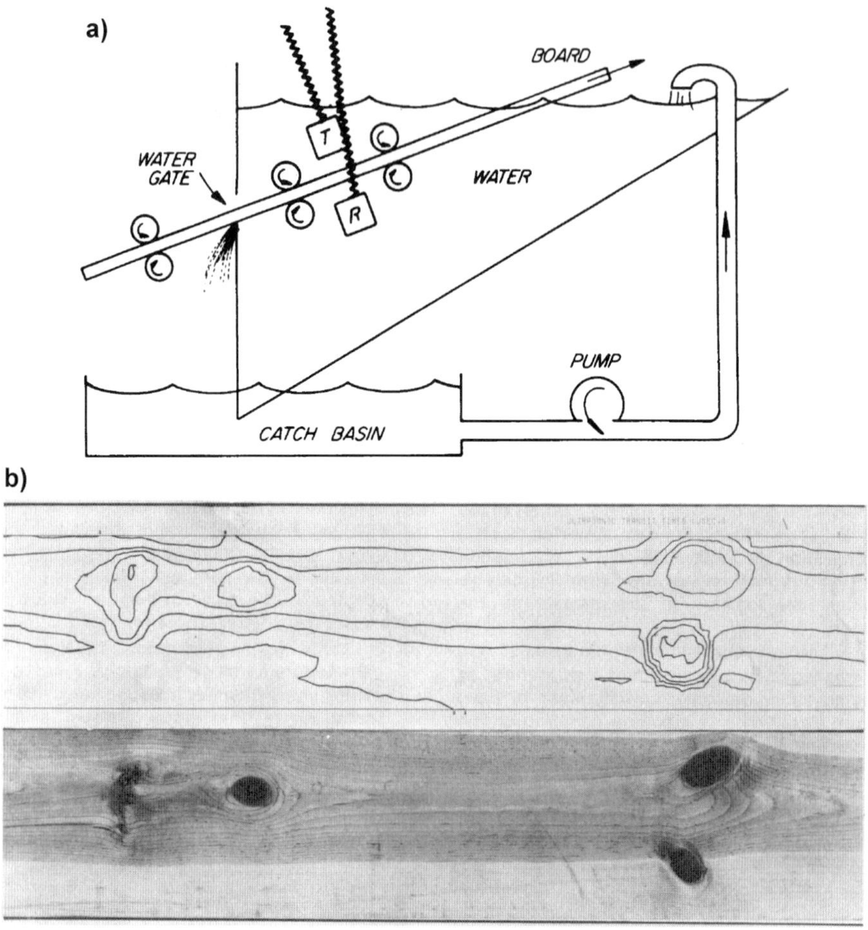

Fig. 5.7a,b. Ultrasonic scanning of lumber by immersion in a water tank.. **a** Device for lumber scanning. **b** Ultrasonic and real images of a lumber board (McDonald 1978; courtesy of Forest Products Society)

collection and the authors noted that "scanning at relatively high industrial speeds is feasible."

For direct transmission measurements, the direct coupling of the transducer requires couplants that can be liquids (Birks 1972), grease, different adhesives, air, etc. The air-coupled transducers were used in line production of particleboard and plywood for the detection of blows and blisters with a speed of 90 m/min. The equipment for ultrasonic tomography available today is only for laboratory measurements and consists of an ultrasonic generator, transducers and an oscilloscope. The ultrasonic signal obtained from field measurements require modern and complex processing (Berndt et al. 2000).

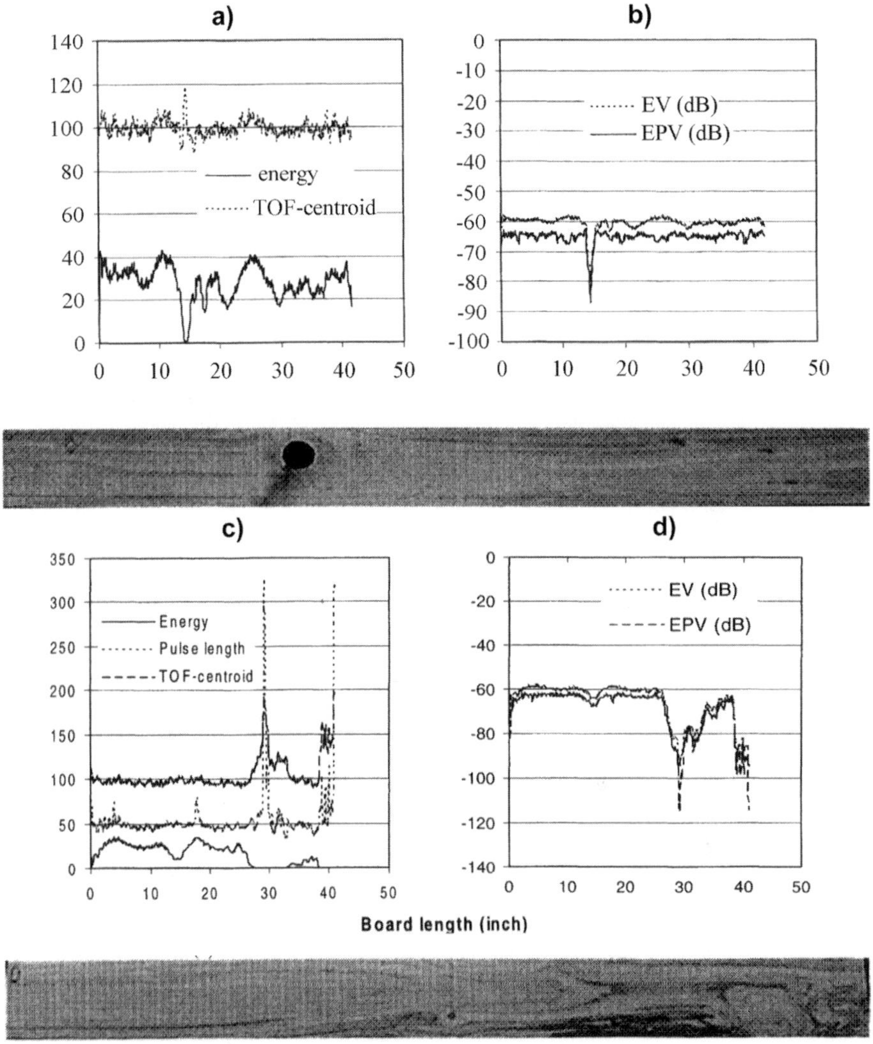

Fig. 5.8a–d. Parameters selected for ultrasonic scanning of board: time-of-flight centroid (*TOF*), time-of-flight energy, time-of-flight amplitude, pulse length energy value (*EV*) and energy/pulse (*EPV*). **a** Measurements of the TOF centroid through a yellow poplar sound knot. **b** Measurements of the energy value and energy/pulse value through a yellow poplar sound knot. **c** Measurements of energy, pulse length and TOF centroid through a decayed red oak board. **d** Measurements of energy and energy /pulse value through a decayed red oak board.

Berndt et al. (1999) reported images obtained with relatively high frequency ultrasonic transducers such as: 1 MHz focused transducers and 5 MHz unfocused transducers using the device depicted in Fig. 5.9a. The line scan in the radial direction of the sample is shown in Fig. 5.9b, in B-scan technique. The

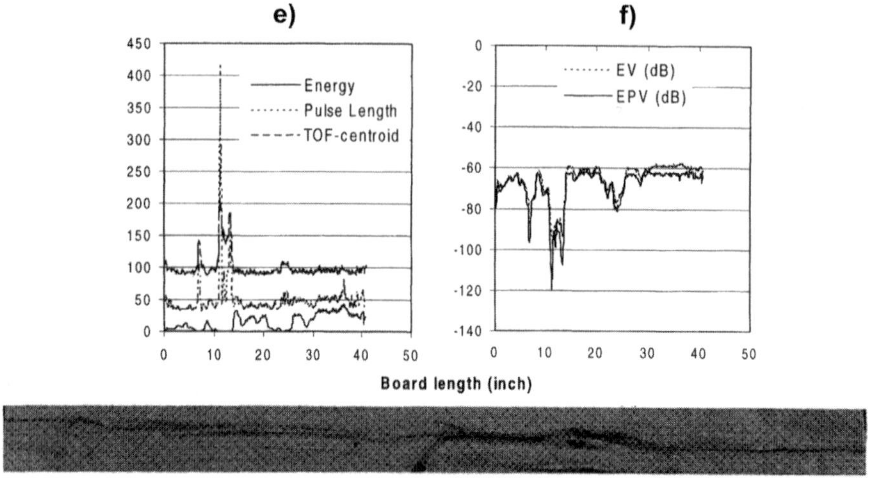

Fig. 5.8e,f. **e** Measurements of energy, pulse length and time-of-flight centroid through a crack in red oak lumber. **f** Measurements of energy and energy /pulse value through a crack in red oak lumber. (Kabir et al. 2001, with permission)

contrast function is given by the instantaneous reflected energy. The overlaying of the total reflected energy with the echo arrival time in radial direction shows that the time is relatively constant, whereas the energy is more sensitive to the structural features. It is generally accepted that the ultrasonic energy incident on wood is split into three components, and that the energy propagates in a direction normal to the slowness surface. The deviation of energy flux can be calculated and can provide useful information about the sample state (Berndt et al. 2000). Collecting signals at multiple locations or using multiple receivers at different positions and frequencies can improve the results of experimental ultrasonic measurements.

5.3.2 Equipment for Noncontact Scanning

The development of equipment for noncontact scanning is related to the progress achieved in producing noncontact probes, for which air is the coupling agent. Noncontact probes are also called as air (or gas) coupled ultrasonic transducers. Coupling efficiency related to energy transfer (Berndt et al 2000) depends on impedance matching. Ultrasonic velocity in air is 330 m/s. Wood exhibits surface acoustic waves ranging from 385 to 1450 m/s (Bucur and Rocaboy 1987). Under these conditions, air can be a perfect coupling medium for the detection of relatively slow acoustic waves.

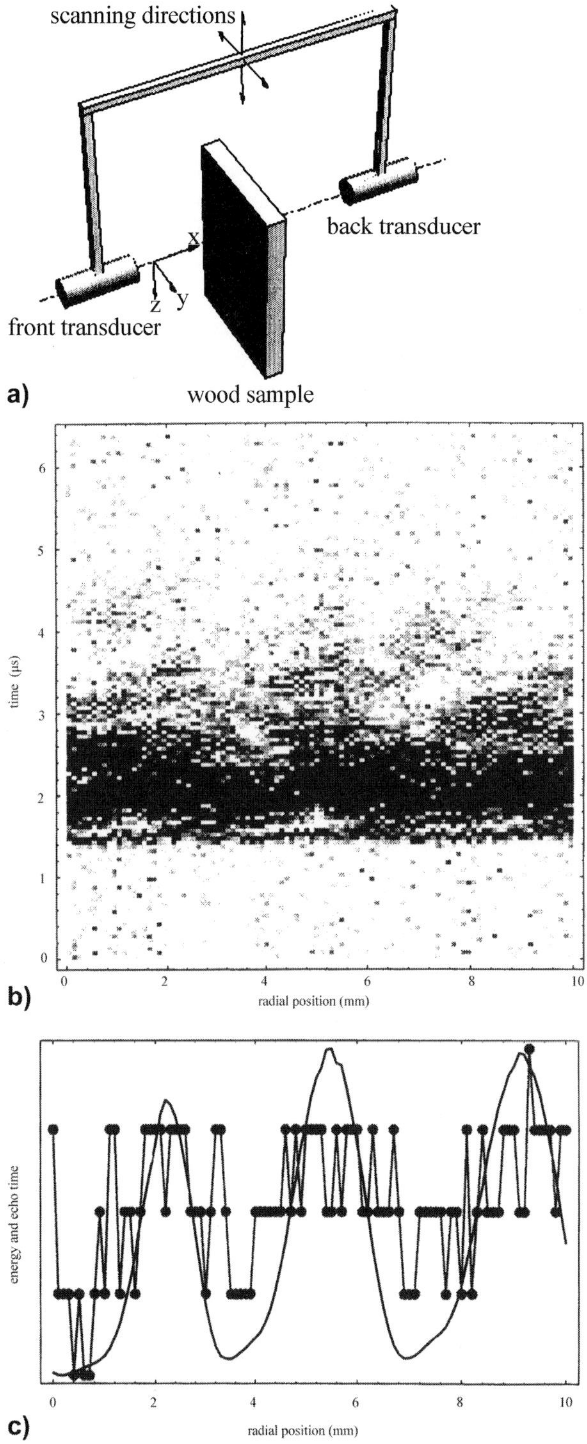

Fig. 5.9a–c. Equipment for contact scanning with immersion technique. **a** Ultrasonic scanning setup. **b** Scan line in radial direction at 5 MHz. **c** Total reflected energy and peak reflection arrival time vs. scan position. (Berndt et al. 1999, with permission)

Several types of transducers are available for noncontact measurements, e.g., piezoelectric transducers unfocused or spherically focused, electromagnetic acoustic transducers (EMAT), capacitive transducers, and laser beam optical generators and detectors. EMAT probes have been used successfully to detect defects in metallic pipes, tubes and plates and for studies of the anisotropy of materials. These probes are inefficient ultrasonic generators and require large power supplies for proper operation. Capacitive probes cannot be used as ultrasonic generators. They require little space between the sample and the probe. Laser beam ultrasound function avoids these difficulties. They operate in both metallic and nonmetallic materials and can be used in hostile environments. The ultrasonic waveform is generated by a pulsed laser incident normal to the sample, and the resulting waveform is detected by a piezoelectric transducer located on the opposite side of the sample. For noncontact characterization of composite materials, laser beam interferometers as well as laser generation and detection of ultrasonic waves have been used (Green 1987).

The use of air as a coupling medium has an important advantage related to the avoidance of permanent damage and decontamination of the specimen tested (Luukkala et al. 1971). On the other hand, because of high acoustic impedance contrast, air-coupled transducers exhibit very high sensitivity to surface roughness.

On paper webbing in motion, air-coupled measurements were performed at 150 kHz (Jarti and Luukkala 1977). Panametrics transducers (type T7) have been commercially available since 1996 for measurements on composites, metal sheets, etc. The frequency of air coupled transducers for industrial applications ranges from 50 to 500 kHz as reported by Lynnworth et al. (1997) and Lynnworth (1989). The internal characteristics of a noncontact probe and its operating modes are shown in Fig. 5.10. The impedance matcher is placed before the piezoelement, and the structure is protected with a thin metal window.

The use of airborne ultrasound is affected by different factors, such as the transmission and reflection coefficients at material boundaries, where a large mechanical impedance mismatch exists, which leads to a large dynamic range between the incident reflection and the echoes generated within the sample. In air, the impedance is 410 $kg\,m^{-2}\,s^{-1}$, while for wood in the fiber direction it is about $2 \times 10^6\,kg\,m^{-2}\,s^{-1}$ and in metals about $20 \times 10^6\,kg\,m^{-2}\,s^{-1}$. The large acoustic impedance mismatch at the transducer/air boundary makes the transmission of energy across the interface difficult. About 99.9% of ultrasonic energy is reflected from material surfaces. Table 5.1 gives the transmission coefficients and energy transfer in various materials. Array transducers for nondestructive application and imaging of solids were developed for the improvement of image quality (Papadakis 1999).

The experimental arrangement for the examination of wooden plates with air-coupled ultrasonic waves and with spherically focused piezoelectric transducers is shown in Fig. 5.11. Two experimental configurations with 1-MHz transducers were used to measure the ultrasonic velocities in wooden plates

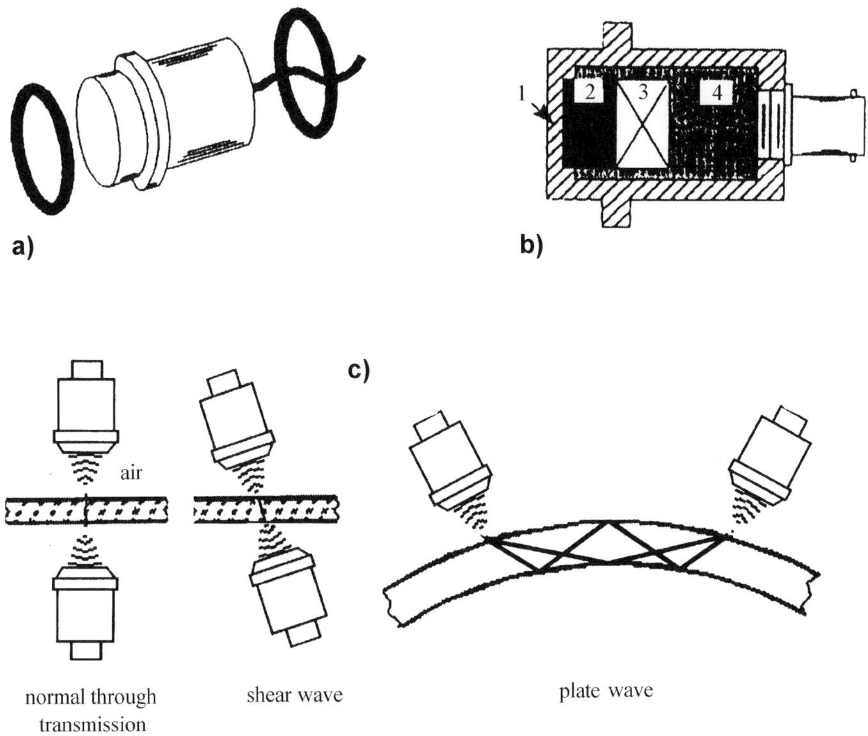

Fig. 5.10a–c. Transducer for noncontact ultrasonic measurements. **a** Transducer's flange is supported between two silicone attenuating rings (Lynnworth and Magory 1999). **b** Internal details: *1* thin metallic window; *2* $\lambda/4$ impedance matcher; *3* 100-kHz piezoelectric element; *4* potting (Lynnworth and Magory 1999). **c** Different experimental arrangements for thorough transmission mode, with longitudinal and shear waves and with Lamb waves. (Buckley 2000; Academic Press, with permission)

Table 5.1. Ultrasonic energy transfer in various materials in noncontact mode. (Bhardwaj 2000; with permission)

Material	Impedance (Mrayl)[a]	Transmission coefficient in air	Transmission coefficient in water	Energy transfer in air (dB)	Energy transfer in water (dB)	Loss energy at interfaces in air (dB)	Loss energy at interfaces in water (dB)
Steel	51	34×10^{-6}	0.11	–89	–19	178	38
Aluminum	17	10^{-4}	0.3	–79	–10	158	20
Acrylic	3.5	5×10^{-4}	0.84	–66	–1.5	132	3
Silicone rubber	1	18×10^{-3}	0.96	–35	–0.35	70	<1

[a] The impedance is expressed in Mrayl (1 Rayl = 1 kg m^{-2} s^{-1}). The impedance of air is 440 Mrayl and the impedance of water is 1.5 Mrayl.

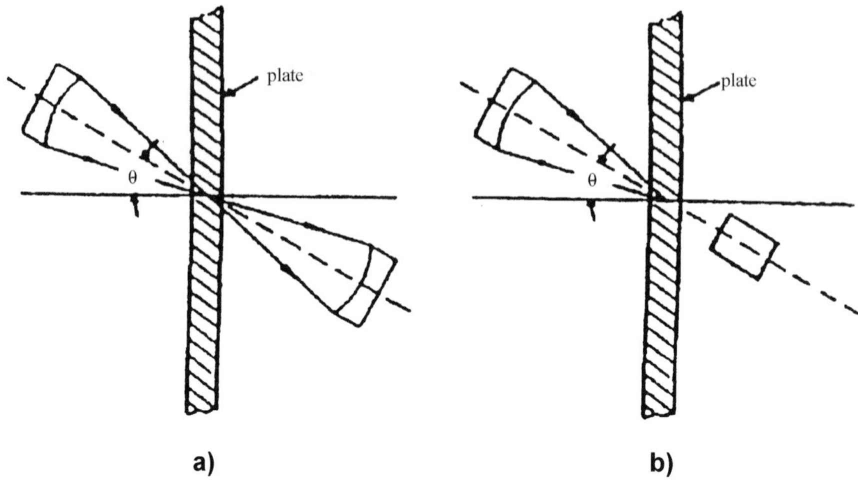

Fig. 5.11a,b. Air-coupled ultrasonic waves in a wooden plate. **a** Two spherically focused piezoelectric transducers coaxially aligned. **b** The focused transducer is coaxially aligned with a flat piezoelectric transducer. (Fortunko et al. 1991; copyright 2001 IEEE)

with a thickness between 3 and 25 mm. The first configuration using two coaxially aligned, spherically focused transducers, having the axes of symmetry parallel, is shown in Fig. 5.11a. The central plane of the sample contains the foci of both transducers. The second configuration (Fig. 5.11b) used only one focused piezoelectric transducer and one unfocused transducer. The focal point of the spherically focused transducer was located at the surface of the wooden plate in front of the unfocused transducer. In both cases, the angle of incidence (θ) is variable to allow the measurement of the time of flight and of the amplitude of the ultrasonic signal as a function of this angle. It is important to note that transmitting transducers must be driven by a high-power, unipolar pulse generator (450 V output voltage, 3Ω impedance and 2000 V peak-to-peak voltage applied to the transducers). A low-noise preamplifier and four precision attenuators (total range 132 dB) were used to minimize the noise factor.

The design of higher frequency noncontact probes (>500 kHz) is difficult because of the short wavelength and high absorption in air, and only two solutions were considered efficient: piezoelectric ceramics elements modified to optimize its transduction in air and capacitive transducers.

Recent advances in high frequency piezoelectric transducers (Bhardwaj et al. 2000), in the frequency range of 200 kHz to 5 MHz, coupled with a high dynamic range–nanosecond accuracy ultrasonic analyzer that is able to provide 100-dB extra gain, 30 dB of which is used to compensate the noncontact mismatch and 70 dB to overcome air-material acoustic impedance mis-

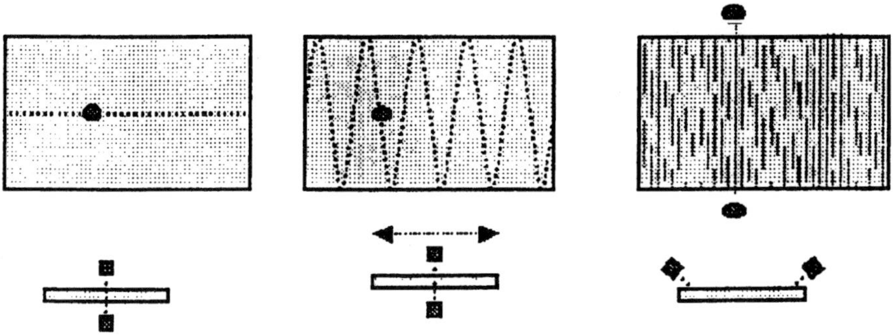

Fig. 5.12. Different ultrasonic scanning modes of sheet products. (Buckley 2000, with permission)

match, made it possible to obtain performances comparable with conventional contact mode transducers. The accuracy of the time of flight measurement was ±50 ns. Tomographic images with this equipment were obtained for an eight-ply graphite fiber re-enforced plastic composite 1.5 mm thick, in pulse-receiver arrangement. The transducers were 5 mm away from the sample. The scanned area was 25 × 25 mm, with 2-MHz frequency and 64-dB gain. The pulse was produced by 4Ω square wave, the scanning was performed with a 3-mm-diameter focused receiver above the sample and with a 25-mm unfocused transmitter underneath it.

Most applications of air-coupled ultrasonic probes for wood composites are for single channel C-scan systems, used for testing flat panels (particleboards, veneer, plywood) less than 20 mm in thickness (Buckley 2000). Multichannel systems were developed for large composite panels, for rapid inspection at the rate of 1 m^2/min. The scanning of sheet products for industrial quality control can be performed as shown in Fig. 5.12 with a bulk wave transmission mode on a central line, in a zigzag course or by inducing Lamb waves with a beam incident angle to the surface of the specimen.

In the lumber industry, air-coupled transducers are used for:

- quality assessment of bulk lumber for internal decay and voids prior to processing. The transducers are used in conjunction with dry pressure coupling for laminated lumber.
- the detection of delaminations in particleboard, veneer and plywood. By using Lamb waves, it is possible to inspect the entire width of the boards.
- the detection of cracks in structural beams and utility poles for in-service inspection of wood products.

Noncontact tomographic imaging of aluminum was reported by Hutchins et al. (1993). They used a pulsed laser to generate Lamb waves in metallic sheets that were detected by electromagnetic acoustic transducers (EMATS).

For wood-based composites, Grimberg et al. (2000) reported the utilization of Lamb waves for imaging of veneer structure. The equipment for Lamb waves launched by Hertzian contact and the operating mode of the transducer are shown in Fig. 5.13. The buffer rods used for emission and reception were identical and were made from an aluminum–magnesium alloy with a radius of 3 mm. Relationships between the amplitudes of P and S waves generated into the plate-inspected specimen and the elastic constants of the wood composite were established. When a delamination between the veneer sheet and the particleboard exists, the Lamb waves split into two beams, one propagating in the veneer layer of thickness h_1 with velocity c_1 and the other one propagating in the particleboard of thickness h_2 and velocity c_2. When delamination exists, the propagation velocity along the y-axis, corresponding to the plate thickness, is smaller than the corresponding velocity in a sound zone. The results presented here were obtained using the assumption that the sheets are homogeneous and isotropic. A better understanding of propagation phenomena can probably be achieved in the future with more complex calculations, based on the assumption that wood in an orthotropic material, with corresponding Young moduli and Poisson ratios.

5.4 Applications

Topics covered in this section include imaging at macroscopic level of the internal structure of standing trees, of lumber, and of defects in wood-based composites. Acoustic microscopy imaging is not discussed in this book.

In considering the application of ultrasonic imaging, it is important to note that for wood the transmission technique seems to be the most appropriate. Since resolution in an ultrasonic image is dependent on frequency, it is important to note that the size of the sample to be inspected, the type of the transducer and the band width must be chosen with care, keeping in mind the propagation phenomena and the anisotropy of the material.

Most ultrasonic inspection and imaging are done at frequencies ranging from 50 kHz to 1 MHz. As the frequency increases above 1 MHz, the image resolution increases but the signal attenuation increases also. Ultrasonic images are degraded by blurring due to practical limitations imposed by beam diameter and pixel size. Digital processing methods may be used to improve the resolution. The spatial resolution, at which the original data were acquired, limits the improved resolution in the processed image. The resolution inherent to acoustic images is basically determined by the beam diameter and by the pixel size. Nyquist's theorem explains the effect of the pixel size on the image resolution. The pixel size is inversely proportional to the resolution. In order to support the spatial resolution, the beam must be spatially sampled at less than half the dimension of the pixel size or, in other words, the pixel size of the image must be one half of the ultrasonic beam diameter.

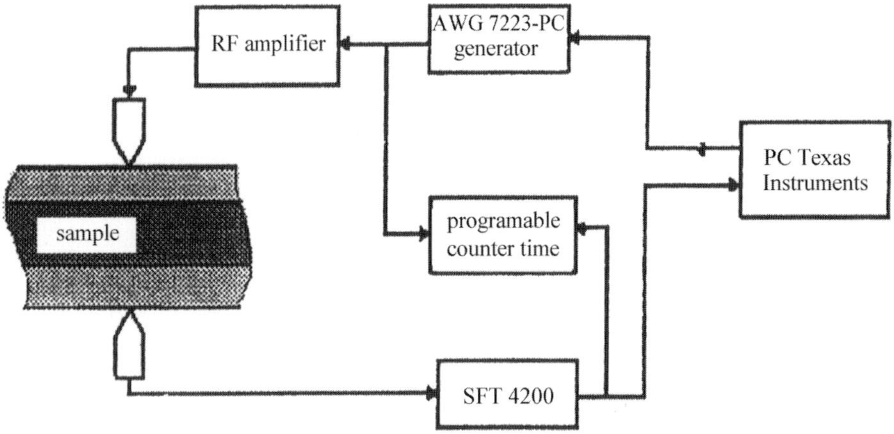

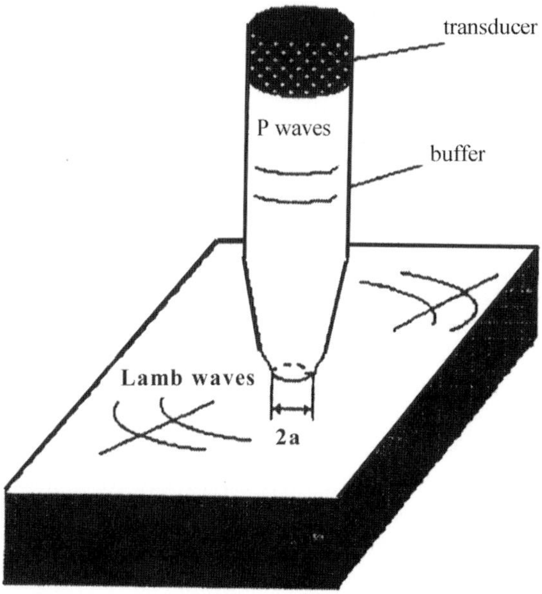

Fig. 5.13. Inspection of a veneer sheet with Lamb waves. *Above* Experimental setup for Lamb waves launched by Herzian contact in a multilayered structure composed of a veneer sheet glued on a particle board. *Below* Transducer for Lamb waves measurements. (Grimberg et al. 2000, with permission)

5.4.1 Imaging of Internal Structure of Standing Trees

Ultrasonic tomographic imaging of the internal structure of standing trees has been reported in several recent articles by the group at Politecnico di Torino, Italy (Comino et al. 2000; Socco et al 2000, 2001; Martinis 2002) using the velocity calculation with SIRT. The ultrasonic velocity was measured by the direct contact transmission method, using exponential probes operating at 54 kHz. The transducers were located on 16 equidistant points around the trunk perimeter selected for the tomography. By changing the reciprocal position of the transmitter and receiver for each measurement, 120 independent acquisitions were obtained. The distance for the tomographic data was calculated assuming a straight path between the two transducers and was determined with an uncertainty of 0.1 mm which was supposed to be equal to the radius of the transducer cross section.

Tomographic image resolution is related to the pixel size and from a mathematical point of view the number of cells cannot exceed the number of measurements, and the size of the cell should not be smaller than the radius of the first Fresnel zone. To define the parameters that rule the iterative processing, the average time uncertainty (0.1 μs) and the spatial resolution (5 cm) were defined by a theoretical approach.

The software used for this purpose was developed for interpretation of seismic borehole data (Friedel et al. 1992). The algorithm is based on the relation:

$$\mathrm{T} = \int_S^R \frac{1}{v} dl = \int_S^R p dl \tag{5.1}$$

where T is the travel time of the ultrasonic wave from the source (S) to the receiver (R), l is the ray path length, **v** is the velocity and p = 1/**v** is the slowness.

The mathematical background of the inversion algorithm is treated in reference books and articles and will not be discussed in this book because of space limitations. As noted by Socco et al. (2001), the procedure can be described as follows: "the velocity calculation was performed with SIRT, starting from an initial arbitrary velocity distribution model, computing the forward model travel times, comparing the calculated travel times with the measured time, and correcting the initial velocity model. These steps are repeated until the fitting between calculated and measured travel times is considered good enough."

The flow chart for data acquisition, analysis, and interpretation is shown in Fig. 5.14. After the processing of ultrasonic data, the tomographic image was compared with the real (photographic) image of the transverse section of the tree. For this purpose, a disk of 40 cm thickness was cut from the tested tree. From this disk, cubic specimens were cut, on which ultrasonic velocities were measured and used to construct an alternative of velocities distribution. This

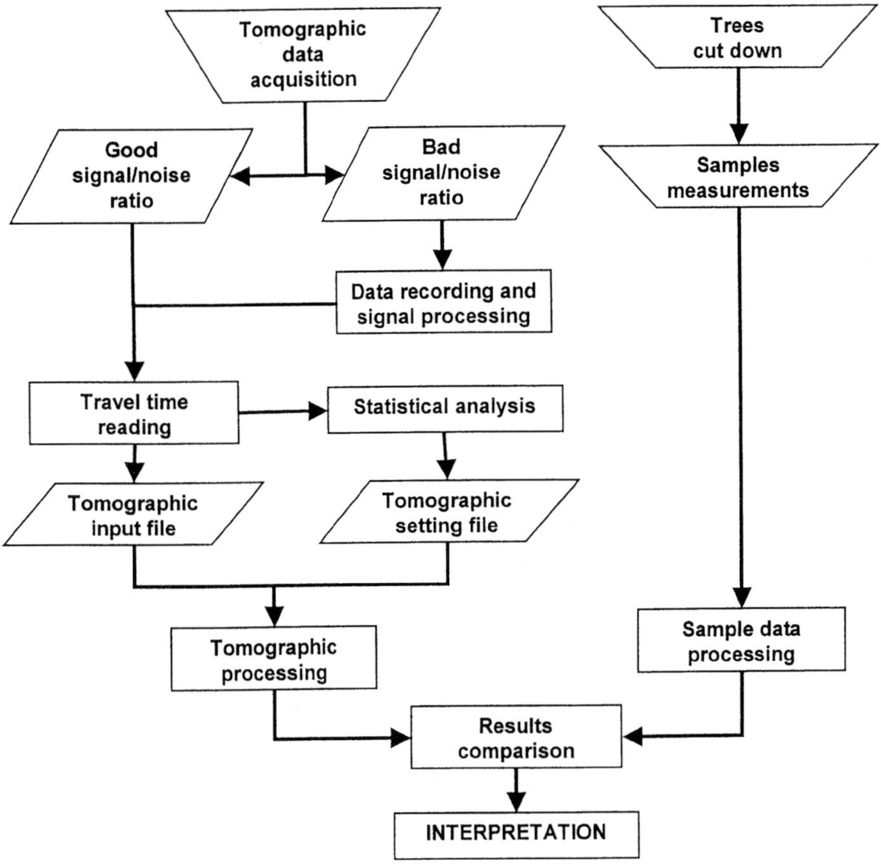

Fig. 5.14. Flowchart for the tomographic imaging of the transverse section of a tree. (Socco et al. 2001, with permission)

reconstruction was compared with the first tomographic image obtained directly on the tree.

For field and laboratory measurements, the signal was observed on an oscilloscope and stored on a personal computer for processing. Spectral analysis was performed with FFT and Hilbert transforms of the signal obtained. The best resolution obtained was 5 cm and was determined taking into account simultaneously the mathematical and the physical resolution, for which the first Fresnel zone (R_F) ray distribution was calculated with Eq. (5.2), which gives the relationship between velocity (v), frequency (f) and distance (d) for the ultrasonic signal

$$R_F = (d\,v/4\,f)^{1/2} \tag{5.2}$$

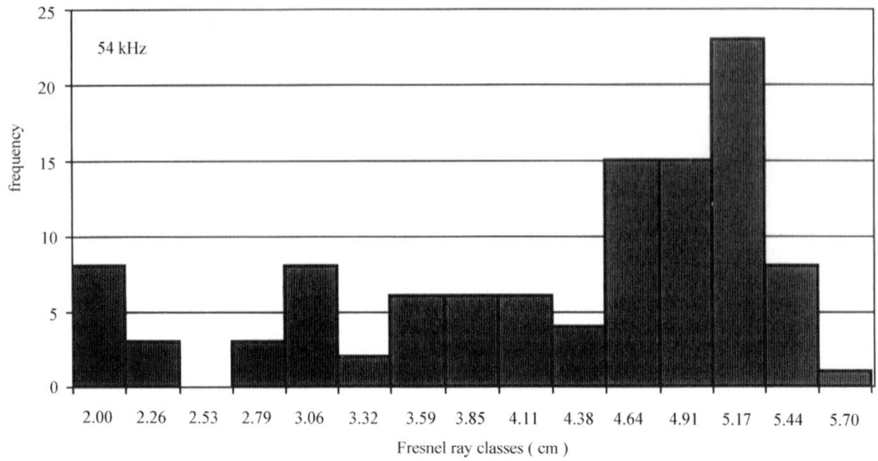

Fig. 5.15. Fresnel ray distribution calculated at 54 kHz on the hackberry trunk (*Celtis australis*) shown in Fig. 5.14. (Socco et al. 2000, with permission)

The Fresnel ray distribution is given in Fig. 5.15 for measurements performed on a 63-year-old hackberry tree (*Celtis australis*) with very complex trunk shape caused by previous wounding and pathological attacks (Fig. 5.16). The most probable Fresnel ray class is at 5 cm, which corresponds to the best image resolution. A good agreement was observed between the three images, that is, the photographic image (Fig. 5.16), the tomographic image of the transverse trunk section obtained from 120 independent velocity measurements (Fig. 5.17), and the reconstructed image with measured velocities on cubic samples (Fig. 5.18a). It should be noted that the sharp velocity contrasts are smoothed by the tomographic inversion.

For the illustration of the simultaneous variation of the velocities in the L, R and T directions, the specimens presented in Fig. 5.18b were used to measure the corresponding velocities. The variation of velocity values versus the position of the specimen is illustrated in Fig. 5.19. Comparing these results and the values of the velocities employed for the tomographic imaging, it can be concluded that the tomographic images match the images constructed from velocity measurements on the individual blocks.

Another example is shown in Fig. 5.20, in which the tomographic image of the transverse section of a standing tree (*Platanus acerifolia*) of 450 mm diameter with an important decay zone is presented. The velocities measured in the RT plane range between 1000 and 1800 m/s. In the central, decayed zone, the velocities are relatively low compared to sound wood. This is probably due to the low density of the tissue affected by the fungal attack. The low velocity peripheral zone observed on the tomographic image is caused by the experimental configuration and is likely to be an artifact of the inversion algorithm, because this zone lies outside all measured ultrasound paths.

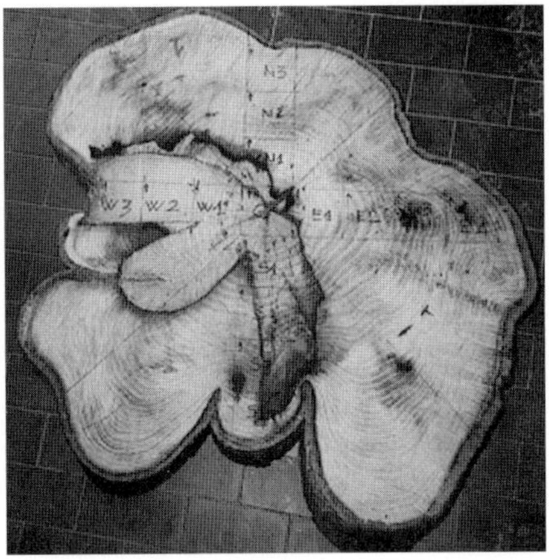

Fig. 5.16. The complex shape of the transverse section of the trunk of hackberry (*Celtis australis*) (Comino et al. 2000, with permission)

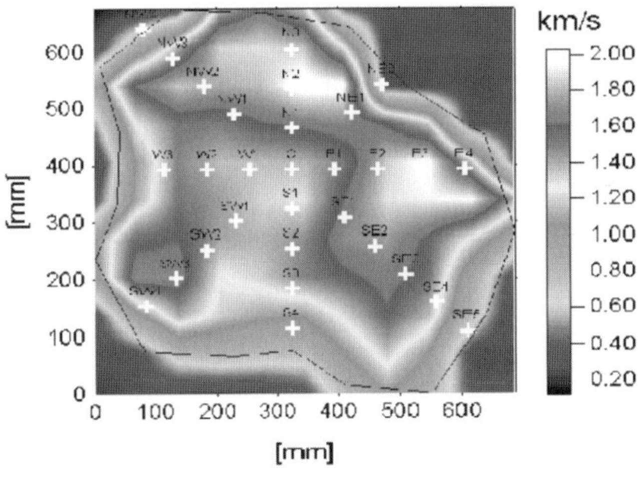

Fig. 5.17. Tomographic images of the transverse section of the trunk in Fig. 5.16 obtained with 120 independent velocity measurements. (Comino et al. 2000, with permission)

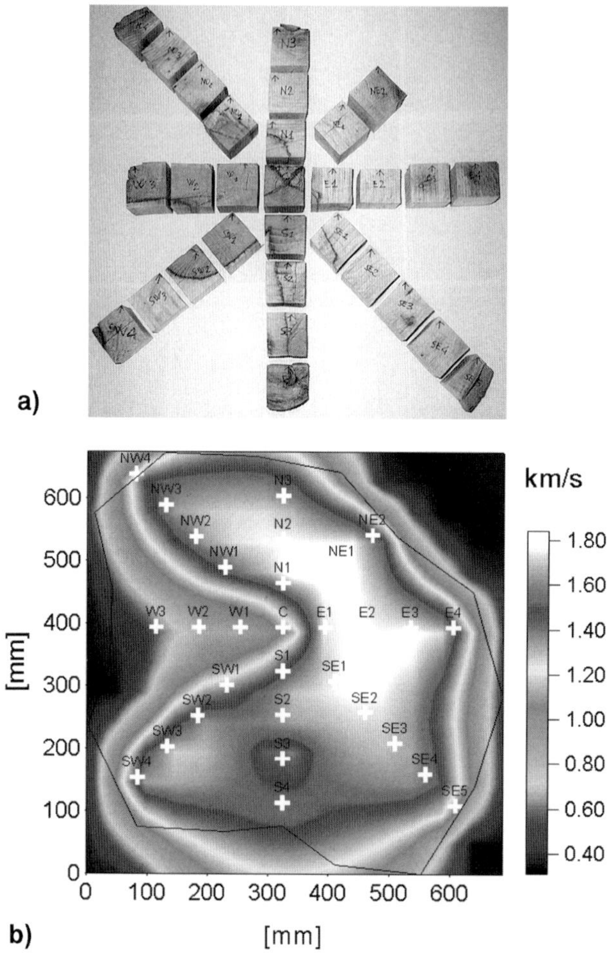

Fig. 5.18a,b. Reconstructed image with velocity values measured on cubic specimens and positions of the specimens in the trunk section. *Above* Reconstructed tomographic image. *Below* Cubic specimens. (Comino et al. 2000, with permission)

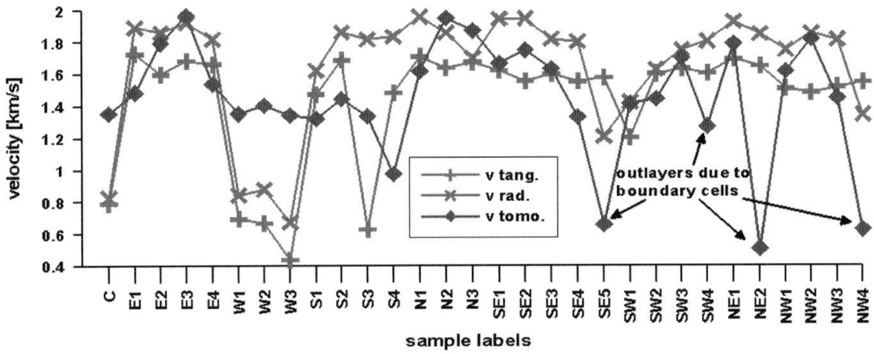

Fig. 5.19. Measured ultrasonic velocities and calculated velocities for two orthotropic directions R and T on selected specimens shown in Fig. 5.18. (Socco et al. 2000, with permission)

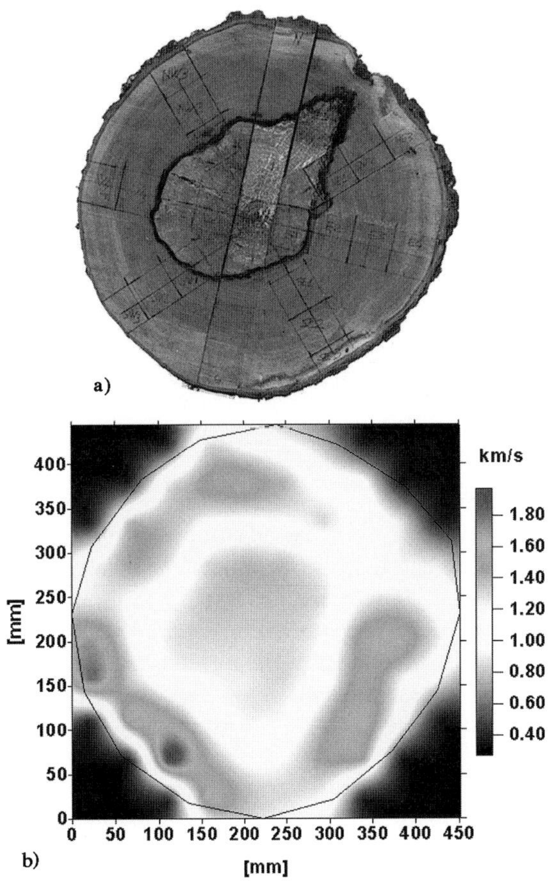

Fig. 5.20a,b. Ultrasonic tomographic reconstruction of the transverse section of a standing tree (*Platanus acerifolia*) 40 cm above ground level. (Martinis 2002, with permission)

5.4.2 Imaging of Lumber Structure

The development of the methodology for the characterization of wood with a direct transmission technique, that is, the calculation of the off-diagonal terms (Bucur and Archer 1984; Bucur 1986) of the stiffness matrix, supported the exact calculation of elastic constants of lumber with knots. Using an array of cubes of $2 \times 2 \times 2$ cm (Chazelas et al. 1988), it was possible to determine the variation of the stiffnesses at each point of the sample. The specimen was divided into numerous cubes, on which ultrasonic velocities were measured and corresponding stiffnesses were calculated.

The mapping (Fig. 5.21) of the variation of elastic constants in three anisotropic planes of wood was possible, and in this way the imaging of knots and of slope of grain deviation was obtained with a resolution of 2 cm (Chazelas et al. 1988).

High resolution imaging of wood, at a millimeter scale, was possible when image reconstruction from projection was developed using very sophisticated signal processing for ultrasonic tomography for materials associated with the development of the corresponding ultrasonic transducers.

Neuenschwander et al. (1997) have published ultrasonic images of lumber, corresponding to a 1-mm resolution of pine lumber knots as shown in Fig. 5.22. The image was obtained in C-scan mode, by an immersion technique in water by reflection with 2.25-MHz broadband transducers. The image of the radial section (Fig. 5.22a) clearly shows the typical pattern of latewood and earlywood as well as the deviation of the slope of grain around the knots. In the transverse section of the specimen (in the TR plane), the circular pattern of the annual rings is well defined. The presence of compression wood can be observed in the central zone of the figure in the 16th and 21st rings. Cracks several millimeters wide are visible in the first annual ring.

5.4.3 Imaging of Defects in Wood-Based Composites

As noted by Maloney (1996) in the family of wood-based composites, the materials described under the label "engineered wood products" (particleboard, medium density fiberboards, different structural panels, etc.) are manufactured annually in millions of tonnes all over the world. Effective internal defect detection for these materials takes full advantage of the development of nondestructive imaging methods for quality assessment. In this section, we present the methodology developed recently for the imaging of the most frequently observed defects: delaminations and voids.

Imaging of delaminations in wood composites was performed with the equipment described by Grimberg et al (2000; Fig. 5.13). The inspected structure was composed of a 20-mm-thick particleboard, on which two beech

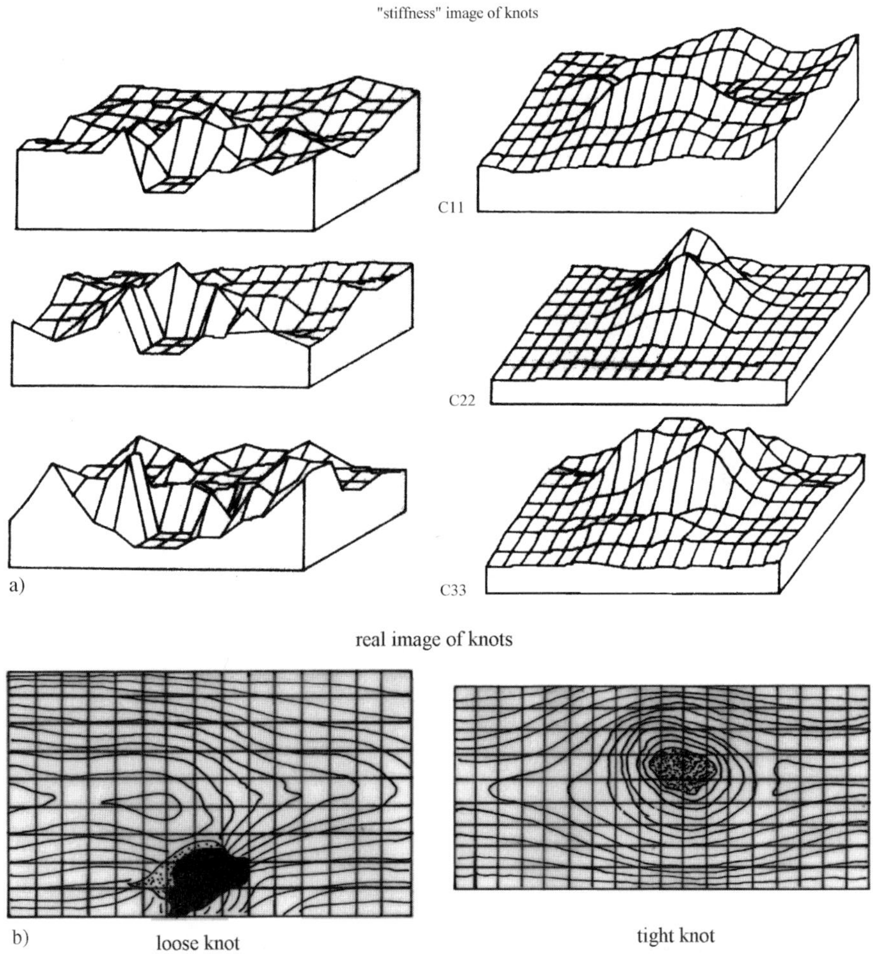

Fig. 5.21a,b. Reconstruction of the image of boards with loose and tight knots in a pine sample, based on the measurements of stiffness $C_{LL} = C_{11}C_{RR} = C_{22}$and $C_{TT} = C_{33}$ and related angles on cubic specimens compared with the real image of boards. (Chazelas et al. 1988, with permission)

veneer sheets of 1-mm thickness were glued onto the two faces. The delaminations were simulated by inserting rayon foils of $50\,\mu m \times 0.5\,mm$ as shown in Fig. 5.23a. The density of the wood composite was $780\,kg/m^3$, the Young's modulus in the longitudinal direction was $1.4 \times 10^{10}\,N/m^2$, and the corresponding Poisson ratio was 0.14. The velocity in sound zones was 4300 ± 90 m/s. In a zone with delamination, the velocity measured in the veneer layer was significantly lower, ranging from 1100 to 2128 m/s. The map of the delaminations is shown in Fig. 5.23b. The location of each delaminated zone is clearly visible.

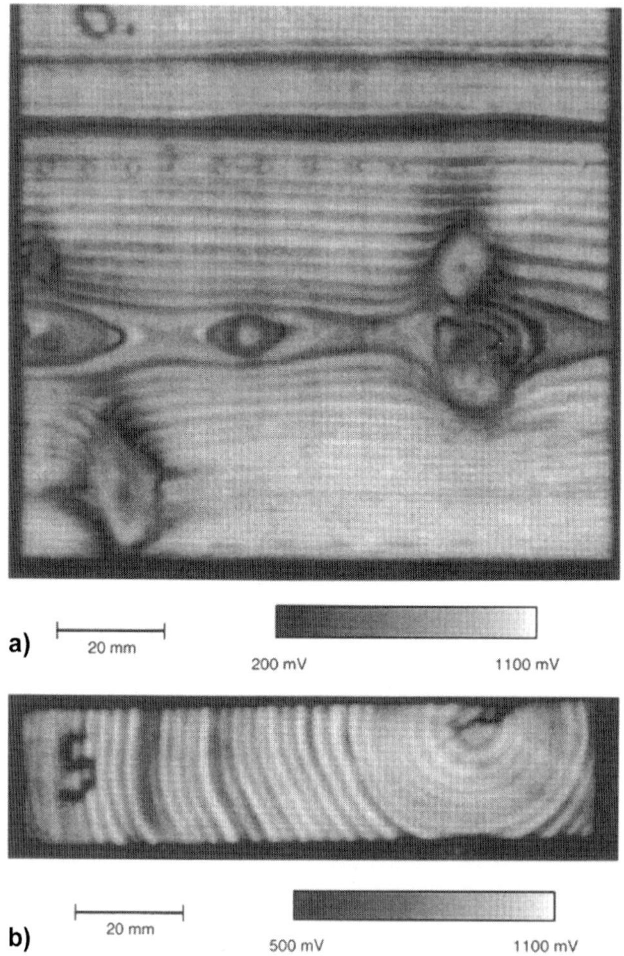

Fig. 5.22a,b. Ultrasonic images obtained in C-scan mode by immersion and in reflection mode of the structure of a pine specimen. **a** Longitudinal–radial section with knots, the pattern of the earlywood and latewood and the slope of grain deviation around the knots. **b** Transverse section with typical pattern of the annual rings and the compression wood. (Neuenschwander et al. 1997, with permission)

For better understanding of the differences between the velocities measured in veneer and in particleboard layers, we suggest that the initial hypothesis should be improved. It was originally suggested by Grimberg et al (2000) who considered the layers as isotropic solids. A better approximation can be obtained with the hypothesis of transverse isotopy for the mechanical properties of the veneer. For industrially produced veneer sheets, the reference anisotropic plane is the plane LT. The same hypothesis can be used for the

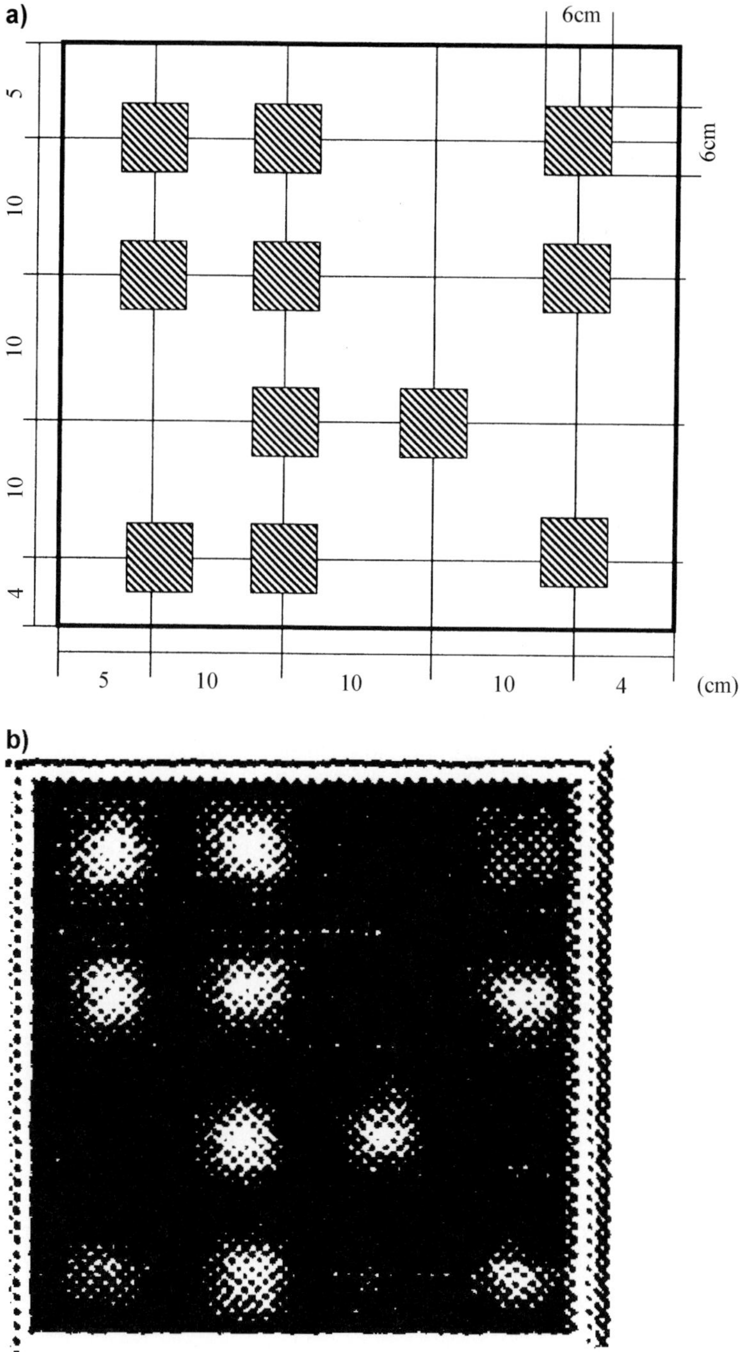

Fig. 5.23a,b. Imaging of delamination in a composite with Lamb waves. The composite was produced with two beech veneer sheets glued to the two faces of a particleboard layer. **a** Position of the delaminated zones. **b** Image of the delaminations obtained with Lamb waves. (Grimberg et al. 2000, with permission)

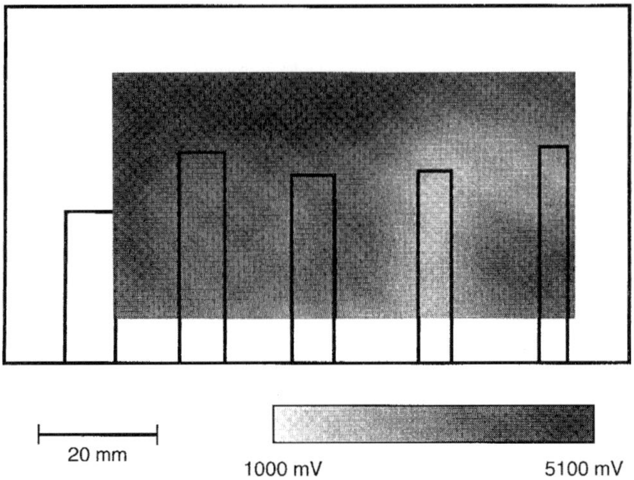

Fig. 5.24. Imaging of holes in medium density fiberboard of 18-mm thickness in C scan mode, by the immersion technique with 0. 5-MHz transducers. The image was superimposed on the geometry of the specimen. The position of the second hole from the right is well defined (Neuenschwander et al. 1997, with permission)

particleboard, where the elastic constants in the direction of the board formation and in the corresponding perpendicular direction are similar, and only the elastic constant thickness direction of the board is relatively low (Bucur 1992).

The ultrasonic imaging of voids in medium density fiberboards of 18-mm thickness is shown in Fig. 5.24. The voids were simulated by holes of different diameters bored in the fiberboard. The image was obtained in C-scan mode, by a transmission technique, with the specimen immersed in water and with broadband transducers of 0.5-MHz central frequency. In the future, for on-line quality assessment, research on the development of dry or noncontact scanning techniques in a large frequency range must be performed.

5.4.4 Imaging of Defects in Wooden Poles

The nondestructive inspection of wooden poles in service for electrical power distribution, for communication and for housing is a very important safety measure. The traditional inspection method is tapping with a hammer to generate a sound. The interpretation and quality of the decision about the service life of the pole depends on the ability and experience of the inspector. X-ray-computed tomography can also be used, but is not easy to handle on a large scale and is hazardous for the operator and the environment. As an alternative, ultrasonic inspection is of great interest for electrical companies all over the world because of its handling ease and safety. The ultrasonic propagation characteristics measured on poles (Arita and Kuratanik 1984) and especially the tomographic imaging (Tomikawa et al. 1985), readily interpreted by the field operators of electrical companies, are excellent alternatives to the old percussive hammer method.

The attempt to reconstruct ultrasonic tomographic images is based on the assumption that the ultrasonic waves propagate in a straight line and that the sound speed greatly decreases in zones where decay is present. The fan-beam capture of ultrasonic time-of-flight was used and 324 points were employed to reconstruct the image using 78-kHz transducers, for which the wavelength was 2 cm. This experimental arrangement was completed with 97 projection data of parallel beams, obtained by interpolation of the fan-beam arrangement. This large number of data was deconvoluted using the technique of filtered back-projection. The computational time for producing an image was about 1 h, which is too long for practical use. To reduce the time requirements, the reconstruction of the images must be done in hardware. It is necessary to develop an on-line system that enables rapid reconstruction for in-situ pole inspection.

5.5 Summary

The main purpose of this chapter has been to identify the breadth of applications of ultrasonic tomographic reconstruction techniques of wood at macroscopic levels. Ultrasonic tomography is a diffraction type tomography that is noninvasive and is safe at low energy levels. Different types of ultrasonic waves can be used for wood imaging, but the most common are bulk longitudinal waves. Ultrasonic images can be reconstructed from all characteristic parameters of the wave: time of flight, amplitude, frequency spectra of the waveform, the phase, etc. The energy distribution and energy flow are important parameters for enhancing the image contrast. In this chapter, we have discussed very recent results related to tomographic imaging with the time of flight of ultrasonic waves.

There are three main types of algorithms that can be used to form tomographic images from ultrasonic data: transform techniques, iterative techniques and direct inversion techniques. For the algebraic reconstruction algorithms (ART), each equation corresponds to a ray projection. The sums of the computed rays are a poor approximation of the measured ones, and the image suffers from significant noise. The SIRT (simultaneous iterative reconstruction technique) reduces the noise of ART and produces better images. The factors that limit the accuracy of the images obtained with diffraction tomographic reconstructions are related, firstly, to the theoretical approach to the approximations in the derivation of the reconstruction process and, secondly, to the experimental limitations. High-resolution images have been obtained for small, clear specimens of different species, for standing trees, for lumber and for wood-based-composites. Two-dimensional imaging requires advanced signal processing and modern digital computers.

The equipment for ultrasonic imaging can operate in mode A, B and C in contact and noncontact (or air-coupled) scanning, using frequencies ranging from 50 kHz to 5 MHz. As the frequency increases above 1 MHz, the image resolution increases, as well as the signal attenuation. The resolution inherent in acoustic images is basically determined by the beam diameter and by the pixel size.

The piezoelectric transducers used for ultrasonic tomographic imaging can operate in direct transmission mode, inducing bulk or surface waves. Conventional piezoelectric transducers in the range of 50 kHz to 1 MHz are commonly used for data acquisition. Air-coupled ultrasonic transducers are mainly used for inspection of wood based composites like veneer or low thickness fiberboard and particleboard. Ultrasonic tomography has a very large field of applications in decay diagnosis of standing trees, in parks and public gardens in the imaging of lumber structure to reveal knots, grain deviation, cracks and compression wood and in imaging of defects in wood composites, such as delaminations between layers and voids.

6 Nuclear Magnetic Resonance

6.1 Introduction

Nuclear magnetic resonance imaging was first developed for medical applications. However, since 1980, an increasing number of researchers have reported applications of this technique to materials science using standard NMR spectrometers, in which modifications had been made, and computing facilities had been adapted to produce images. Every measurement in conventional NMR spectroscopy, such as chemical shift, nuclear spin relaxation times, dipolar couplings and the spectrum of translational motions can, in principle, be translated into a spatial resolved contrast and used to produce images. The major impediment to the extension of this technique was and still is the price of the equipment.

Nuclear magnetic resonance imaging offers to wood science and wood processing a nondestructive, noninvasive and noncontact technique to study wood structure characteristics, such as growth rings, the sapwood/heartwood boundary, rays, reaction wood, resin canals, knots (Bucur 1990), wounds, wetwood, decay by fungi, etc. Images obtained from magnetic resonance properties of ^{1}H, ^{13}C or other nuclei provide a tool for in vivo analysis of trees. This technique is useful in evaluating normal events in the life of trees as well as different perturbations induced by external factors, such as climate, diseases, etc. Tree vitality as indicated by the sapwood/heartwood ratio can also be studied. Water distribution within the growth ring, in roots, branches and leaves, diurnal or annual periodicity of the water content in the stem can be determined in vivo. Wood drying is an important field of application of magnetic resonance imaging. Another research aspect is related to the distribution of preservatives in wood or to the decomposition of wood by fungi and/or other pathologic agents able to induce a modification of the moisture content of wood (Birkeland 1990; Kucera 1990). When combined with localized spectroscopy, nuclear magnetic resonance imaging provides a unique experimental tool for the study of the structure of wood.

Knowledge of the fundamentals of nuclear magnetic resonance technique is critical to the productive use of this technology for wood, which is a porous material. The physics of magnetic resonance phenomena is well understood and is described in various reference books. Amongst these books and review articles, one can mention the books written by Abragam (1961), Talpe (1971), Slichter (1978), Gadian (1982), Mansfield and Morris (1982), Schau (1984),

Callaghan (1991), Gil and Neto (1999), and Lambert et al. (2000). The principal phenomena are identified in the literature by their initial letters such as: NMR for nuclear magnetic resonance and MRI for magnetic resonance imaging. This chapter reviews methods and applications of the NMR technique to wood considered here as an anisotropic porous material, with emphasis on imaging and spectroscopy techniques developed specifically for wood.

6.2 Basic Aspects of NMR Imaging

NMR is an important technique for measurement of wood properties that provides a number of bulk parameters (moisture content distribution, wood porosity, proton diffusion and absorption as a function of time, chemical composition, chemical shifts that gives information about the position of protons in the samples, etc), that cannot be provided by any other methods. In addition, by using pulse NMR techniques, one can obtain additional important information about the diffusion and flow of moisture in wood.

A detailed presentation of the theoretical background of the NMR technique is far beyond the scope of this book. In this chapter, we shall give a brief description of the basic properties of atomic nuclei, explaining their interaction with the applied magnetic field.

In this book, the term *nuclei* is used synonymously with *nuclear magnetic moment.*

The NMR properties of the nuclei commonly used in wood science are given in Table 6.1. High resolution experiments utilize ^{1}H, ^{13}C and ^{31}P which have a spin quantum number equal to 1/2. The distribution of hydrogen or other atoms and their relaxation time in a strong magnetic field lead to differences in the NMR signal used for the reconstruction of the images.

Some atomic nuclei, such as ^{1}H, ^{13}C and ^{31}P, possess an angular momentum or spin and an associated magnetic moment. The spin can be visualized as the spinning motion of the nuclear magnetic moment about its own axis. The magnetic moment determines the couple on the nuclei that tends to align it with

Table 6.1. NMR properties of some nuclei (Gadian 1982) (© DG Gadian 1982, reprint from Nuclear Magnetic Resonance and its applications to living system by David G Gadian (1982) by permission of Oxford University Press)

Nucleus	Spin-quantum number	Resonance frequency (MHz) at 5 tesla	Relative sensitivity at constant field
^{1}H	1/2	213	100
^{13}C	1/2	53.5	1.6×10^{-2}
^{31}P	1/2	86.2	6.6

the applied magnetic field. In the absence of a magnetic field, the average sample magnetic moment produced by the nuclear magnetic moments is zero because the spins are randomly oriented.

Quantum mechanically, a spin 1/2 nuclei can only align parallel or anti-parallel to the field. Boltzmann statistics gives the fraction of nuclei in each state with more nuclei in the lower energy state, with the nuclei magnetic moment parallel to the field, and anti-parallel to the field direction There will therefore be a net sample magnetic moment in the direction of the field. Because of their angular momentum, the nuclei are subjected to a torque and they precess like a gyroscope around the direction of the field. The frequency of precession is the Larmor frequency ω_o and is proportional to the gyromagnetic ratio of the nucleus γ and field B_o as defined by:

$$\omega_o = \gamma B_o \tag{6.1}$$

The frequencies associated with conventional NMR techniques are usually in the range of 100 to 600 MHz. NMR frequencies can be identified with the absorption of microwave photons of energy $\hbar\ \omega$ (where $\hbar$ is the Plank's constant $= 6.6255 \times 10^{-27}$ erg. s $= 6.6255 \times 10^{-34}$ N ms), causing transitions between the energy levels associated with the quantum mechanical states of the nucleus in a magnetic field as:

$$\Delta E = h\ \gamma B_O \tag{6.2}$$

where ΔE is the difference in energy between the two states.

The nuclei resonance frequency is therefore directly proportional to the local magnetic field experienced by the nuclei. However, the applied B_o field also induces electronic currents in the surrounding atoms, producing a further small shielding or screening field σ_{Bo}. The effective field B_{eff} at the nucleus is therefore given by:

$$B_{eff} = B_O(1 - \sigma) \tag{6.3}$$

σ is known as the shielding or screening constant and depends on the atomic environment of the individual nuclei in a chemical compound.

The NMR frequency for an individual nuclei is therefore

$$\omega_O = \gamma\ B_O(1 - \sigma) \tag{6.4}$$

where σ has typical values from 10^{-6} to 10^{-3}.

Different nuclei in a chemical structure will therefore have slightly different NMR frequencies, depending on their chemical environment.

Resonant frequency is defined by the chemical shift δ (in ppm = part per million) and is expressed as a fractional change in the frequency measured. For example, the NMR spectrum of acetic acid CH_3COOH gives rise to two distinct frequencies for the protons in the CH_3 and COOH parts of the molecule. The zero value of the chemical shift for ^1H and ^{13}C spectra is defined by comparison with measurements on tetramethylsilane. The chemical shift is independent of the magnetic field strength. The application of a small radio frequency field at the Larmor frequency, perpendicular to the static, applied field causes the spin to be excited between the energy levels of the nuclei in the static field. This leads to the absorption of power at the Larmor frequency. For ^1H nuclei in a field Bo of 9.3 tesla an ac B_1 field at a Larmor frequency of 400 MHz is required to produce a resonance spectrum.

As can be seen from Fig. 6.1, in a stationary coordinate system, the momentum M_o has two components, M_z and M_{xy} rotating in the XY plane at frequency ω. If a new magnetic field is applied in a direction perpendicular to M_{xy}, the whole system will rotate and will have the new coordinates X′, Y′ and Z′. The rotating magnetic field is produced by a radio frequency (RF) oscillator. No effect is seen on M_{xy}, but a torque is exerted on M_z that will thus precess about the X′-axis and about the applied field B_o direction. Rotation of M_z will induce a voltage in a detector coil situated along the Y-axis, and resonance will be detected.

The maximum induction signal, induced in the detector coil, will occur when the pulse of RF energy at the Larmor frequency is sufficiently long to rotate M_z at 90° into the XY plane. On removal of the B_1 field, the nuclei will precess with their spin perpendicular to the applied static field B_o. However, the nuclei will experience small local fields that alter B_o. These local fields are both positive and negative, so that some spins precess faster and some slower and are no longer in phase. Their magnitude reduces to zero as $\exp(-t/T_2)$, where T_2 is the transverse relaxation time.

The thermally induced fluctuation in the local field associated with the motion of atoms also produces components of the local field at the Larmor frequency. This causes net magnetization. M_z returns to its equilibrium value as $M_z = M_o[1 - e^{(-t/T_1)}]$, where T_1 is referred to as the longitudinal relaxation or the spin-lattice relaxation time.

The characteristic relaxation times for a decay of the Z and XY components of the magnetization T_1 and T_2 are strongly dependent on the environment of the excited spins. The effects of relaxation and precession are combined in the Bloch equations describing NMR (Abragam 1961). In general, T_1 differs from T_2 because T_1 involves a transfer of energy from the spin system to the surroundings, while T_2 describes the energy concerning loss in phase coherence within of the spin system.

Fig. 6.1a–d. Macroscopic description of magnetic moment of the nuclear spin system. **a** Nuclei are in equilibrium with the static magnetic field Bo. *Mo* is the overall magnetic field vector; *Mz* is the static magnetic field along the axis Z; *Mxy* is the rotating magnetic field in the XY plane; θ is the angle of precession. **b** Rotation of *Mz* in the presence the of B_1 magnetic field generated by the Larmor frequency. *Mz* is tipped away from Z direction by B_1 field. **c** Loss of phase coherence of spins after the extinction of the B_1 field, leading to a reduction in precessing Mxy as $Mxy_{(o)}e^{-t/T2}$. **d** In practice, decay is observed because nuclei experience different fields in different molecules and shift the effect that can be measured. The local variation in the field B_1 is observed due to the spin–spin interaction. T_2 is generated by sample properties (pore size, chemical bonding of H_2 molecules)

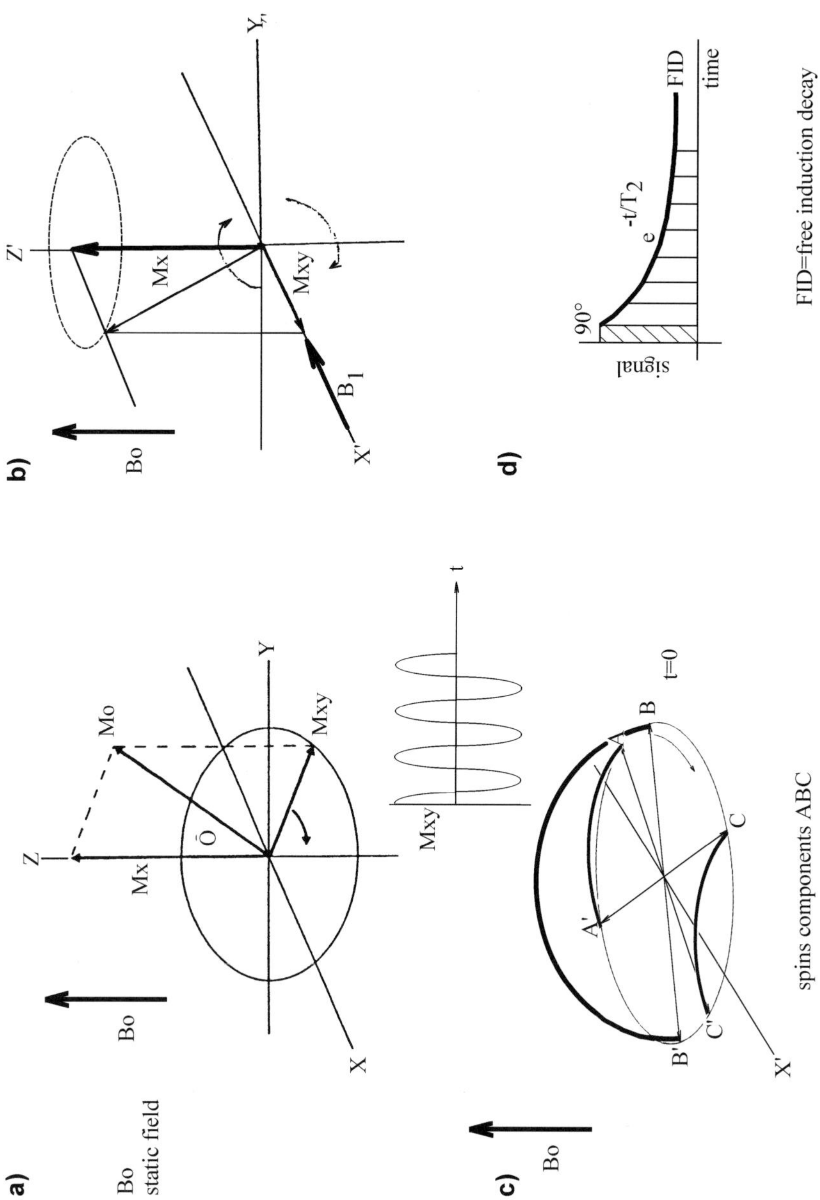
a)
Bo
static field
Bo
Z
Mx
Mo
Ō
Mxy
Y
X
Mxy
t
b)
Bo
Z'
Mx
Mxy
B1
X'
Y'
c)
Bo
A'
B
B'
C
C'
X'
t=0
spins components ABC
d)
signal
90°
e-t/T2
FID
time
FID=free induction decay

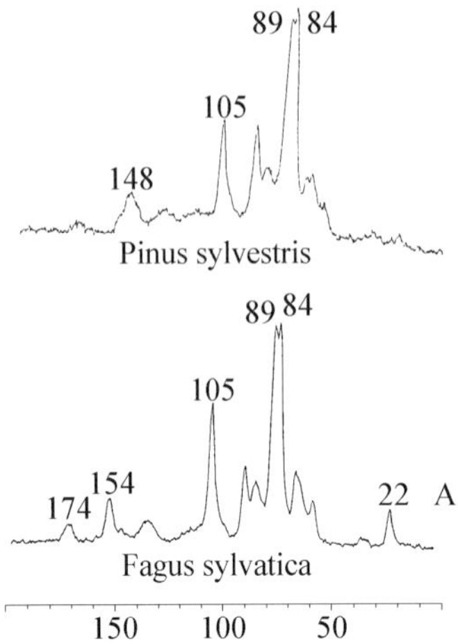

Fig. 6.2. ^{13}Carbon cross-polarization/magic angle spinning (CP/MAS) NMR spectroscopy spectra for pine (*Pinus sylvestris*) and beech (*Fagus sylvatica*), determined at 75.47 MHz ^{13}C frequency, 4.5 kHz rotor spinning frequency and 1.5-ms contact time for cross-polarization. (Gilardi et al. 1995, with permission)

In the case of NMR spectroscopy, the investigation of NMR relaxation mechanisms in inhomogeneous solids involves the characterization of decay curves as a sum of exponential functions. The estimation of T_1 and T_2 is relatively accurate for two or three exponentials (Menon et al. 1987, 1989; Flibote et al. 1990).

Figure 6.2 shows a typical NMR spectrum of pine and beech with major differences in the lignin and hemicellulose components. The peaks corresponding to chemical shifts δ of 105 ppm correspond to proton resonances in cellulose and are present in both spectra. The peaks of higher intensity at 84 and 89 ppm show that in pine the cellulose crystallinity is higher than in beech. The lignin peaks are at shifts of 148 ppm (phenolic carbon, with methoxyl groups attached) in pine and in beech at 154 ppm (with methoxylated phenolics). The acetate groups present in the xylan in beech give rise to peaks at 22 and 174 ppm.

NMR techniques can be used to determine T_1 – the spin-lattice relaxation time – and T_2 – the spin–spin relaxation time in wood. The factors that influence T_1 and T_2 are the physical state of the wood sample and the presence of water molecules in the cell wall and lumina (Araujo et al. 1992). The quality of the NMR signal for wood samples is strongly dependent on the amount of water present. A significant difference can be seen between free water, water in the cell lumina, and the bound water, in the cell wall. The contribution of each subsystem to the total magnetization is called abundance and is proportional to the amount of protons in the governed subsystem. The abundance increases with the moisture content in wood and is 100% at the fiber saturation point. The relaxation time T_1 of protons in pure bulk water is 3 s while electrically bonded protons have relaxation times of about 10^{-3} s.

The spin-lattice relaxation time T_1 and the spin–spin relaxation time T_2 are determined by fluctuating magnetic fields and by molecular motions. The time

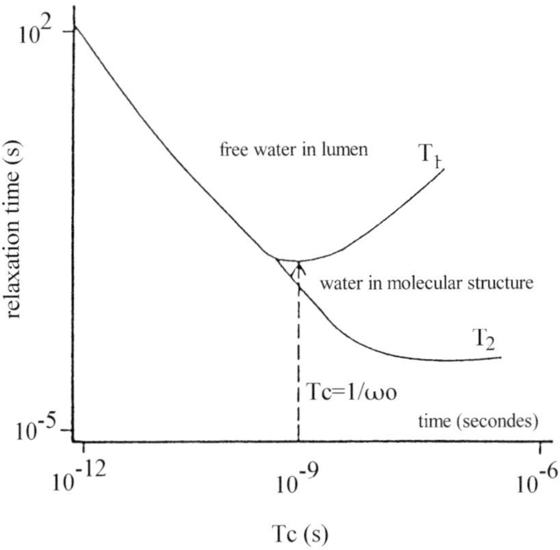

Fig. 6.3. Relaxation time as a function of molecular correlation time T_c at typical frequencies associated with spin motion. (Pfeffer and Gerasimowicz 1989, CRC Press, with permission)

scale, as shown in Fig. 6.3 for these relaxation processes, can be quite different. Slow motions in the kHz range cause the spin to lose the precessional phase coherence determining T_2, while protons are associated with fluctuations of the magnetic field, at the Larmor frequency (typically many hundreds of MHz) are required to restore the spin population to equilibrium along the B_o direction determining T_1.

Polymers of molecular weight between 10,000 to 25,000 Da may have relaxation times on either side of the minimum of the curve, depending on the Larmor experimental frequency. For molecules having relaxation times Tc $>10^{-9}$ s the relaxation times T_1 and T_2 are different. In contrast to the relaxation of proton spin magnetization in wood samples, on which one finds a multi-exponential decay time constants ranging between less than 1 ms to several tens of ms, relaxation in bulk water is a single exponential process: $T_1 = T_2 = 2.4 \times 10^{-9}\,m^2/s$ at 25 °C. Diffusion measurements are of particular interest in wood science and industry. For this purpose, we have to study the effects of field gradients and of static inhomogeneities in field on spin–spin fluctuations.

By using field gradients perpendicular to the X, Y or Z axes, one can make measurements on slices in all directions, using familiar computed tomographic techniques. Three-dimensional information can be obtained from all points of the space available for object imaging.

The field gradient is also important in determining the diffusion time because during the measurements the water molecules diffuse in and out of the measured section. Only molecules remaining with the section contribute to the magnitude of the pulse echo in a 90°–180° sequence. This enables the diffusion coefficient in the direction of the field gradient to be measured. In practice, for imaging improvement, Chang et al. (1999) demonstrated the utility of the sophisticated sequence of switching of pulse gradients in all directions.

Interestingly, any reduction in precessing the M_{xy} component due to field inhomogeneities can be reversed by applying a 180° pulse as indicated in Fig. 6.4. As the spins represented by the components ABC continue to precess relative to each echo by the same amount, they all come together at a time 2t, to give what is called a spin echo signal. However, alternatively, the effects of a static inhomogeneity in the field can be reversed. Random processes such as the spin–spin fluctuation affecting T_2 and any diffusion of spins from one region to another in an inhomogeneous field cannot be reversed. In a homogeneous field, this technique enables T_2 to be measured, while in a field gradient the decay enables the diffusion content to be measured along the field.

For spins diffusing in a field gradient, the pulse echo sequence is determined by the spin diffusion length $(2\ D\ t)^{1/2}$. If the length is known from the structural features, we can measure $T_2{}^*$ to determine the diffusion coefficient D, or, if D is known, the diffusion length, which can be related to the dimensions of cell structure.

The NMR technique can also be used to investigate the diffusion of water both along and perpendicular to the fibers. This is also achieved by NMR measurements in a field gradient. In practice, during the sequence, the water molecules can diffuse out of the structural region. If an interaction takes place between the water and the cell wall, the alignment of spins can be destroyed. This limits the time the spin can remain in a coherent state (the T_2 time) so that the diffusion time can be deduced and the pore size can be determined. Moreover, a long diffusion time can be related to the diffusion measurement along the fibers and can be used for the calculation of porosity.

Brownstein (1980), using NMR data published by Hsi et al. (1977) for wood specimens at 38% moisture content of *Thuja occidentalis*, explained the observed nuclear relaxation rates in terms of a strong relaxation. The deduced relaxation times are shown in Fig. 6.5. For these specimens, the relaxation time T_2 has two components, a slow relaxation time T_{2s} of the order of a few tens of ms and a fast component T_{2f} of the order of a few ms. T_{2f} is identified to be the same as T_1. Similar data were reported by Byrne et al. (1986) for *Pinus radiata*, with two values of T_2, 0.8 and 10 ms measured at 60 MHz. When, the specimens were saturated with water, T_{2s} increased considerably, while T_{2f} leveled off at about three times its initial value. To explain the phenomenon, Brownstein (1980) proposed a model entailing (we suppose) diffusion of free water present in the lumina and a monomolecular layer of water near the cell wall (corresponding to a 6% wood moisture content), which is illustrated schematically in Fig. 6.6.

The model consists of two water layers of thickness *a* and *b* with diffusivities D_I and D_{II}, where $D_I \ll D_{II}$. The strong magnetic relaxation is supposed to be present only in region I. Calculations demonstrated that in region II the

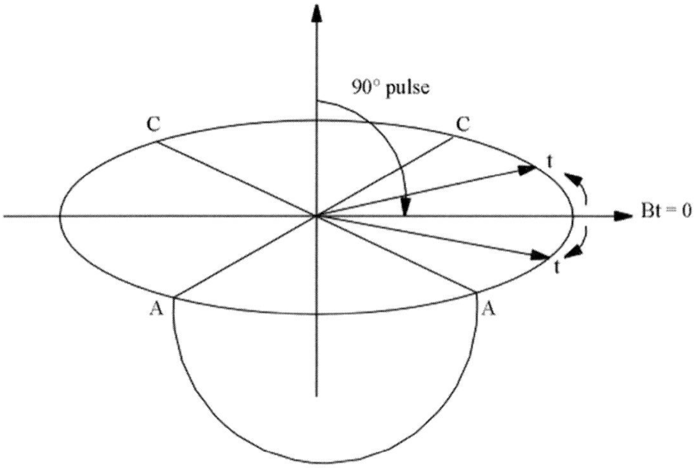

a

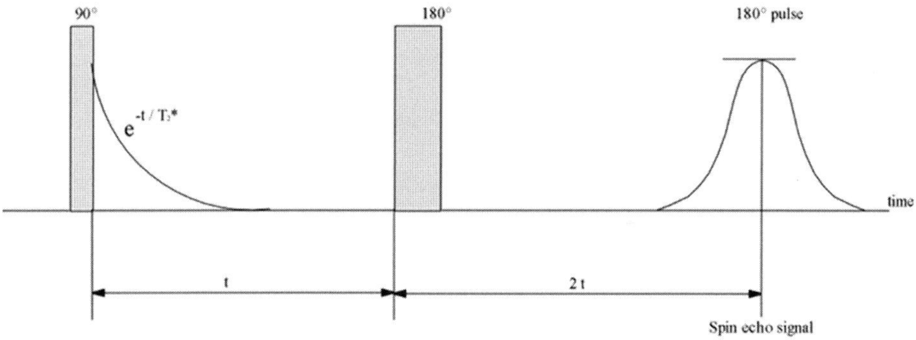

b

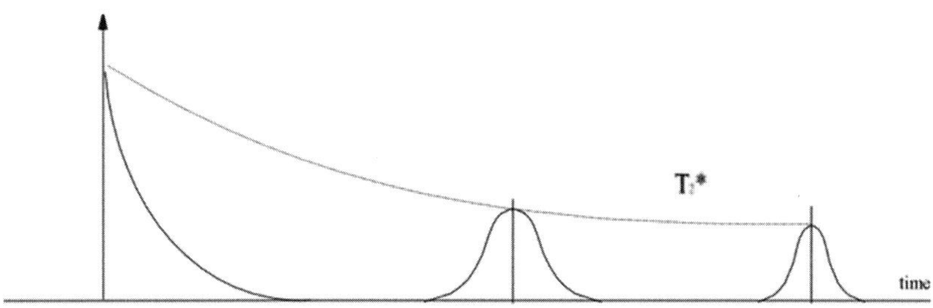

c

Fig. 6.4a–c. Relaxation time as a function of pulse sequence time. **a** Reduction of precessing M_{xy} component due to field inhomogeneity can be reversed by applying a 180° pulse. **b** The spins represented by the components ABC continue to precess, and they come together at a time 2t, corresponding to the spin-echo signal. **c** For spins diffusing in a field gradient, the pulse-echo sequence is determined by the spin diffusion length. If the diffusion length is known from geometric structural features, one can determine the diffusion coefficient from T_2* measurements

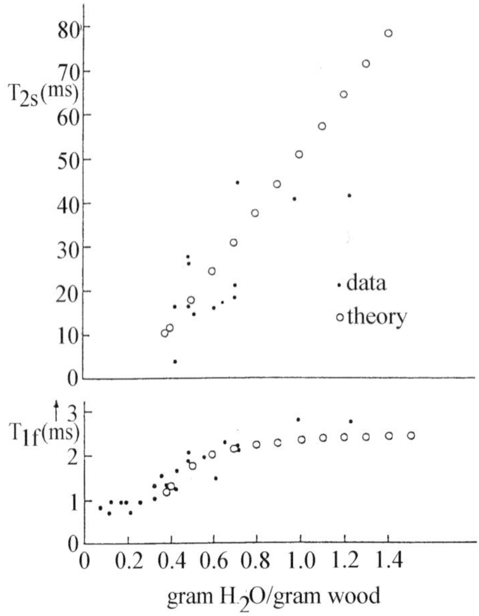

Fig. 6.5. Theoretical and experimental data of T_{1f} and T_{2s} as a function of moisture content. (Brownstein 1980, Academic Press, with permission)

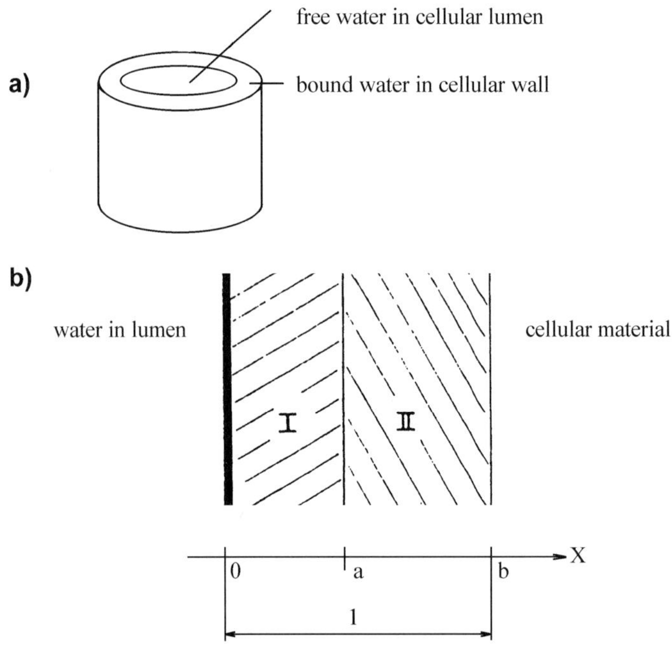

Fig. 6.6a,b. Physical model proposed by Brownstein (1980) for a wood cell, to explain the diffusivity of water in wood, with 1-D geometry of the density of proton magnetic moment. **a** Free water and bound water in a cell. **b** Diffusion model. *Region I* is defined by the thickness x = a and diffusivity D_I and corresponds to free water in the wood, which corresponds to 0.38 g H_2O/g wood. When only a monomolecular water layer is present, we have $D_I \ll D_{II}$; $a^2/D_I = 25.8$ ms and a $\leq 0.26\,\mu$m. The diffusion process is inhibited within this layer. *Region II* is defined by the thickness x = b and the diffusivity D_{II} and corresponds to the bound water in the cell wall. The proton distribution is constant and $D_{II} \gg D_I$ and $D_{II} = 2.4 \times 10^{-5}$ cm^2/s. (Academic Press, with permission)

Table 6.2. Diffusivity (D) in some species determined with the NMR technique, compared with classic techniques

Species	Units	D parallel	D Perpendicular	Ratio	Reference
Douglas	$10^{-9}\,m^2/s$	1.7	0.88	1.93	
Maple	$10^{-9}\,m^2/s$	1.2	0.36	3.3	MacGregor et al. (1983)
Fir	$10^{-10}\,m^2/s$	3.0	0.28	10.7	Choong (1963)
Theoretical	$10^{-10}\,m^2/s$	5–40	–	–	Byrne et al. (1986)

proton density distribution was a constant for each mode under consideration and that the relaxation time was dependent on a^2/D_I, corresponding to the diffusion term for water molecules across the thickness **a**. This parameter determines the time scale and "does not affect the form in which the relaxation time varies with the ratio b/a." The model requires that the ratio of the two diffusivities $D_I/D_{II} = 1/900$. If D_{II} is taken to be $2.4 \times 10^{-9}\,m^2/s$, for water, the parameter a^2/D_I has the value of 26 ms and consequently $a < 0.26\,\mu m$. This means that region I consists of rather a thin layer of water molecules.

Like all physical parameters of wood, the diffusion of water is anisotropic. MacGregor et al. (1983) calculated the diffusion coefficient of water molecules under equilibrium conditions (static or self-diffusion) as a function of the angle Θ between the longitudinal anisotropic axis corresponding to the cell axis and the field gradient (Fig. 6.7).

The coefficients parallel and perpendicular to the cell axis are given in Table 6.2. The ratio between the two coefficients reflects the anisotropy of wood. Water diffusion in sugar maple is more anisotropic than in Douglas fir. In the direction parallel to the cell axis, the diffusion coefficient is smaller than that of bulk water, probably because the water in wood (sap) contains different magnetic ions (Mn, Fe, Cu) and is different from bulk water and also because the cell walls restrict diffusion.

Lindgren (1994) summarized in a very concise manner, the principles of the NMR method:

1. the random orientation of the magnetic moment of hydrogen nuclei in the absence of a field;
2. the lining up of nuclei when a magnetic field is applied;
3. the 90° tips on magnetization on applying a radiowave field;
4. the removal of the radiofrequency signal, and subsequent generation of a radiowave by the precessing tipped magnetization frequency and amplitude.

Several methods for inversion recovery to measure T_1 and T_2 are used, including saturation recovery and progressive saturation. The following describes the inversion recovery pulse sequence as given by Hailey et al. (1985).

It was noted previously that the magnetic moment precesses at a frequency proportional to the local magnetic field strength. The tipping at 90° and at 180° depends on the strength and duration of the pulsed radio frequency signal. A

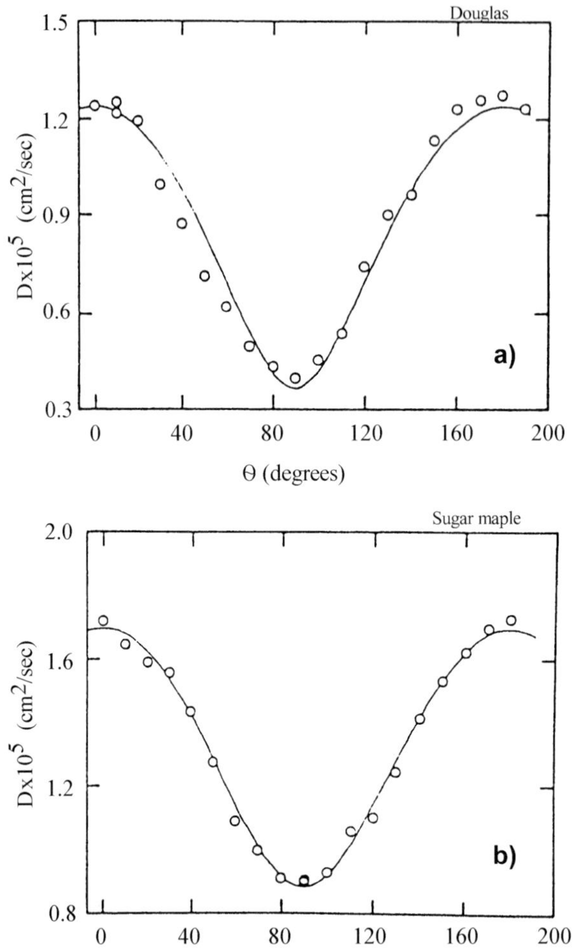

Fig. 6.7a,b. Diffusion coefficient vs. the orientation angle of the specimen for **a** Douglas fir and **b** sugar maple. The theoretical values were deduced for Douglas fir with $D_{//} = (1.7\pm0.1)\ 10^{-5}\ cm^2/s$. $D_{\perp} = (0.88\pm0.05)\ 10^{-5}\ cm^2/s$ and for sugar maple with $D_{//} = (1.2\pm0.1)\ 10^{-5}\ cm^2/s$ $D_{\perp} = (0.36\pm0.05)\ 10^{-5}\ cm^2/s$. (MacGregor et al. 1983, with permission)

particularly important sequence of pulses CPMG is named after the inventors Carr and Purcell (1954) and Meiboom and Gill (1958), cited by Abragam (1961). This sequence (Fig. 6.8) consists of a 90° RF pulse, tipping the M_z into the XY plane (Fig. 6.8b) magnetization precess, around the field direction (Fig. 6.8c) growing an "induction signal" in an RF pick-up coil with its axis in the plane. However, the fluctuations in the local field result in small local changes in the local field, leading to a slightly different time, varying precessional frequency in different microscopic regions, so that the individual components lose their phase coherence resulting in a reduction in amplitude amounting to

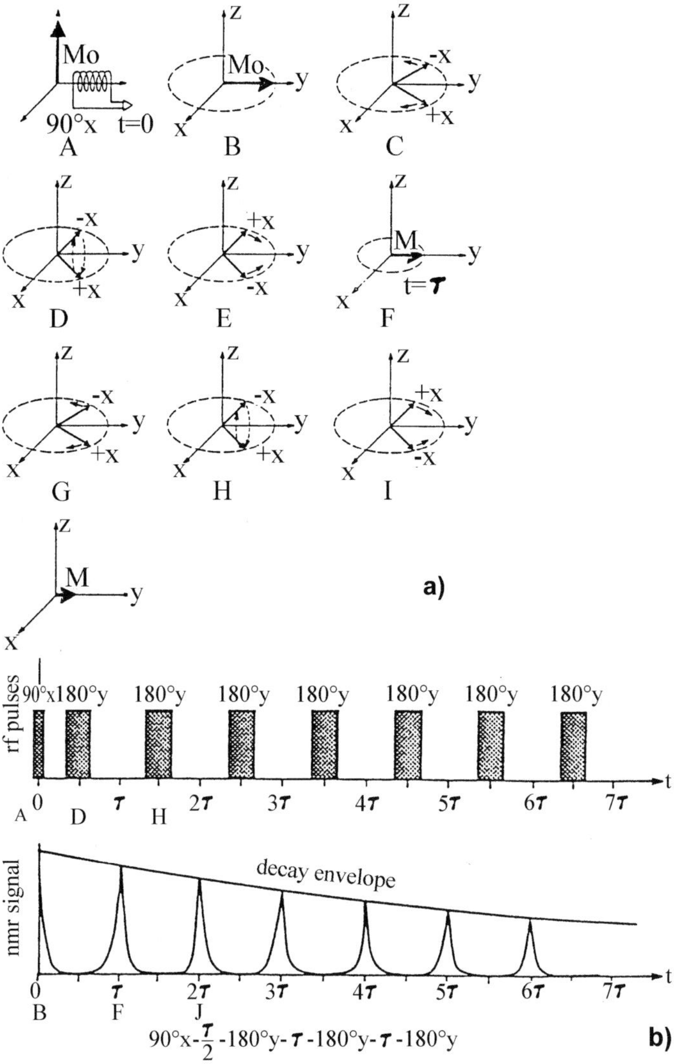

Fig. 6.8a,b. Carr-Purcell/Meiboom Gill sequence. **a** Spin isochromatic representation. **b** Pulse sequence and NMR signal for T_2 measurements. (Hailey et al. 1985; courtesy of Forest Products Society)

$e^{(-t/T_2)}$. This is a random process and cannot be reversed. However, in practice, the applied fields are never completely homogeneous and therefore result in dephasy, giving an induced signal recovery as $e^{(-t/T^*_2)}$ where T^*_2 depends on the homogeneity of the field.

The free induction decay (FID) curve describes the decay of the induced signal arising from free precession of the nuclei in the field B_o. The Fourier

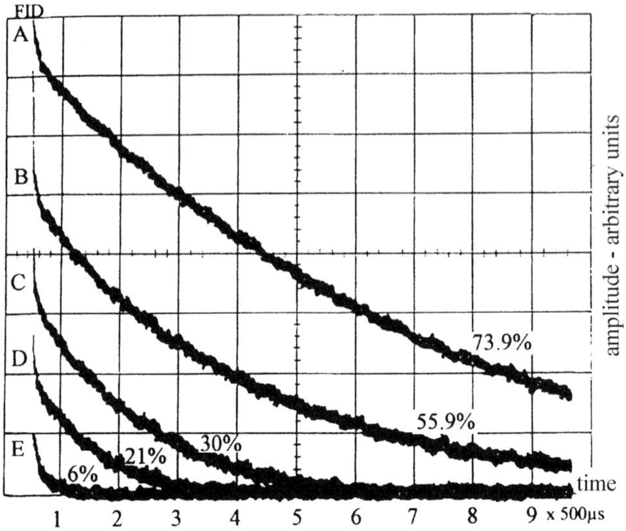

Fig. 6.9. Free induction decay (FID) amplitude of ^1H nuclei in maple with different moisture content as a function of time following a 90° pulse. Note that the signal increases in amplitude and the relaxation rate decreases with moisture content. (Sharp et al. 1978; courtesy of Wood and Fiber Science)

transformation of this decay gives the conventional NMR spectrum, in which the signal amplitude is plotted as a function of frequency. If the nuclei are present in different environments, the resulting FID contains contributions of different frequencies, each component having its own characteristic time constant.

An example of free induction decay versus time is given in Fig. 6.9, for maple at different moisture contents above and below the fiber saturation point. Figure 6.9a shows a very short relaxation time <20 μs for the sample at 6% moisture content and a long relaxation time >500 μs for wood at 30% moisture content. Different moisture contents give signals of different sizes and different relaxation times can be measured. The wood moisture content therefore can be related to the FID signal. Sharp et al. (1978) showed that the wood moisture was proportional to the FID, as shown in Fig. 6.10. (It should be noted that the amplitude of the free induction signal just after a pulse is simply proportional to the number of protons.)

The amplitude of the free induction decay signal can be used to obtain data relating to the mobile and less mobile components of the protons. An example is given for western red cedar, for which the FIDs were acquired with a 90° RF pulse sequence and with 5 s between pulses. Note that the measurements depend on the time between pulses. The free induction decay curve for three levels of moisture content in western red cedar is shown in Fig. 6.11. One can describe these results in terms of the FID from protons directly associated with

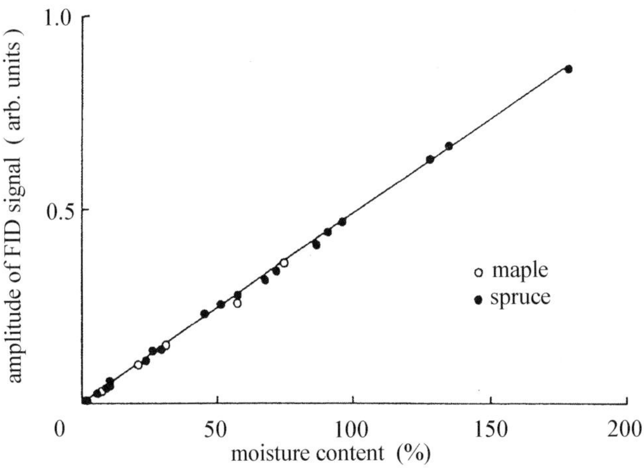

Fig. 6.10. Amplitude of FID signal measured at 50 μs as a function of moisture content. (Sharp et al. 1978; courtesy of Wood and Fiber Science)

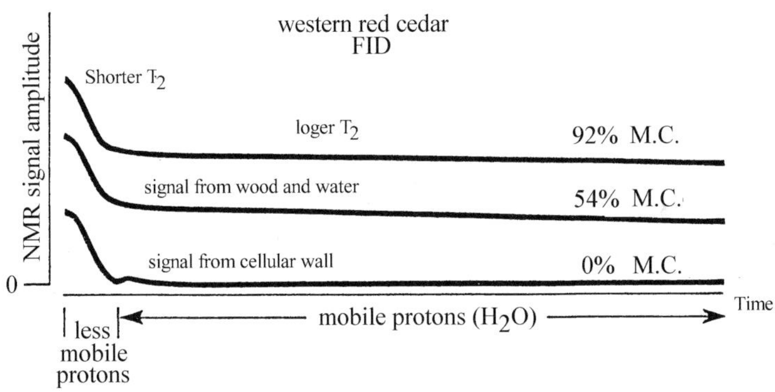

Fig. 6.11. Free induction decay at three moisture content levels. (Hailey et al. 1985; courtesy of Forest Products Society)

the wood structure with a superimposed additional signal from the absorbed water. No mobile protons were detected on the signal corresponding to 0% moisture content. The mobile protons of water in a sample of western red cedar of 25% moisture content can be quantified on the graph as the intersection of the profile of the mobile protons with the axis of the signal amplitude.

Menon et al.(1987) measured the ^{1}H NMR free induction signal for western red cedar sapwood at three moisture contents: 216, 92 and 54%, and observed three relaxation times (Fig. 6.12). The very rapid component at 54% is related to T_1 and is <10 μs. Two components were measured for T_2, a fast one, decaying at $T_{2\ fast} < 100$ μs and a slow one decaying at $T_{2\ slow} < 200$ μs. The intensities

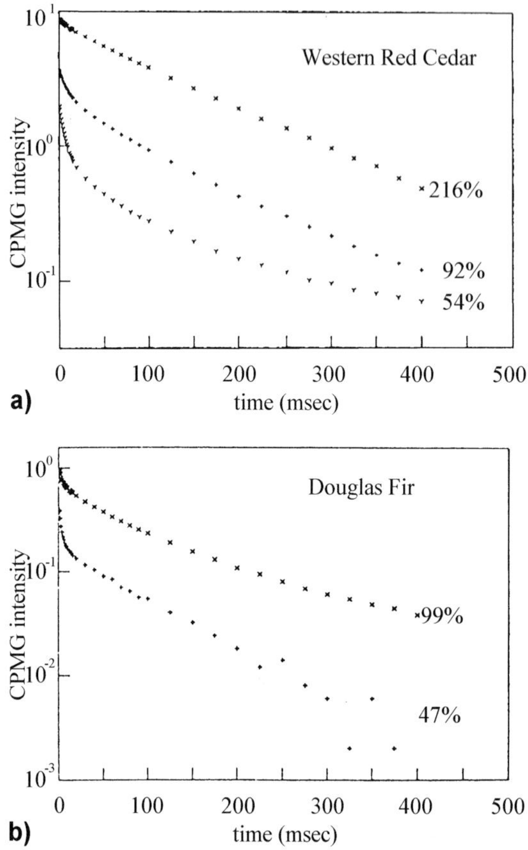

Fig. 6.12a,b. CPMG decay curves. **a** Western red cedar at 216, 92, 54% moisture contents. **b** Douglas fir at 99 and 47% moisture contents. N.B. The logarithmic intensity scale that could give a straight line for a single component exponential decay. (Menon et al. 1987, Academic Press, with permission)

of CPMG data normalized versus the solid proton signal from FID at each moisture content were plotted versus time for western red cedar and for Douglas fir. The authors detected three components of the relaxation time on visual inspection of the slope of the regression line. The spin lattice relaxation curves are decaying nonexponentially except at the highest moisture content. It should be noted that the protons near the surface interact with the cell walls. T_2 becomes a measure of pore size.

The spin-lattice relaxation time T_1 can be determined from an inversion sequence illustrated in Fig. 6.13. Immediately following the magnetization a 180° pulse is reversed. The magnetization then reverts to its equilibrium value as $M_{z(t)} = M_o[1-2\ e^{(-t/T_1)}]$. This defines what is known as the longitudinal relax-

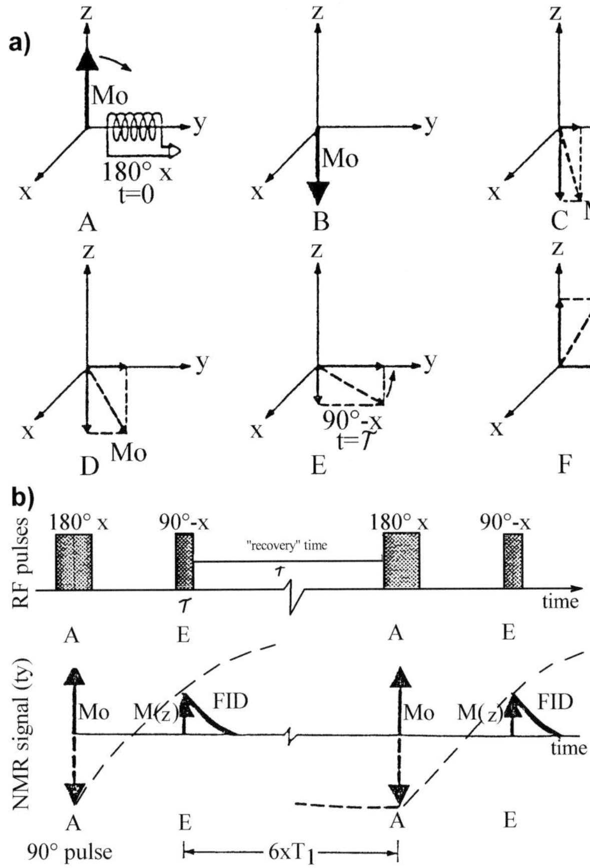

Fig. 6.13a,b. Inversion recovery pulse sequence and NMR signal for T_1 measurement. **a** Inversion recovery spin isochromatic representation. **b** Pulse sequence. (Hailey et al. 1985; courtesy of Forest Products Society)

ation time T_1. This time can range from ms in metals to several hours in insulators at low temperature and is determined by local fluctuation in the field experienced by the individual nuclei but at the Larmor frequency. This decay depends on the nuclei concerned with larger fluctuations in a liquid than in a solid. For example, the nuclei in the water of wood comes into equilibrium faster than nuclei associated with protons in the cellular structure.

Observation of the amplitude immediately following a subsequent 90° pulse, enables T_1 to be measured and is convenient for varying the recovery time τ. This assumes that at the start of the sequence $M_{z(t)}$ equals M equilibrium, but if T_1 is very long this may not be true. To make measurements on a specific nuclei subsystem, the time between pulse sequences τ must be longer than the time T_1 ($T_1 > \tau$). If the time between pulse sequences is shorter than the relaxation time ($\tau < T_1$), no signal from the nuclei is observed.

For the measurement of T_2, more sophisticated pulse sequences (90°, 180°, 180°, etc.) than for $T_1$are used. The presence of the component relaxation times T_1 and T_2 is attributed to the different mobility of water molecules in different structural regions, as for example the free water in the lumen and the bound water in the cell wall. Molecular processes and the total signal intensity depend on these two times as well as on spin density and proton mobility.

Since the FID signal is proportional to proton density, the moisture content, MC, in wood can be deduced using

$$MC = [\rho_{Hwood}/\rho_{Hwater}] \times M_O(S_O - M_O) \times 100\% \tag{6.5}$$

where MC is the moisture content of wood measured with the NMR technique; ρ_{Hwood} is the proton density in wood calculated from the amount of cellulose and lignin ; $\rho_{H\ water}$ is the proton density in water; M_o is the zero intercept of the FID curve of the mobile component S_o is the zero intercept of FID of the total signal component. The proton density is expressed as the number of protons per gram. The ratio $\rho_{Hwood}/\rho_{H\ water}$ is 0.549 for western red cedar and 0.538 for Douglas fir.

The agreement between the NMR moisture content measurements [MC] and routine oven-dried moisture measurements [*MC*] has been shown with the regression equation calculated for the sapwood and the heartwood of western red cedar and of Douglas fir, which is $R^2 = 0.9916$

$$MC = 0.9791\,\mathit{MC} - 0.7561 \tag{6.6}$$

This agreement is a milestone in the further application of NMR techniques in wood science since it demonstrates that it is possible to measure directly the moisture content of wood without oven-drying the sample. From such data an upper limit on the difference in moisture content between the heartwood and the sapwood of the same species can be established.

Because nuclei in different environments decay with different characteristic time constants, Olek et al. (1994) used NMR protons to distinguish bound water from free water in birch. The spin–spin relaxation time T_2 was measured with a pulse spectrometer at 30 MHz and with the CPMG method. The time between π pulses was 200 μs for samples with a moisture contents close to the fiber saturation point and 5000 μs for samples with an even higher moisture content. The repetition time between pulse sequences was 5 s. The multi-exponential spin echo decays were calculated as a function of a two-component relaxation functions, such as:

$$A_{(t)} = A_A \exp(-t/T_{2A}) + A_B \exp(-t/T_2B) \tag{6.7}$$

where $A_{(t)}$ is the amplitude of spin echo at time t; T_{2A}, T_{2B} are the spin–spin relaxation times of the spin subsystems and A_A and A_B their amplitude contributions (abundances) of the subsystems.

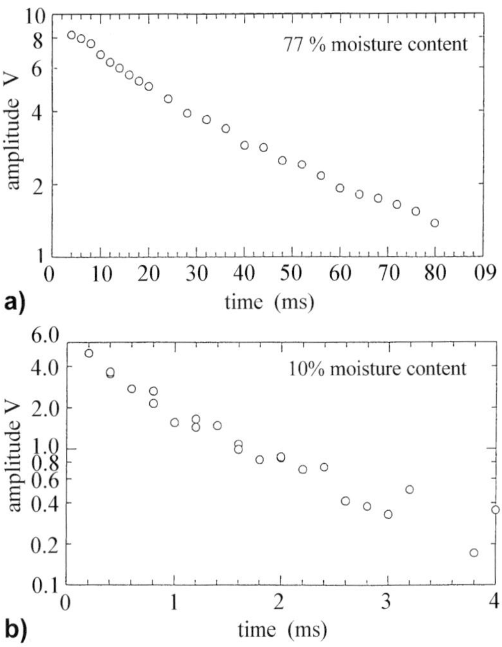

Fig. 6.14a,b. Spin-echo amplitudes versus time in birch above (**a** 77%) and below (**b** 10.4%) the fiber saturation point. (Olek et al. 1994, with permission)

The spin echo amplitudes versus time for a saturated specimen (77%) and dry specimen (10%) are shown in Fig. 6.14. The spin echo's decay with time is in the range 0–90 ms. The sample with 77% moisture content clearly shows two distinct relaxation components corresponding to T_{2A}, 60 ms and to T_{2B}, 100 ms. For the dry specimen the decay is also nonexponential and characterized by a relaxation time (<4 ms).

The moisture content *MC* above and below the fiber saturation point can be expressed as a function of the corresponding relaxation time T_2 (respectively, T_{2B} and T_{2A}) and is then calculated with:

$$T_2 = MC/(\mathrm{a} + \mathrm{b}\, MC) \tag{6.8}$$

where a and b are empirical constants determined for birch as: a = 2.481 and b = –0.0124. above the fiber saturation point and a = 18.67 and b = –0.1503 below the fiber saturation point and *MC* is the gravimetric moisture content of wood, determined by routine measurements.

The relationships between the abundances A_B and A_A and the moisture content above the fiber saturation point are given in Fig. 6.15. The abundance

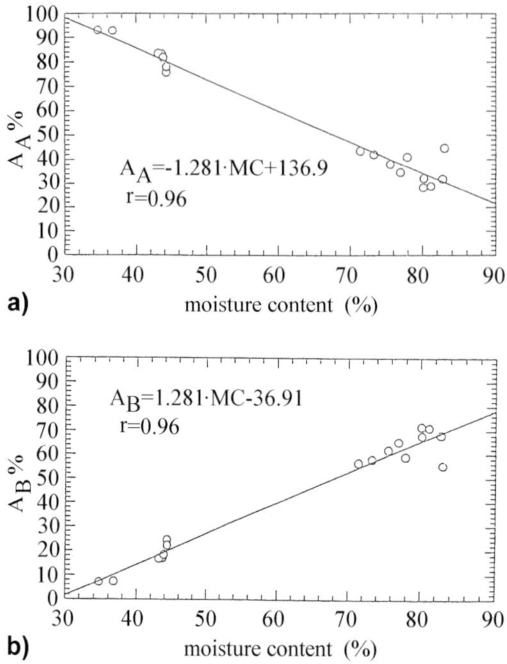

Fig. 6.15a,b. The abundance A_A of the subsystem A and the abundance A_B of the subsystem B vs. the moisture content in the range 30 to 90% moisture content. **a** A_A of the subsystem A vs. moisture content. **b** A_B of the subsystem B vs. moisture content. (Olek et al. 1994, with permission)

A_B deduced from the fast decaying component of spin-echoes decay, T_{2B}, increases with the moisture content (the correlation coefficient is $R = 0.96$), and the regression line passes through 0. The abundance A_A deduced from T_A decreases with moisture content. By extrapolation, the maximum moisture content of birch can be associated with a relaxation time T_{2B} of 75 ms. This is much less than the relaxation time of bulk water, which is several seconds. The abundance A_A increases while A_B decreases with increasing moisture content. Above the fiber saturation point, the NMR signal is therefore produced by the protons of bound water.

Above the fiber saturation point, T_{2A} (Fig. 6.16) and T_{2B} increase with moisture content. Water molecules of the subsystem with relaxation time T_{2B} ($>T_A$) have a restricted mobility compared with the molecules of bulk water. The same subsystem has also the longest observed spin–spin relaxation time and it is expected that this system corresponds to magnetic relaxation of the water molecules filling the cell lumina (free water) albeit with a reduced relaxation time due to the cell wall. Probably, the capillary forces restrict the mobility of the water molecules with the relaxation time T_{2B}. The time T_{2A} is shorter than

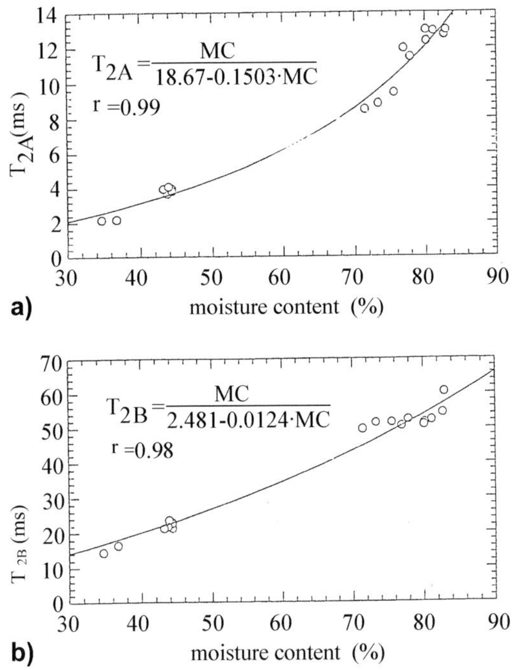

Fig. 6.16a,b. The spin–spin relaxation time T_{2A} and T_{2B} vs. the moisture content above the fiber saturation point. **a** T_{2A} vs. the moisture content. **b** T_{2B} vs. the moisture content. (Olek et al. 1994, with permission)

Table 6.3. Spin–spin relaxation time (ms) expressed by fast, slow and intermediate components in birch wood as a function of the moisture content. (Data from Olek et al. 1994, with permission)

Moisture content	Fast component	Slow component	Intermediate component
Above fiber saturation point	2–14	15–60	–
Under fiber saturation point	0.4–2.8	–	2–10

T_{2B} and varies in the range 2 to 17 ms and also increases with the moisture content above the fiber saturation point (Table 6.3). The low spin–spin relaxation time T_{2A} indicates that the very low mobility of the water molecules associated with this system, is mainly related to bound water.

Below the fiber saturation point, we have seen that the decay of spin echoes is nonexponential. Two relaxation times were measured, the first T_{2A}, ranging between 0.4 and 4 ms and a relatively slow time called T_{2C}, ranging from 2 to 10 ms. This time is shorter than the time involving capillary water. Both T_{2A} (Fig. 6.17) and T_{2C} (Fig. 6.18) increase with the moisture content. For the rela-

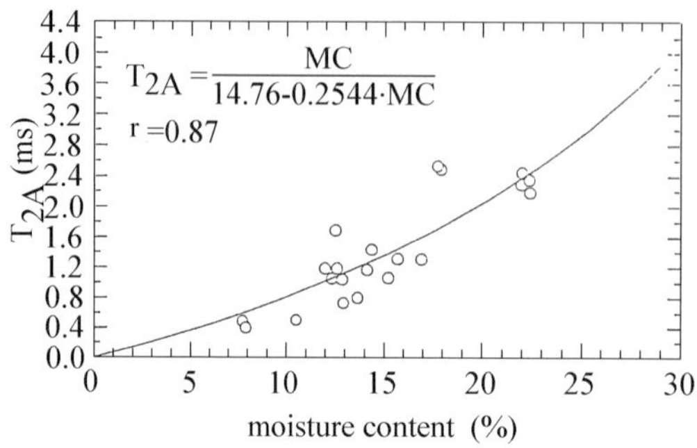

Fig. 6.17. The spin–spin relaxation time T_{2A} vs. the moisture content below the fiber saturation point. (Olek et al. 1994, with permission)

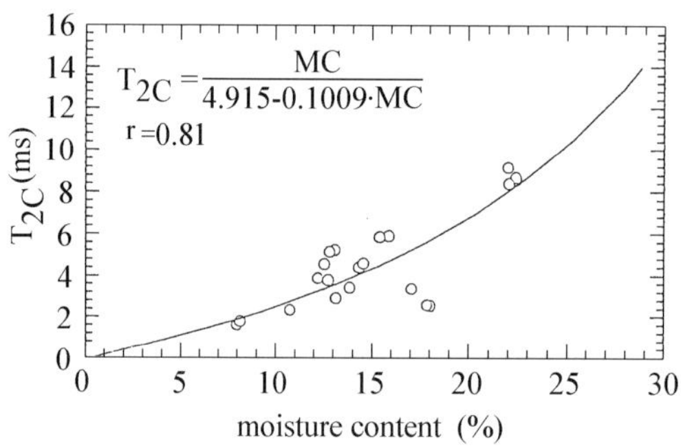

Fig. 6.18. The spin–spin relaxation time T_2C vs. the moisture content below the fiber saturation point. (Olek et al. 1994, with permission)

tionship between T_{2C} and the moisture content, the correlation coefficient is positive and relatively high $r = 0.81$. At zero moisture content, the relaxation time was estimated to be zero. The relationships between different relaxation times are: $T_{2C} < T_{2B}$ and $T_{2C} > T_{2A}$, and for all terms, we have $T_{2B} > T_{2C} > T_{2A}$. Because $T_{2C} > T_{2A}$, it was supposed that the mobility of the molecules of the system is much higher than that of the molecules of bound water and consequently, the corresponding NMR signals could be produced by condensed water molecules in very thin anatomic elements of wood. Another explanation was advanced by the authors who suggested the existence of a very thin monomolecular layer of water on the cell wall. This water determined the increase in all mechanical properties of wood in the vicinity of 5% moisture content. From all experimental data ($T_{2B} > T_{2C} > T_{2A}$) presented by Olek et al. (1994), the weight of evidence suggests that bound and free water relax separately and that the corresponding relaxation times can be precisely measured.

6.3 The NMR Imaging Technique

In this section, we will outline the NMR imaging technique, with special stress on what the images represent and on the algorithms used to identify and isolate particular structural features of wood.

6.3.1 Techniques for Imaging

The magnetic resonance technique can image nuclear spin density, NMR relaxation times and diffusion of water molecules at various points throughout the wood sample. The values of relaxation time T_1 and T_2 are important in determining the quality of an image and can be used in selective imaging of bound and free water molecules in a wood sample under certain conditions. The intensity can be proportional to the product of the proton density and the ratio of the relaxation times. High resolution NMR imaging is very sensitive to the more mobile protons. The dependence of the image intensity on relaxation times is very important in differentiating between wood tissues with the same proton density but different values of the ratio of the relaxation times. In general, the quality of an image depends on many parameters. It is also possible to vary the experimental conditions in such a way that the image is particularly sensitive to a specific chosen parameter. The spatial discrimination is dependent on the magnetic gradient field.

The techniques currently used in NMR image reconstruction are: back projection, similar techniques used in X-ray computed tomography, rotational frame, echo planar for two-dimensional images and two-dimensional Fourier transform techniques for three-dimensional imaging of the object. The Fourier transform technique, which encodes the spatial information in phase with the NMR signal, is optimal with respect to signal-to-noise ratio. The imaging of porous solids like wood requires particular considerations. The solid matrix material has an important effect on the relaxation, which can limit the resolution of standard imaging methods. The different water phases in wood must be observed separately if possible. If not, the proportion of measured signals corresponding to various phases must be distinguished.

For all imaging methods, there is a strong dependence of imaging time on spatial resolution. The imaging time depends on spatial resolution as $(\text{resolution})^4$. However, the spatial resolution tends to decrease as the size of the object increases. Typically, resolution is currently of the order of 1–2 mm. In the future, improvements of the technology will be directly related to the improvement of acquisition time. The usual approach to obtain 3-D images is to use a 2-D image of a selected thin slice of a 3-D object, and then to move to the next slice and so on until the whole object is reconstructed A thin slice is defined by using frequency-selective radio frequency pulses, or a time-dependent field

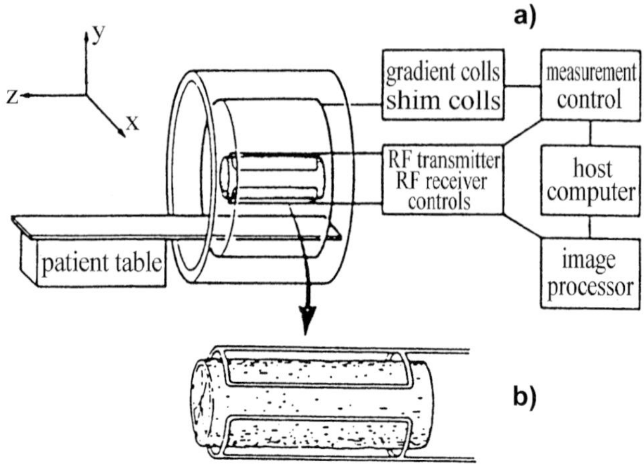

Fig. 6.19a,b. Outline of the NMR system. **a** Block diagram. **b** Position of the log in the center of the magnetic field. (Wang et al. 1990, with permission)

gradients. The back-projection reconstruction method is the most popular method for NMR imaging using scanners.

The major components of an NMR scanner are the magnet, three gradient coils (for X, Y and Z directions), an RF transmission and receiving coil and computers for control of pulse sequencing, data analysis and 3-D reconstruction (Fig. 6.19). The magnet provides a uniform static field. Both electromagnets and superconducting magnets are available for imaging. To avoid problems with attenuation or phase shifts of the radiofrequency fields within the sample, it is usual to work at relatively low fields of a 0.1 tesla, which corresponds to a ^1H frequency, which is about 4 MHz.

Three sets of gradient coils modulate the magnetic field to spatially encode the object. Two-dimensional images can be obtained by choosing one of the three orientations of the three gradient coils as the slice selection. The other two coils can then be used for readout and phase encoding gradients to map out information on the plane. The gradient provided by these coils has a direct effect on the spatial resolution of the image. The stronger the gradient, the better the spatial resolution. The RF coil delivers a radiofrequency power to flip the magnetization. The same coil picks up the NMR signals as a receiver. The signal detected by the receiver coil is amplified, filtered, digitized and fed into an array processor or a computer for image reconstruction. The computer also controls the execution of the RF pulse sequence, the field gradients, data acquisition, image reconstruction and display function.

An NMR image of the water content in a log (Fig. 6.20) can be obtained with the following views: coronal from top to bottom, sagittal from left to right and cross-sectional. The pixel is usually 128×128 or 256×256. The size of the

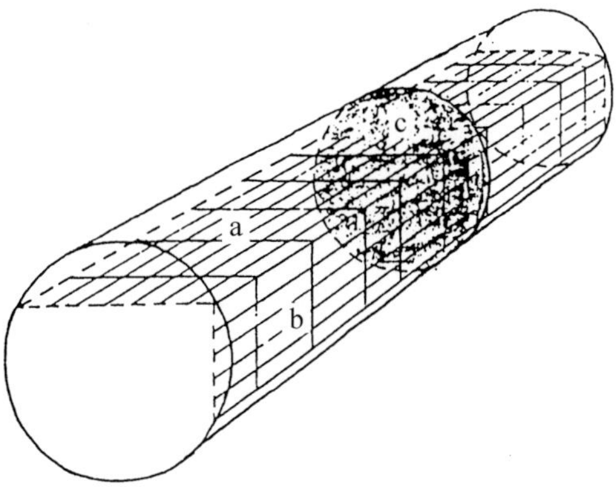

Fig. 6.20. Three NMR imaging views of a log. *a* Coronal view, from top to bottom. *b* Sagittal view, from left to right. *c* Transverse view. (Chang et al. 1989; courtesy of Forest Products J)

sample imaged can typically be about 50 × 50 cm^2 and can be imaged with a spatial resolution of order 1 mm. The imaging time depends on the pulse 90°, 180° sequence repetition or "recovery" time, the number of phase encoding steps and the number of averages in the sequence. Initially, NMR systems used repetition times of 500–800 ms. Faster repetition times are required for industrial applications, but there are fundamental limits in repetition times set by physical parameters.

In modern imaging facilitation, "clever tricks" are used to maximize the rate with which the data can be obtained by sophisticated pulse sequences and switched into field gradients in all three directions. Typically, Chang et al. (1989) using modern scanners in a saw mill, were able to acquire ten images per second for log inspections.

Another interesting imaging technique called "current density imaging (CDI)" involves the passing of a current to the object to be imaged. This current produces a local magnetic field reflecting the current distribution through the sample. In this way, it is possible to distinguish regions of high electrical conductivity from those of poor conductivity. An interesting example of this technique is shown in Fig. 6.21, which provides additional quality information to that obtained from conventional imaging alone. This imaging technique, which generates maps of the spatial distribution of electric currents through different wood tissues, superimposed on a conventional spin-echo technique, was reported by Beravs et al. (1998) in a study of 2-year-old trees. One current pulse is applied between a radiofrequency 90° pulse and a 180° pulse and a second between a 180° pulse and a signal pulse echo. Both have the same magnitude and duration, but are in opposite directions. During imaging, the electric pulses produce phase shifts in the proton image, proportional to the magnetic field produced by the current and to the duration of such pulses.

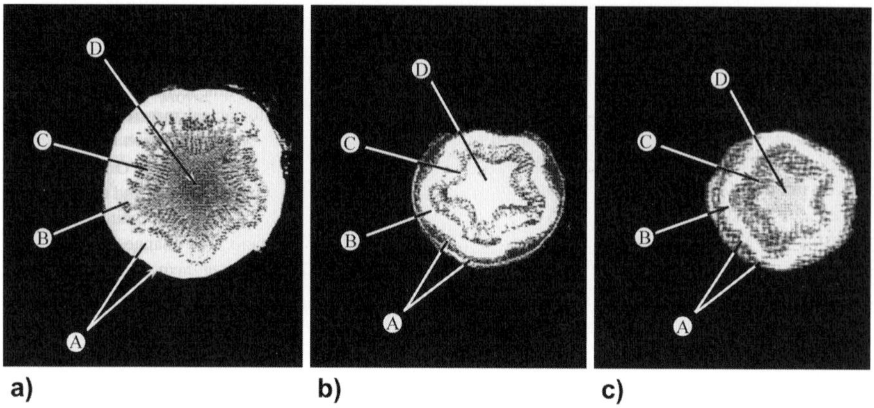

Fig. 6.21a–c. Electric current density imaging of a young oak tree. **a** Cross section in light microscopy. **b** Nuclear magnetic resonance image. **c** CDI (current density image). *A* Cambial zone + differentiating xylem + complete phloem; *B* differentiated secondary xylem of the current growth ring; *C* first annual ring; *D* pith. (Beravs et al. 1998; courtesy of Holzforschung)

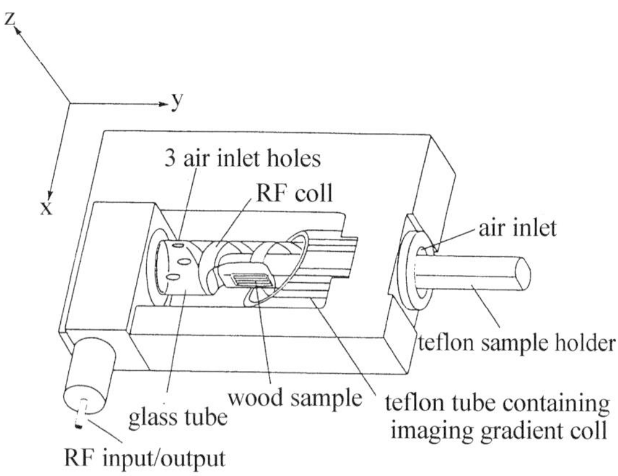

Fig. 6.22. The NMR probe containing the wood sample. (Quick et al.1990; courtesy of Wood Fiber Science)

Two-dimensional images of sapwood and heartwood, growth rings, knots, reaction wood, resin pockets and other features have been reported by Chang et al. (1989b), Wang et al. (1990), Niemz et al. (1998a, 2000b,c).

One-dimensional images can be obtained more rapidly and more easily than 2-D images, using a modified solid-state spectrometer, as noted by Quick et al. (1990). The NMR probe containing the sample is shown in Fig. 6.22. This device was designed for measurement of the radial profile of the moisture content during drying of western red cedar wood above the fiber saturation point. One-dimensional images for pine specimens was reported by Lindgren

(1994). The drying of a $10 \times 10 \times 10$ mm pine sample was imaged over 25–11% moisture contents.

6.3.2 Algorithms

Specific algorithms were developed for data interpretation in NMR spectroscopy and nuclear magnetic imaging. The most common method is to fit the decay curves with the smaller number of discrete exponential terms, which provide a satisfactory representation of the experimental data. This approach fails when more than three exponential components are considered. A second approach is to analyze relaxation times in terms of their discrete or continuous distribution (Whittall and MacKay 1989). This procedure requires no a priori assumption about the relaxation behavior and provides a more stable fit to the experimental data. The application of linear inverse techniques to characterize decay curves in terms of discrete or continuous distributions of relaxation times is presented in Fig. 6.23. The amplitude of the nonnegative least-squares spectra is plotted versus T_2. Three peaks were observed: one at 0.006 s due to the bound water in the cell wall, and two peaks at 0.030 and 0.200 s due to the free water present in cells of smaller and larger diameters respectively, probably in earlywood and latewood. As a comment on this exper-

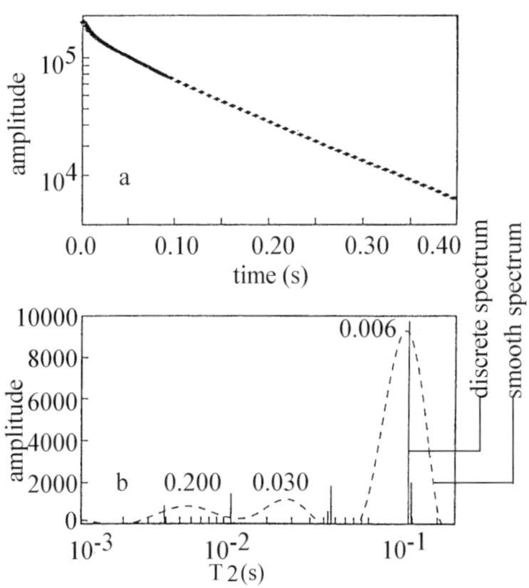

Fig. 6.23a,b. Linear inverse technique to characterize decay curves in terms of discrete or continuous distributions of relaxation times. **a** Amplitude vs. time. **b** Amplitude vs. T_2 with discrete and smooth spectra. (Whittall and MacKay 1989, Academic Press, with permission)

imental approach, we note that an essentially continuous spectrum introduces more adjustable parameters to fit the results than does a discrete spectrum.

A fine example of the power of algorithms for the NMR technique to image the cross section of a whole log is shown in Fig. 6.24, in which one can clearly see the individual rings, the core region and, in diametrically opposed positions, the first indication of the growth of side branches (top of the figure) and the more fully developed side branches at the bottom of the figure.

6.3.3 Deduction of Relaxation Times from Measurements

For simple systems involving protons in water, the relaxation of M_z and M_{xy} are given by simple exponentials with relaxation times T_1 and T_2. However, wood is a naturally occurring composite material, where the water molecules in different macroscopic regions have very different relaxation times, where the values of relaxation times from different regions are significantly different.

In some circumstances it is possible to extract T_1 and T_2 for the protons in different regions of the sample by fitting the relaxation to two exponential terms. Some authors (Menon et al. 1987) have even attempted to describe the relaxation process with three separate relaxation times. However, to obtain reliable information in such a way with any confidence would require an extremely high signal to noise ratio and very long relaxation times to enable the various regimes to be differentiated. For example, with an observation value for T_2 of 6 and 200 ms for water in the wood, a wide range of permeability measurements is possible for small and large anatomic elements (fibers, vessels, etc.) which are in fact capillary tubes.

Another important feature of NMR imaging is its ability to identify wood structure and defects. An example of this is given by the pulsed NMR measurements of black oak (*Quercus velutina*) samples used by Coates et al. (1998). An example of the identification of major structural areas is shown in Fig. 6.25. The spatial variations observed are probably due to variations of the moisture content in the sample and associated variations of density in growth rings and rays.

In all defective regions, the gray level of some pixels widely deviate above and below the gray levels of the clear wood region. In clear wood areas, the variations in the gray levels are also important due to growth rings and rays also because of random noise. Constant threshold limits are not able to give the degree of variability of gray levels found in a particular log. A more effective approach is to introduce threshold limits that follow the contour of the average moisture content. The largest gray level deviations are associated with the defect area. Using values obtained for clear wood zones and subtracting the largest deviations associated with defect areas, it is possible to obtain a clear map of the defects. Computer algorithms have been developed by Vincent (1993) and Peters (1995) to enhance the imaging of defect regions (Tsai 1985).

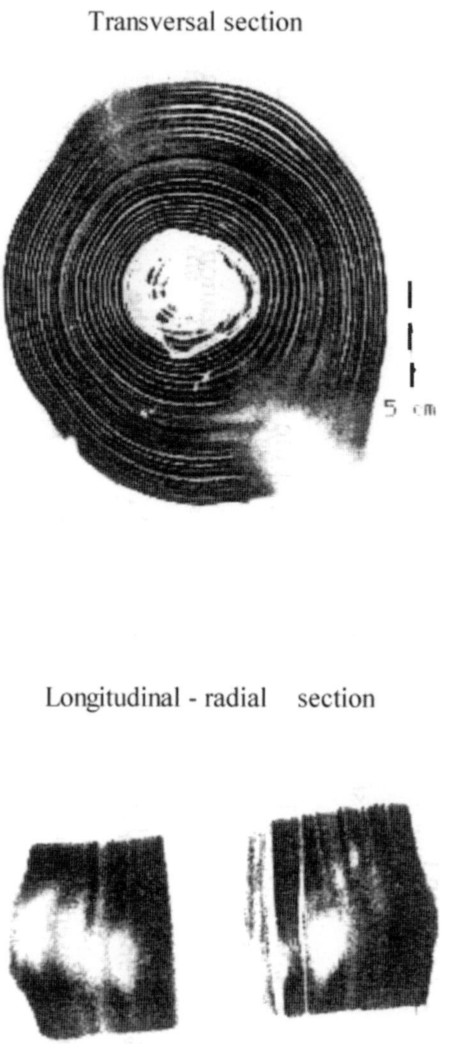

Fig. 6.24. Imaging of transverse and longitudinal radial section of a beech. (Photo Escanyé, in Bucur 1990, with permission)

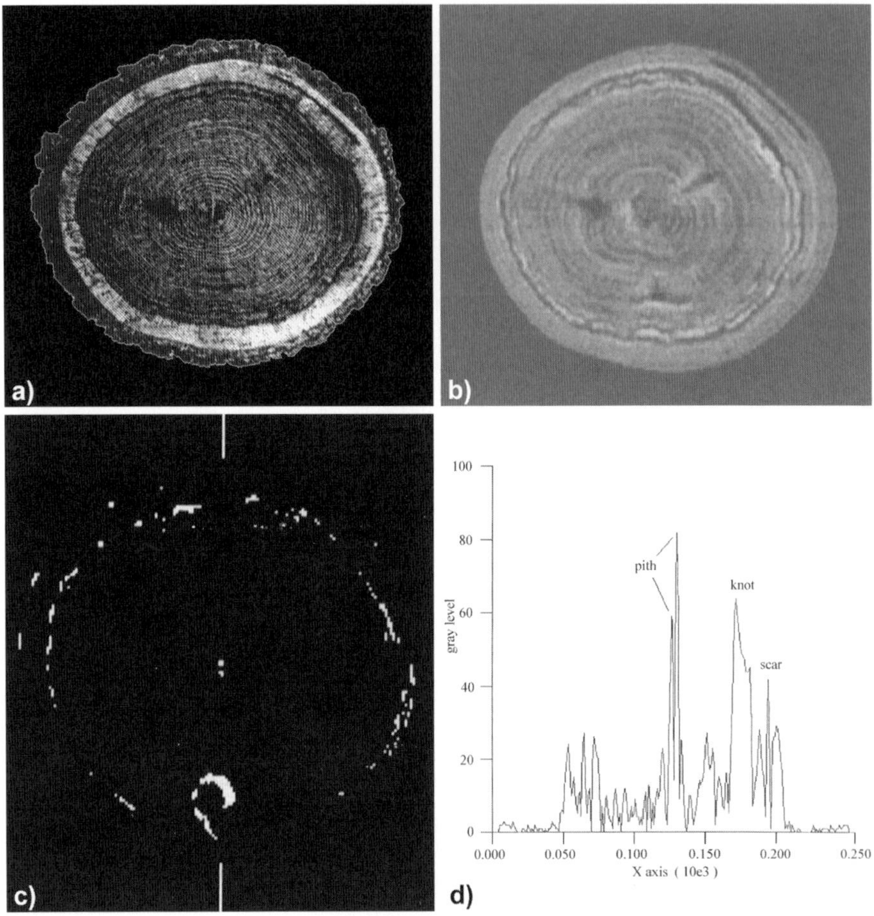

Fig. 6.25a–d. Comparison between the photographic image and the automatically segmented image of oak, using the algorithm proposed by Coates et al. (1998). **a** Photographic image of the cross section. **b** Magnetic resonance image of the section. **c** Deviation of 40 gray levels or more between original pixel and median axis. **d** Central profile of **c** where pith, knot and scar are visible. (Courtesy of Forest Products J)

6.4 Applications

In this section, the physical and chemical characterization of wood using NMR imaging techniques are discussed in relation to the inspection of trees, logs and lumber and the possible industrial control of drying, impregnation of wood products with preservatives and adhesion. For all such applications, it is important to be able to quantify the water content, and the corresponding moisture content in wood, and, on the other hand, the structural features at macroscopic and microscopic level. Different aspects of wood processing include the control of the moisture content in lumber during drying, the control of the moisture content in wood-based composites, the control of impregnation processes with preservatives and the kinetics of curing during adhesion.

NMR imaging can also be used to image bulk physical properties of wood, such as electrical conductivity. For many applications, it would also be useful to be able to image the elastic properties of wood in situ. A very interesting field of potential application is the simultaneous measurement of elastic constants of bulk objects. Until now, this approach has not been used for wood materials. Nevertheless, we present a theoretical background for this application, which seems to be very promising for the mechanical quality estimation of green lumber, trees and logs.

6.4.1 Inspection of Living Trees, Logs and Lumber

Examples of NMR imaging for the in situ inspection of living plants have been given by Halloin et al. (1994) and Nakanishi and Matsubayashi (1997). Examples illustrating distribution and movement of water in roots and meristems during development, in leaves or needles, and seeds have been reported. MacFall et al. (1990) investigated the water depletion region in loblolly pine roots. During the first stages of growth, the roots were easily visible and the formation of a distinct water-depletion region around the roots was evident. Later, this region extended around the lateral roots and clusters of mycorrhizal short roots. The development of roots of Douglas fir seedlings, grown in containers at normal temperature, has also been investigated (Brown et al.1991). The lengths of the roots determined from NMR images was in agreement with the measured lengths of excavated roots. Southon et al. (1992) studied the influence of freezing on the roots of Scots pine and Norway spruce seedlings over a large temperature range from +1to −20 °C. The amount of root material detectable by NMR imaging decreased with decreasing temperature, presumably because of a significant increase in T_1 and T_2 in the material as the polar mobility of water molecules, particularly in root material, decreases with the temperature.

Several aspects of leaf and needle anatomy and morphology have been discussed by Masuch et al. (1991), Veres et al. (1993) and Millard and Chudek (1993). In vivo measurements of growth were made as a function of the water content of the soil. The NMR signals of needles from declining forests showed a larger NMR signal and a different distribution of free water from that of healthy needles. NMR imaging has been also used for the discrimination of the germinability of seeds as noted by Vozzo et al (1996).

6.4.1.1 Water Content in Living Trees

In this section, we will discuss techniques for water content imaging in living trees, in fresh logs and in green timber. Previously, it was noted that the nature of NMR signals is strongly dependent on the amount of water present in different wood tissues. The presence of magnetic ions (Cu, Fe, Mn) in the sap and in the cell walls causes a strong reduction in the nuclear relaxation times T_1 and T_2. For example at 1% of copper sulfate solution in water reduces T_1 and T_2 from about 3 s to 20 ms. NMR imaging and spectroscopy can also provide information about the anatomic structure of healthy and infected wood of living trees which will be discussed in this section.

6.4.1.1.1 Water Content in Healthy Wood Tissues

It is generally agreed that increment cores bored into the trunk of living trees are representative of the quality of the tree. Byrne et al. (1986) pioneered the measurement of the water content of trees by using an NMR spectroscopic technique on 5-mm-diameter increment cores bored from healthy young *Pinus radiata* and *Eucaliptus grandis* trees, using NMR at 60 and 90 MHz at two temperatures (20 and 30 °C), to investigate their frequency and temperature dependence, as a noninvasive technique for measurement of water flow in young trees. NMR measurements of T_1 and T_2 were made using two spectrometers. T_1 was measured by the 180°, 90° sequence inversion recovery pulse sequence. T_2 was measured from the decay of spin-echo signals as a function of pulse spacing in both species. The two measured relaxation times T_1 and T_2 had two constituents, one long and one short, and were determined from logarithmic plots of the NMR decay envelope at ambient pressure and temperature. The moisture content of the two specimens ranged from 39 to 43%. To identify any changes between the freshly taken samples and those used for laboratory experiments, specimens were tested in their fresh state and after 3-day storage in a sealed container.

The results are given in Table 6.4. The scarcity of experimental data permits only quantitative conclusions:

- for both species, T_1 long decreases with increasing frequency.
- for pine, T_1 short also decreases with increasing frequency.
- for both species, T_2 long and T_2 short at 90 MHz are greater than the values at 60 MHz.
- for fresh eucalyptus specimens, all measured constants are higher than after storage for 3 days.

The data reported in this paper suffer from some inconsistencies (reference to fresh and old eucalyptus samples). The presence of the two-component relaxation times can be attributed to the different mobility of water molecules

Table 6.4. Some values of the relaxation times T_1 and T_2 (ms) measured for *Pinus radiata* and *Eucalyptus grandis* at 40% moisture content (Byrne et al. 1986, 2002, with permission of Elsevier Science from Agricultural and Forest Meteorology, vol. 38: 307–317, Bynne GF, Fenn MD, Burgar MI 1986, reprinted from Nuclear magnetic resonance studies of water in tree section)

Frequency	Time	*Pinus radiata*	*Eucalyptus* old	*Eucalyptus* 3 days	*Eucalyptus* fresh
60 MHz	T_1 long	140	205	310	–
	T_1 short	22	–	–	–
	T_2 long	10	7	7	–
	T_2 short	0.8	0.9	0.7	–
90 MHz	T_1 long	115	120	230	330
	T_1 short	15	22	26	80
	T_2 long	22	21	33	46
	T_2 short	1.6	3.6	3.0	1.0

in different regions. The differences in T_1 measurements at different frequencies can be due to the mobility of the water molecules as well as to different experimental conditions. More experimental data are needed to clarify this point. The longer values of T_2 on storage could be related to the mobility or diffusion length of water molecules. Short T_2 values indicate a highly saturated wood structure and the presence of small capillary channels (T_2 is proportional to the ratio $\mathbf{a}^2$/diffusion coefficient, where **a** is the size of a capillary). Short T_2 values will also be associated with bound water present in the cell walls. T_1 and T_2 long values can be associated with bulk water in the xylem vessels, fibers and tracheids, while T_1 short and T_2short could arise from water trapped in the wall material. The water in fresh trees, more exactly in the sap, has a T_2 of about 46 ms and is in a different physical state from free water, for which T_2 is only 3 s. Byrne et al. (1986) demonstrated that NMR can be used for measurement of the water flow in trees, and that only the cost of the equipment is a serious impediment for field applications.

Another interesting imaging technique involves the passing of current to the object to be imaged. This current produces a local magnetic field, reflecting the current distribution through the sample. In this way, it is possible to distinguish regions of high electrical conductivity from those of poor conductivity. An interesting example of this technique was shown previously in Fig. 6.21, which provides additional high quality image information to that obtained by conventional imaging alone. The aim of Beravs et al. (1998) was to create an image in very young oak trees (2 years old) of the cambial zone, the xylem, the phloem, the first annual ring and the pith. The cambial zone is located between the xylem and the secondary phloem and both originate from the cambium. The imaging conditions were as follows: frequency 100 MHz, repetition time 600 ms, echo time 25 ms, field view 2 cm, slice thickness 4 mm,

Table 6.5. Average values of electric current density, of electric conductivity as a function of wood density and wood moisture content in a young oak tree .(Beravs et al. 1998, courtesy of Holzforschung)

Zone	Tissues	Electric current density [A/m^2]	Electric conductivity (s/m)	Density (kg/m^3)	Moisture content (%)
A	Cambium, xylem, phloem	1582	0.070	610	119
B	Second annual ring	1314	0.058	550	63
C	First annual ring	842	0.037	430	52
D	Pith	1205	0.053	610	74

pixel grid 256 × 256. The imaging time was 15 min to average six scans. The voltage applied to the sample was 180 V, with a current pulse length of 10 ms. The 3-D images were obtained from two perpendicular orientations of the sample. As is evident from the image, current density imaging offers a higher contrast between the different tissues than conventional NMI imaging.

The current density image shows a conductivity contrast in four different regions of the oak sample. The corresponding values of electric current densities, electric conductivity, and density and moisture content of the wood are shown in Table 6.5. The increase in the moisture content from the first annual ring to the cambial zone results in an increase in all measured parameters.

All reported experimental data in this section were obtained at room temperature. The influence of low temperatures on relaxation times was studied qualitatively by Johansson (1985) with frozen logs in Sweden. No NMR signals were obtained in this case, probably because of the solid state of frozen water. The mobility of water molecules decreases with decreasing temperature so that there are fewer higher frequency components of the local fluctuating magnetic fields to restore M_z to its equilibrium value (hence T_1 increases) with similar effects on T_2.

6.4.1.1.2 Water Content in Infected Wood Tissues

The second part of this section is devoted to the imaging of fungal colonization of young living trees. The imaging of fungal attacks with different pathological agents of living sapwood of sycamore has been reported by Pearce et al. (1994, 1997). The infection agents for sooty bark disease were *Ustulina deusta, Chondrostereum purpurem, Cryptostroma corticale,* and *Ganoderma adspersum.* The development of fungal lesions in the sapwood of living trees is restricted by an active defense mechanism. The pathological anatomy of sooty bark disease lesions can easily be seen by imaging shown in Fig. 6.26. The inoculated zone is surrounded by a reaction zone located between the

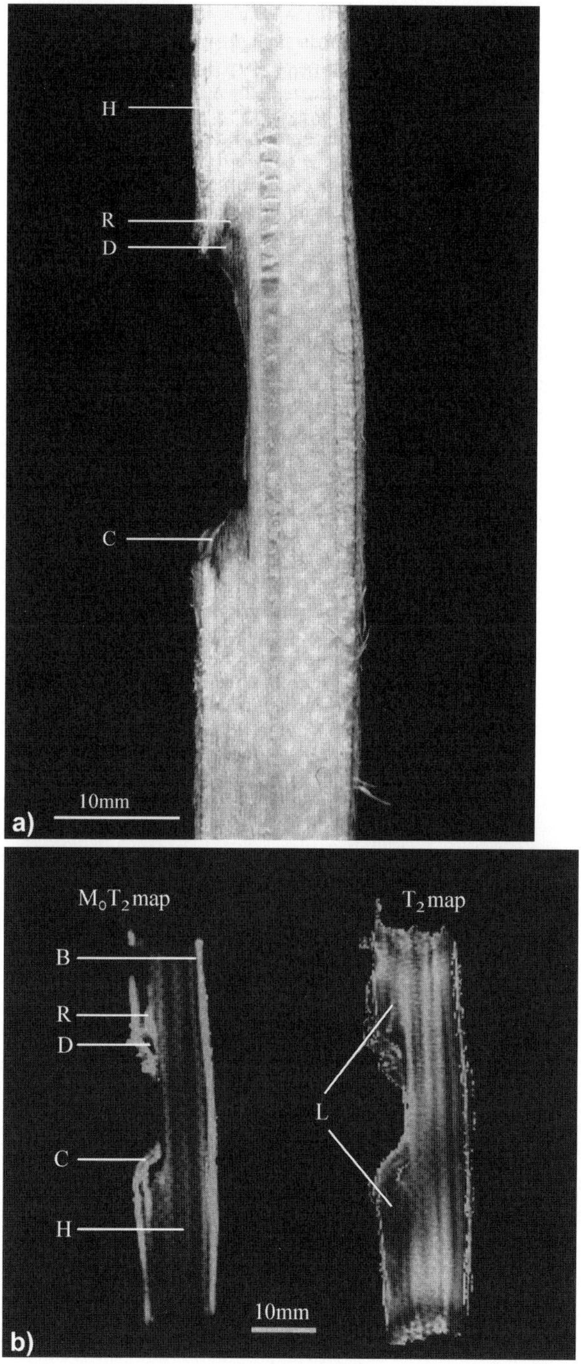

Fig. 6.26a,b. Different sections through a stem of young *Acer pseudoplatanus* tree wound inoculated with *Ustulina deusta*. Longitudinal–radial section inoculated with *Ustulina deusta* where one can see *D* decayed zone; *H* healthy zone; *R* reaction zone, which separates the healthy and decayed zones; *C* callus beginning to grow over the wound faces. *Scale bar* 10 mm. **a** NMI obtained with a spin-echo image sequence for a 2-mm slice thickness, with a resolution higher than 0.47 mm. **b** Comparison of pixel intensities in specific areas of M_0 maps and maps of measured T_1 and T_2 have shown that both techniques are similar for the illustration of different lesions (*L*) in different tree regions. Reaction zone had a high contrast. *Scale bar* 10 mm. (Pearce et al. 1997, Blackwell Science LTD, with permission)

decayed and the healthy wood. It was demonstrated that changes in moisture content are associated with the formation of decay lesions in living trees. Spin-lattice T_1 maps showed little contrast compared with proton density maps. Measured relaxation times were within the range of 700 to 1000 ms.

The spin–spin relaxation time T_2 for healthy wood varied between 35 and 127 ms. In reaction wood, T_2 was reduced by as much as 40%. Neither T_1 nor T_2 was altered in the drier wood adjacent to naturally occurring *Cryptostroma corticale,* but both were reduced in the infected wood. The moisture content was higher in the reaction zone than in the healthy area. Calculated proton density images showed the delimitation of the reaction zones. The contrast of the images of lesions was attributed to proton density variations and to the decrease in spin–spin relaxation time T_2 in the vicinity of the lesions. In the reaction zone, the values of spin-lattice relaxation time T_1 were less sensitive to the pathological modification of the tissue than T_2.

6.4.2 Growth Rate and Other Structural Features in Logs and Lumber

6.4.2.1 Structural Features at the Macroscopic Level

The detection of annual rings, knots, reaction wood, wet wood, decay and other structural features with the NMR technique is a basic approach and is a challenging task for image reconstruction and processing. Pattern recognition techniques to identify, classify and quantify structural features are needed for the on-line internal inspection of logs before sawing at normal production speeds in saw mills and other wood processing. NMR image processing techniques are similar to those used for X-ray computed tomography, but the information content of the NMR images is higher than that obtained from X-ray computed tomography and ultrasonic imaging. Because the NMR signal intensity is directly proportional to the moisture content, the higher the moisture, the better the image. However, even for samples with a low moisture content, a reasonable image can still be obtained by averaging signals over a long time.

Hall et al. (1986a,b), using a relatively low field of 0.14 tesla, obtained clear images of the macroscopic structure of fresh-cut bolts of aspen and spruce of 25 cm diameter. Annual ring patterns, buried knots and decayed regions were clearly observed. Similar images have been reported by Wang and Chang (1986) and Swanson and Hailey (1987).

In this section, we present images from a beech bolt of irregular shape and elliptical cross section (75 cm long and maximum diameter 15 cm), cut from a tree at a height of 20 m from the ground, on which bolt branches were formed. In this sample, the wood structure is very complex, with normal wood, reaction wood, knots and other features. Beech was chosen for imaging because of its commercial importance in Europe. The internal structure of the bolt is given

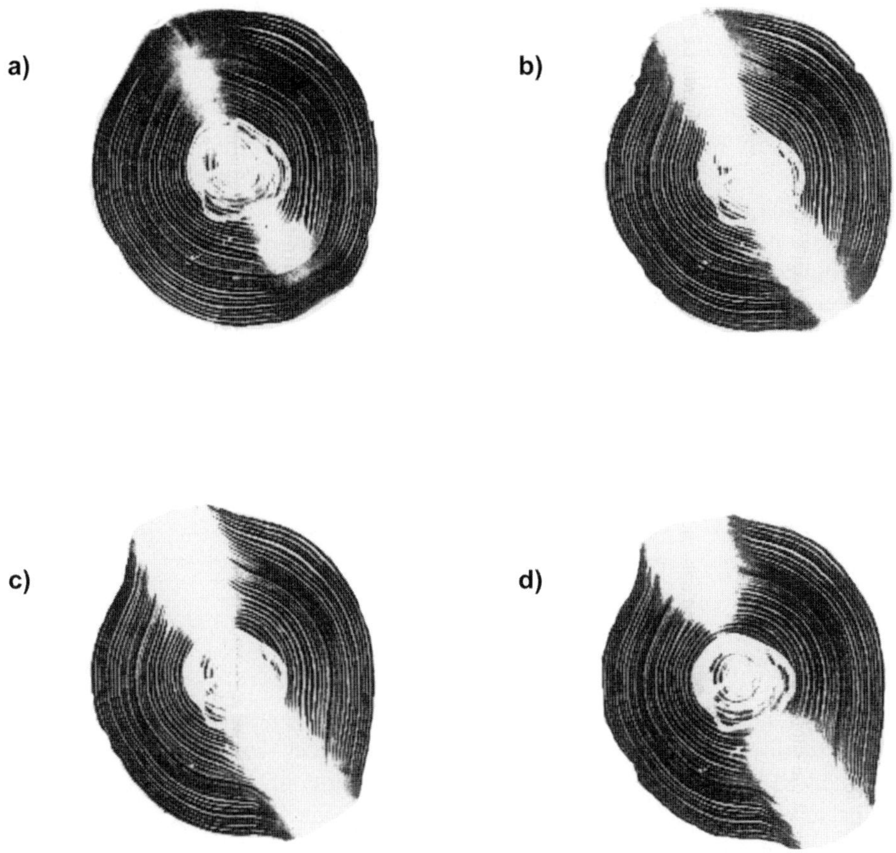

Fig. 6.27a–d. Views of the internal structure of a beech bolt. Images in the transverse plane at different heights (74, 33, 21, 15 cm) noted **a**, **b**, **c**, and **d**, respectively. NB: The images were taken on a medical scanner for 0.5 tesla, with a spatial resolution of 1.2 × 1.2 mm. The echo time (30 ms) permitted imaging only of the distribution of mobile, free water. (Photos Escanyé in Bucur 1990, with permission)

in Fig. 6.27 with four transverse sections at different heights (15, 21, 33, 74 cm) and sagittal views. Well-imaged features are the annual rings, the pith, the red heart zone, the sound wood, the tension wood, and even the bark. The image clearly shows branch formation, which corresponds to the physical state of the bolt. The bark is seen as a very clear zone. Loss of image intensity is due to the low water content. The oval-shaped bright areas correspond to branch formation with decreased water content and reveal the existence of buried knots.

The abnormally bright area around the pith corresponds to red heart. The development of red heart is different at different heights. The dark protuberances of the red heart contain less water than the surrounding area, probably because of to the presence of tyloses.

The diameter of the pith is 3 mm, and the first annual ring has a diameter of 4 mm. In the healthy zone, the earlywood of the annual rings shows a lighter color than the latewood of the same ring, which implies either that it contains more water or that this water has a longer relaxation time. The latewood zone in all rings is always dark, indicating low water content. The natural distribution of water during the life of the tree can be seen. The influence of climatic changes, or of physical and biological damage can be determined from such NMR images. By looking at the image carefully, one ring of the tension wood with a darker latewood zone can be seen in transverse image at 25 mm from the periphery of the bolt. In the sagittal plane image, the same dark vertical projection of the ring can be observed. This ring is therefore likely to be tension wood with less moisture content than the surrounding rings and a shorter relaxation time. White irregular small bright stain checks between the annual rings signify a high level of water content

6.4.2.2 Structural Features at the Microscopic Level

It is not the aim of this section to review the NMR techniques that reveal the porous nature of solids in general, but rather to select the studies on those materials of most relevance for the understanding of wood. Cohen and Mendelson (1982) studied the influence of the size and geometry of pores in rocks on NMR signals. Assuming the validity of the Bloch equations with magnetic relaxation on the surface of the pore voids in a uniform field, T_1 and T_2 could be defined as a single relaxation rate that was related to the surface area and the volume of the pore spaces. The NMR measurements of T_1 and T_2 were made to determine the spin-lattice relaxation and the diffusional attenuation, using a spin-echo signal. A strong magnetic field was used to polarize the water molecules in the pores of the sedimentary rocks. When the magnetic field is reduced, the proton magnetization relaxes to a new equilibrium value. The decay in the amplitude, at a rate determined by T_1, is sensitive to the geometry of the pore space of the rock. Similar approaches can be extended to the specific case of porous woods, and a new method for porosity measurement could be developed in the future.

In this section, we consider the sensitivity of the nuclear magnetic imaging technique to the structural features of wood on a microscopic scale, as a function of cell size and the dimensions of earlywood and latewood in the annual ring.

Wycoff et al. (2000) estimated the cell size in four softwood species (eastern red cedar, redwood, eastern white pine and sugar pine) using NMR measurements. The diameter of the cells in these species is between 25 and 65 μm and the lengths between 2 and 5 mm. Measurements of the translational diffusion coefficients in a field gradient spin reveal barriers to diffusion along the length of the cells and indirectly provide a method to measure boundary dimensions and geometry. Calculation of the apparent diffusion coefficients was performed with a model proposed by Callaghan (1995), in which the signals from both free and bound water were considered. Fields along the diameter in a tangential anisotropic direction were chosen to measure the cell dimensions, because in this direction the cellular wall has the smallest effect on the mobility of the absorbed water. Comparative measurements were performed also in the longitudinal direction to establish the expected anisotropy.

For moisture contents above the fiber saturation point, the NMR signal is primarily due to the "free water" associated with the cell lumina. Only a small portion of the signal is contributed by water molecules "bound" to the cell walls by strong hydrogen bonding. The diffusion of the bound water is also very small and the corresponding diffusion time is very long relative to the total diffusion time. For convenience, the cells are assumed to have a rectangular cross section so that they can be described by a parallel plane geometry. For a short field gradient, the attenuation of a spin in a 90–180° pulse sequence can be considered in terms of relaxation of spin the diffusion by the cell walls

For example in Fig. 6.28, the variation of the diffusion coefficient of water is plotted as a function of time diffusion in eastern white pine. The values mea-

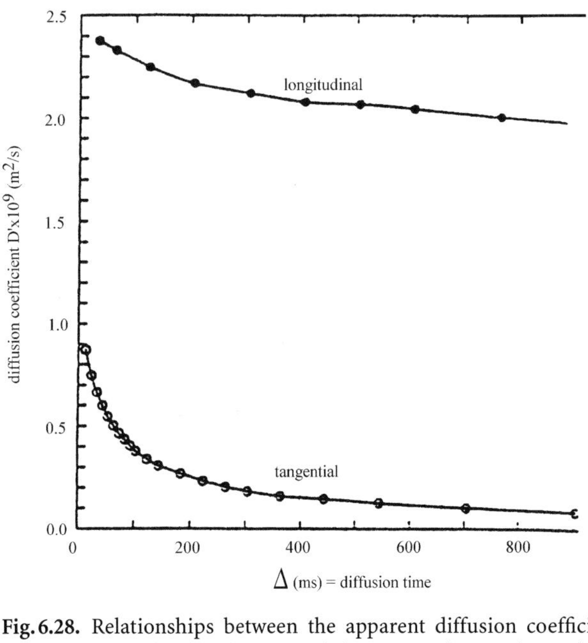

Fig. 6.28. Relationships between the apparent diffusion coefficient and the diffusion time of water in longitudinal and tangential directions of a wood sample of eastern white pine. (Wycoff et al. 2000; courtesy of Wood Fiber Science)

Table 6.6. Diameters of cells in the tangential direction and corresponding standard deviation, measured with the NMR technique and with optical microscopy. (Wycoff et al 2000, courtesy of Wood Fiber Science)

Species	Cell size with microscopy (μm)	Cell size with NMR technique (μm)
Sugar pine	42 ± 3	45 ± 3
Eastern red cedar	17 ± 1	17 ± 2
Eastern white pine	31 ± 1	35 ± 3
Redwood	40 ± 1	43 ± 3

sured for the field gradient in the longitudinal direction are very close to the values for bulk water ($2.2 \times 10^{-9}\,m^2/s$), demonstrating that there is no significant barrier to diffusion in this direction. The deduced diffusion coefficients were between 2.1×10^{-9} and $2.4 \times 10^{-9}\,m^2/s$. The diffusion length along the cells corresponding to a diffusion time of 903 ms corresponded to a diffusion length of 150 μm. For pulsed gradient fields in the tangential direction the barrier walls limit the diffusion length.

Good agreement between the dimensions of cells measured by optical microscopy and NMR is evident from the data shown in Table 6.6. The values measured with the NMR diffusion technique, using water molecules as probes, were only slightly higher than those obtained by microscopy. Wycoff et al. (2000) demonstrated that the NMR provides accurate measurements for cell sizes at moisture contents above the fiber saturation point. However, for dry wood, it would be more difficult to separate the signals for free and bound water.

One-dimensional NMR microscopy can be used to produce images at the millimeter and submillimeter level corresponding to regions with different T_2 values. Menon et al. (1987) demonstrated that NMR signals from water in western red cedar (*Thuja plicata*) can be separated into three components with different T_2 values such as:

- values of a few milliseconds corresponding to water bound within the cell walls,
- values of about 50 ms corresponding to water located in latewood tracheids and ray lumina,
- values of over 100 ms corresponding to water in the lumina of the earlywood tracheids.

It was also found that the relaxation time corresponded to lumen water T_2 scales with lumen diameter (D) as D^2.

The image in Fig. 6.29 was obtained with a Carr-Purcell-Meiboom-Gill (CPMG) pulse sequence in a constant field gradient of 19.4 Gtesla/cm. To isolate an image of each T_2 component of water in wood, it is assumed that

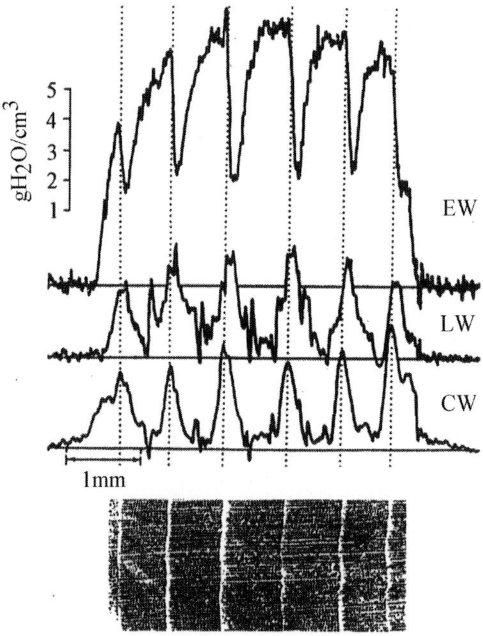

Fig. 6.29. Images of free water in the lumina of earlywood tracheids (*EW*), of the lumina of latewood tracheids (*LW*), and from bound water in the cell wall (*CW*) superimposed by the scanning electron micrograph of western red cedar. (Menon et al. 1989, Academic Press, with permission)

$$S_{(t)} = 0.11e^{-t/5.6} + 0.12e^{-t/46.8} + 0.77e^{-t/184.1} \tag{6.9}$$

where t is the time (ms) from the start of the CPMG sequence and $S_{(t)}$ is the amplitude of the CPMG echo sequence.

The three terms of the equation correspond to the three T_2 components of water content. When the water content in latewood, earlywood and cell wall components were superimposed on the corresponding scanning electron micrograph, near perfect agreement was observed. The authors also demonstrated that the variation of bound water obtained from one-dimensional images across the annual rings on a green specimen is perfectly matched by that of the oven-dry wood density in the radial direction when measured with an X-ray microdensitometric technique on the same specimen, as can be seen from Fig. 6.30. The major contribution of Menon et al. (1989) has been to demonstrate that NMR imaging can be used to separate images on the basis of spin–spin relaxation time T_2, so that images on a submillimeter scale can be achieved. Such studies will enable the role of the bound water to be better correlated with other physical properties, such as dimensional stability, thermal and electrical conductivity and mechanical properties.

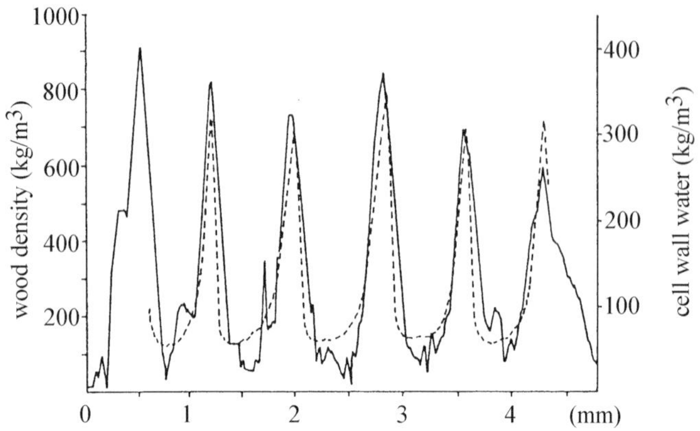

Fig. 6.30. The profile of the bound water in the cell wall compared with the X-ray densitometric profile of annual rings in western red cedar. (Menon et al. 1989, Academic Press, with permission)

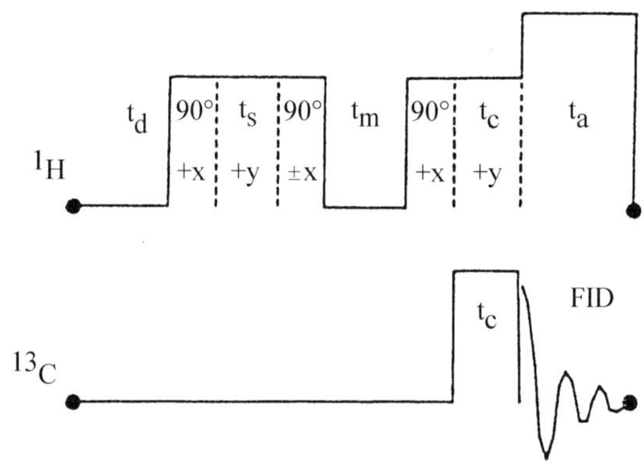

Fig. 6.31. Representation of the pulse sequence MOPS to monitor ^{1}H and ^{13}C spin diffusion. t_c Contact time; t_a data acquisition time; t_d recovery delay time; t_s spin locked time of 8 ms; t_m remaining magnetization time; –x and +x pulses phased along *x*-axis. (Newman 1992; courtesy of Holzforschung)

6.4.2.3 Spatial Distribution of Chemical Components in Cell Walls

Because the Larmor frequencies are specific to different nuclei, NMR methods can be used to investigate the variation of the chemical composition of a wood sample. The spatial variation of different chemical components in the wood cell wall was first studied by Newman (1992) using NMR. He found that natural differences in the proton spin relaxation time constants governed different T_2 values corresponding to ordered and less ordered cellulose and amorphous lignin. The cellulose signals with chemical shift $\delta = 89$ ppm were assigned to ^{13}C in cellulose and $\delta = 56$ ppm ^{13}C at nuclei in lignin. T_1 and T_2 for ^{13}C can be obtained from the same experiment at 50.3 MHz as described in Fig. 6.31 for

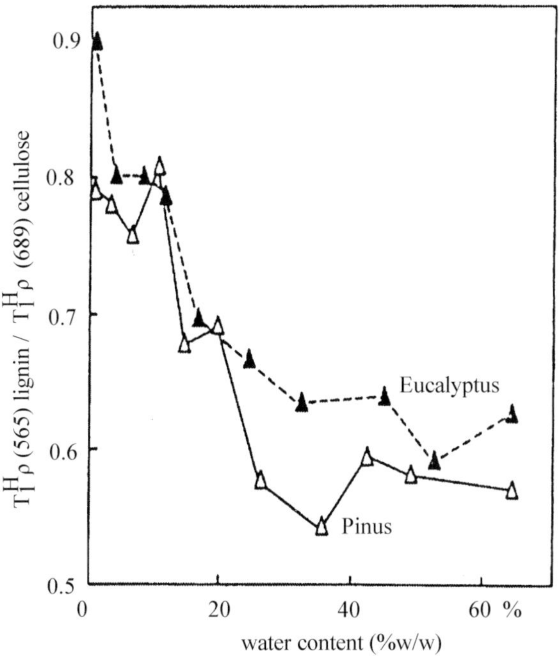

Fig. 6.32. Proton rotating frame spin relaxation time signals vs. the water content for *E. delegatensis* and *Pinus radiata* and for cellulose and lignin. (Newman 1992; courtesy of Holzforschung)

Eucalyptus delegatensis and *Pinus radiata* specimens. The spin locked time was ts = 8 ms. The value of tm, the time needed for the remaining magnetization to return to an orientation parallel to the static field, was variable. Proton magnetization was monitored as a function of tm, with a standard cross-polarization sequence. During the time tm, the phase of the second 90° pulse in the MOPS sequence was alternated to suppress interference from T_1 relaxation. An important experimental detail to note is that D_2O was used in moisturizing the specimen (with 35% by weight), which results in a longer value of T_1. Figure 6.32 shows the variation of the ratio of lignin to cellulose relaxation times $T_{1\rho}^H$ in *Eucalyptus delegatensis* and *Pinus radiata* versus water content. Up to 20% of water content, the curves are superimposed. However, above 30%, corresponding to the fiber saturation point, the values for eucalyptus are slightly higher than those for pine. The differences between the spin-lattice relaxation time, T_1^H, in cellulose and lignin is again evidenced in Fig. 6.33. The increase in $T_{1\rho}^H$ suggests that the spin-lattice relaxation is dominated by mechanisms involving water molecules or hydroxyl groups. The influence of moisturizing on *Pinus radiata* is more important than on *Eucalyptus delegatensis* and can be explained by the difference in strength of bonding of water molecules to the different chemical components of each wood species. The average

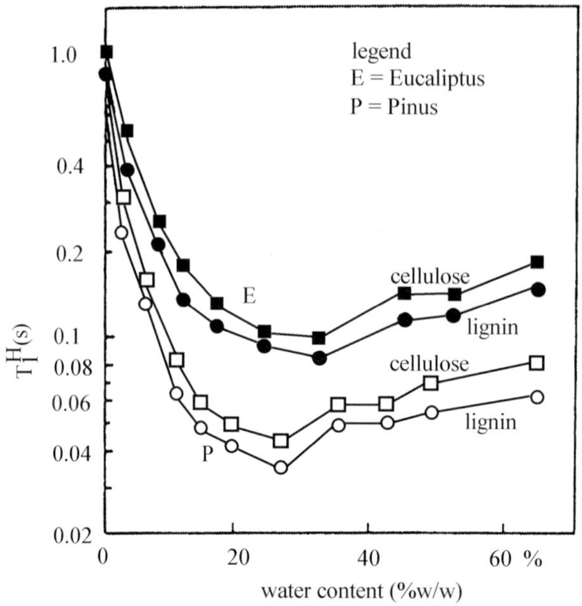

Fig. 6.33. Proton spin-lattice relaxation time vs. water content for cellulose and lignin in eucalyptus and pine. (Newman 1992; courtesy of Holzforschung)

Table 6.7. Proton rotating-frame spin relaxation times $T_{1\rho}^{H}$ (ms) for *Pinus radiata* moisturized with H_2O and with D_2O to 35% of total weight. (Newman 1992, courtesy of Holzforschung)

Specification	Values	H_2O	D_2O	Differences (%)
Cellulose (δ = 56 ppm)	Experimental	3.9	6.2	37
	Theoretical	2.8	5.0	44
Lignin (δ = 89 ppm)	Experimental	7.3	9.8	25
	Theoretical	12.4	14.2	13

difference between cellulose and lignin is 22%. To enhance this difference, the increase in $T_{1\rho}^{H}$ was demonstrated for *Pinus radiata* at 35% moisture content, with the H_2O replaced by D_2O, as can be seen in Table 6.7.

The experimental values were influenced by diffusion between domains. Theoretical values were deduced by computing simulation, using the model cited before with which spatial segregation of chemical components was observed, as can be seen in Fig. 6.34, where the signal height differences between cellulose and lignin were plotted versus the time tm. The perturbation of initial magnetization disappeared after about 20 ms of mixing and no difference between cellulose and lignin was observed. The difference between the theoretical and the experimental values for cellulose are in the range of experimental uncertainties, but for lignin these differences are affected by a

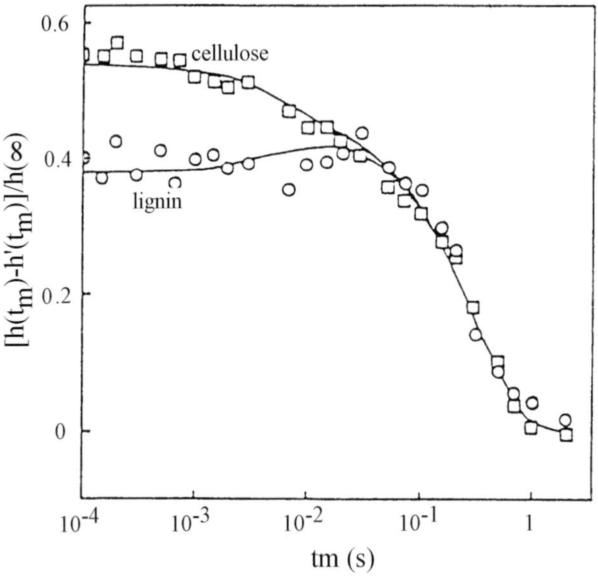

Fig. 6.34. Cellulose and lignin signal height ratios from MOPS experiments in *Pinus radiata* moistened with D_2O at 35% vs. the magnetization time (*tm*). (Newman 1992; courtesy of Holzforschung)

factor of 1.7. It can be concluded that the rotating-frame spin relaxation mechanisms involving water are important in lignin but not in cellulose. Mixing of cellulose and lignin as wood constituents is illustrated by the similarities in $T_{1\rho}^{H}$.

6.4.3 NMR Imaging in Wood Processing

This section describes the potential use of nuclear magnetic imaging in wood processing, and particularly its applications to the control of the moisture content during lumber drying, quality control of wood-based composites and their moisture content at different phases of processing, control of adhesion between adhesives and wood, and the control of impregnation processing of wood with preservatives. The last part of this section is devoted to a study of archeological wood.

6.4.3.1 Moisture Content in Lumber During Drying

Over the last 30 years, extensive literature has been published relating NMR spectroscopy to the measurement of the moisture content in wood. For the forestry industries and for wood science, the most important application of NMR is undoubtedly the measurement of water flow and diffusion in wood, both in situ and in vivo specimens. As noted by Nanassy (1973), the physicists Shaw, Elsken and Rubin in the 1950s were the first to draw attention to the possibility of using NMR to measure accurately the moisture content of wood.

A serious limitation of this technique, at the time, was the large expense of the apparatus required. Using a laboratory spectrometer, Nanassy (1973) proposed a graphic method for measurement of the relative intensity distribution of the narrow component spectral absorption. This approach permitted the moisture content in yellow birch and white spruce to be derived in the range of 5 to 100%, on specimens of about 2 cm^3 volume. The accuracy was around ±3%. A year later, the same author (Nanassy 1974) used a broad-spectrum component system to determine the fiber saturation point, called by the author "critical moisture content" defined in relation to "the phases of the sorbed water, and the ratio of wood protons participating in chemical exchange between wood and water protons." The fiber saturation points at 25 °C for white birch and white spruce were both found to be 38% of the protons in specimens. This value was higher than the generally accepted value of about 28% moisture content for wood species from temperate zones. Nanassy (1974) explained that the diffusion of water in wood over the cell wall surfaces is dependent on the bonding energy of the water molecules to the cell walls, and that such water is "strongly bound to the rigid lattice of wood." The additional free water, called by the author "mobile water," is only weakly bonded to the cell walls. Experiments using improved electronics and corresponding NMR systems during the last 10 years have validated Nanassy's statements, as shown in many articles published in recent years.

As seen in Section 6.2, the diffusion of water in wood is anisotropic (MacGregor et al. 1983). Measurements have been performed with two species, Douglas fir and sugar maple, using a pulsed gradient NMR technique. These measurements confirmed that diffusion along the fibers is higher than diffusion in the transverse directions (Table 6.8). Water diffusion in maple is more

Table 6.8. Effective diffusion coefficients D′ (10^{-5} cm^2/s) in Douglas fir and sugar maple, determined with the pulsed field gradient NMR technique, as reported by MacGregor et al. (1983, with permission)

Species	D′ parallel[a]	D′ perpendicular[b]	Ratio
Douglas fir	1.7 ± 0.1	0.88 ± 0.05	1.9
Sugar maple	1.2 ± 0.1	0.36 ± 0.05	3.3

[a] The coefficient D′ parallel was measured in longitudinal anisotropic direction of wood direction corresponding to the axis of the cylindrical specimen.
[b] The coefficient D′ perpendicular was measured in the perpendicular direction (radial or tangential anisotropic direction) not specified explicitly by the authors.

anisotropic than in Douglas fir, which is in agreement with the anatomical structure of these species. It was also suggested that further experiments should be made using three different cylindrical specimens. Each specimen must have the long axis oriented in one anisotropic direction, L, R or T. In this way, the influence of the specimen geometry on the diffusion phenomena can be avoided.

During the 1980s, many ingenious methods of spin imaging were developed, making possible enhanced one-, two- or three-dimensional NMR imaging. It is undoubtedly of great interest to compare the information obtained from NMR imaging methods with those obtained from high resolution NMR spectroscopy in order to evaluate the state of different wood tissues within the sample under inspection. For clarity, in the following pages, the distribution of water during drying will be studied in small, clear specimens and in boards of industrial size.

6.4.3.1.1 Water During Drying of Small Clear Specimens

The distribution of water in a small, clear specimen can be characterized by one-dimensional radial imaging across the growth rings. This may be achieved with only a single field gradient, which involves only the slice perpendicular to the field gradient as demonstrated by Quick et al. (1990), for western red cedar, and by Araujo et al. (1992) for white spruce.

Lindgren (1994) presented one-dimensional images of pine in the radial direction during drying from 27% moisture content to 11% (Fig. 6.35). One-dimensional images were obtained from the Fourier transform of the FID.

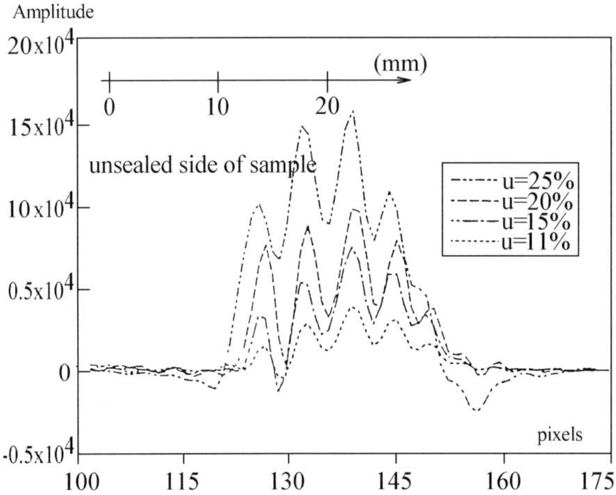

Fig. 6.35. One-dimensional imaging with proton gradient during drying of small, clear pine specimens from 25 to 11% moisture contents. (Lindgren 1994, with permission)

They represent the radial moisture profile with amplitudes proportional to the spatial variation of the water content. For the small, clear specimens an accurate analysis of water distribution above and below the fiber saturation point can be made.

Since the aim of this section is to relate the 1-D distribution of water protons and the structural features of wood, it was therefore natural to superimpose the SEM optical image of the corresponding anatomic structure on the proton profile (Fig. 6.36). To illustrate this statement, we refer to the results on white spruce published by Araujo et al. (1992) and to the definition of moisture content proton density in Eq. (6.6). The drying of white spruce sapwood from 100 to 17% moisture content proton density was followed in four steps, corresponding to 100, 86, 59 and 17%. The radial moisture density profile for total water content, for free water and for bound water was obtained and superimposed on the SEM image. One-dimensional images were obtained from the Fourier transform of the T_2 signals. They showed that the total area under the profile of the 1-D image is proportional to the mass of water in the sample.

One-dimensional images for the representation of the amount of total water (Fig. 6.36a) can be analyzed for each moisture density level as:

- for 100%, the moisture density is higher in the earlywood zone of all rings, because of the larger tracheid lumina in earlywood.
- for 86 and 59%, the rate of drying is more important in earlywood than in latewood.
- for 17%, the distribution of water is quasi-uniform in earlywood and in latewood. It was stated that this curve corresponds to the bound water.

From Fig. 6.36b, representing the 1-D images of the free water present in the lumina, it can be seen that the profiles are very similar to those discussed previously. For the 17% water content sample, the central component indicates the presence of a small amount of water inside the lumina.

In Fig. 6.36c, showing 1-D images of bound water, we observe a superimposition of the bound water profile on the growth ring profile. This allows us to conclude that above the fiber saturation point the bound water distribution is in agreement with the density of solid wood, which is higher in latewood than in earlywood. Below the fiber saturation point, the distribution of the relative moisture density is uniform in the earlywood and latewood. With such evidence, it was possible to suggest that the bound water profile was unrelated to the wood density.

Another possible interpretation of data produced from T_2 spectra is to relate histograms of the lumen radius to that of the proton volume and to the cell number (Fig. 6.37). Both histograms in white spruce sapwood and in compression wood show similar distributions.

This allows us to conclude that the lumen radius is related to the proton volume and that the peaks in the continuous spectra can be attributed to the anatomical location of the water. Different tissues, for example juvenile wood,

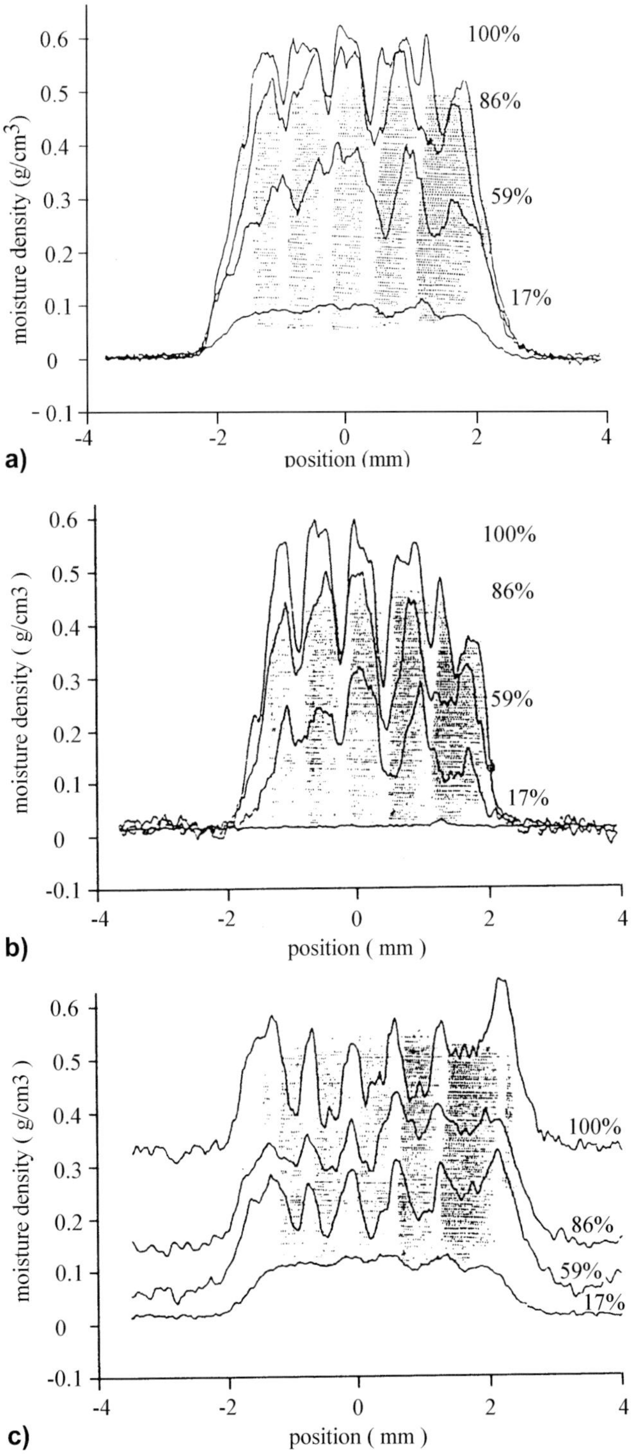

Fig. 6.36a–c. One-dimensional radial moisture profile of protons superimposed on the anatomic structure of white spruce at 100, 86, 59 and 17% moisture contents. The radial moisture density profile corresponds to **a** total water distribution; **b** free water and **c** bound water distribution. (Araujа et al. 1992; courtesy of Wood Science Technology)

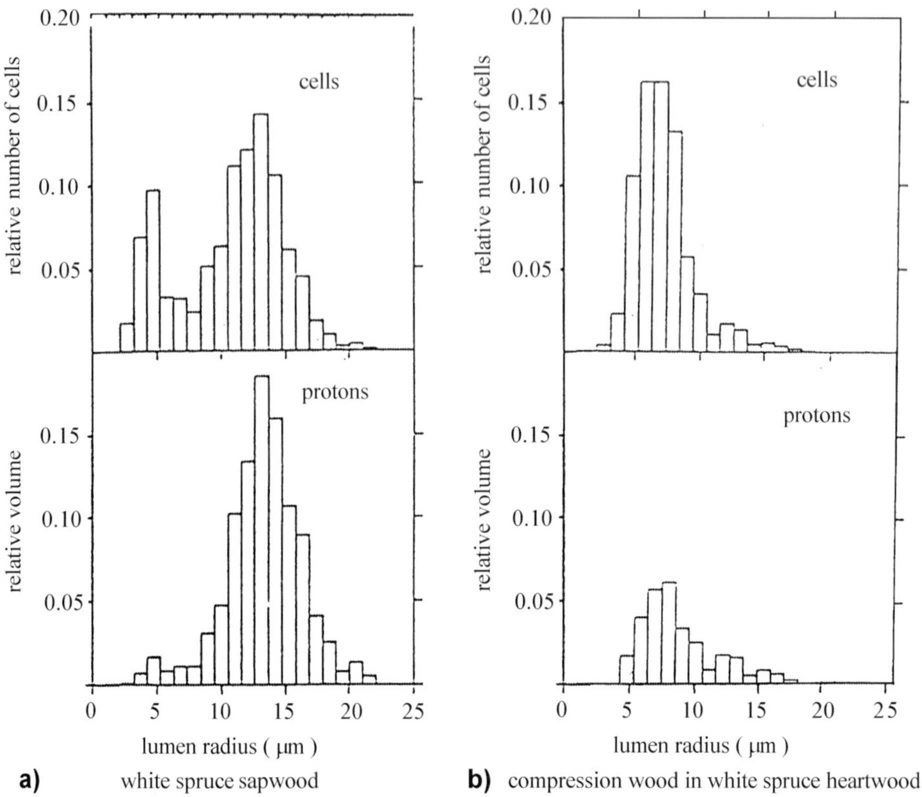

Fig. 6.37a,b. Histograms of numbers of cells and of protons as a function of lumen radius in **a** white spruce sapwood samples; **b** white spruce heartwood compression wood. (Arauja et al. 1992; courtesy of Wood Science Technology)

compression wood or heartwood, are characterized by different cell sizes and can be distinguished by the shape of their T_2 spectra, as can be seen in Fig. 6.38.

Flibotte et al. (1990) used small, clear specimens of $10 \times 5 \times 5$ mm, measuring free induction decay curves (FID) for a series of specimens including undried sapwood, heartwood, juvenile wood and decayed wood in western red cedar. The corresponding hydrogen contents were 6.28, 6.14, 6.13 and 5.27%, respectively. In these specimens, the moisture content was between 30% in juvenile wood and 235% in decayed wood. To monitor the drying of the samples, a pulse sequence was used. The gradient of moisture content during drying is shown in Fig. 6.39. In all signals three distinct components were observed; one corresponding to a rapidly decaying signal, induced by the immobile protons in wood and two other, slowly decaying signals from mobile water molecules, corresponding to water in the lumina of earlywood and late-

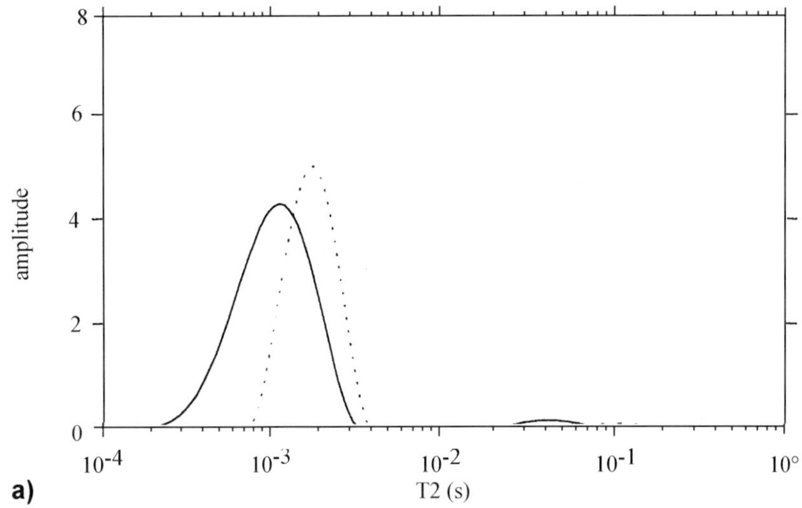

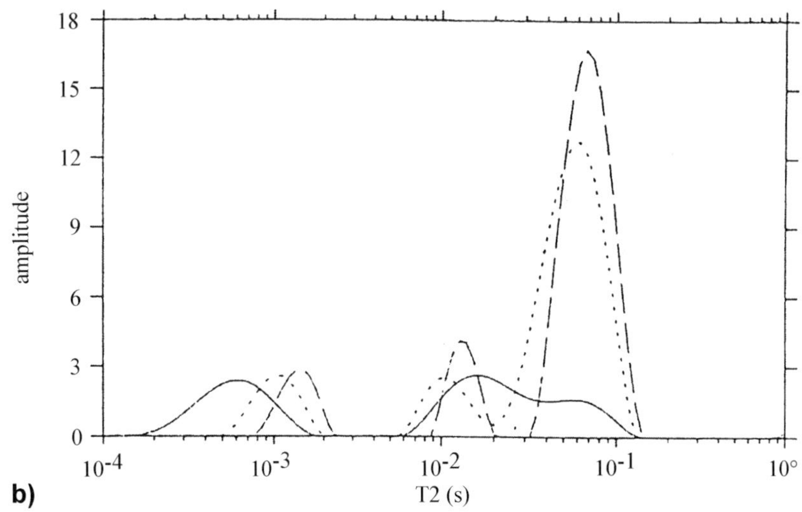

Fig. 6.38a,b. Continuous spectra T_2 in white spruce. **a** Juvenile wood (*solid line*) and heartwood (*dashed line*); **b** in rehydrated samples of heartwood compression wood (*solid line*), heartwood (*long dashed line*) and heartwood with incipient decay (*short dashed line*). (Arauja et al. 1992; courtesy of Wood Science Technology)

wood tracheids and in ray lumina. The moisture content deduced from NMR signals was obtained by "multiplying the ratio of water to wood signal intensities by the ratio of the hydrogen content of wood to that of water." Compared to values for oven-dried wood, the NMR values are overestimated at higher moisture contents and underestimated at lower moisture contents by several

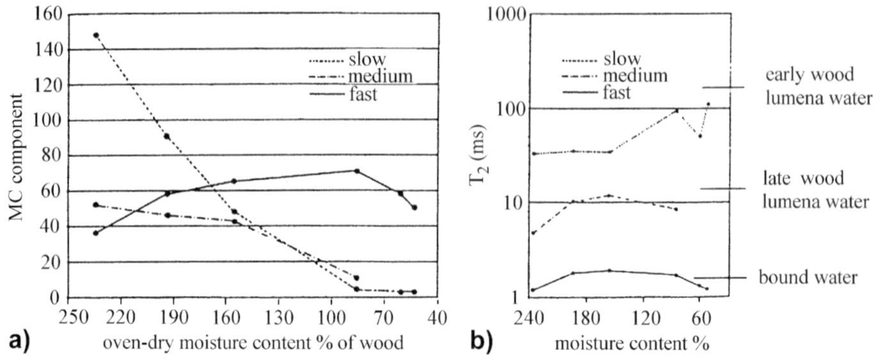

Fig. 6.39a,b. Moisture content determined with NMR technique and the relaxation time T_2 vs. the gravimetric moisture content in cedar. **a** Juvenile wood; **b** decayed cedar wood. N.B. The moisture content deduced from NMR signals was obtained by "multiplying the ratio of water to wood signal intensities by the ratio of the hydrogen content of wood to that of water." (Flibotte et al. 1990; courtesy of Wood Fiber Science)

percent. Nevertheless, bearing in mind the experimental errors, this method can be considered valid for the measurement of moisture content of wood above and below the fiber saturation point.

6.4.3.1.2 Water During Drying of Boards

We will now consider nuclear magnetic imaging in boards of industrial size. For 3-D imaging, only the presence of the fluid in the pore space is observed during drying. Olson et al. (1990) noted that NMR imaging is more effective than NMR spectroscopy in the determination of moisture gradients within a large wood sample. Three-dimensional imaging was used by Olson et al. (1990) to monitor the moisture distribution in oak boards during a 14-day period. Boards of three thicknesses were selected (25, 38 and 50 mm) with a moisture content ranging between 68% at the beginning and 44% at the end of the experiment. A clinical NMR scanner was used for imaging the internal structure of the boards, giving an image intensity proportional to the free water present. The image area was 30 × 30 cm with a spatial resolution of 1.2 × 1.2 mm.

Figure 6.40 shows multislice spin-echo images for the transverse section of the board. The images exhibit the variation of moisture content during the drying time. The quantitative estimation of the brightness of the images can be obtained with a routine optical density technique. The brighter core areas represent higher moisture content in contrast to the darker shell area around the surface. The rays and the annual ring are well resolved. The rays have a

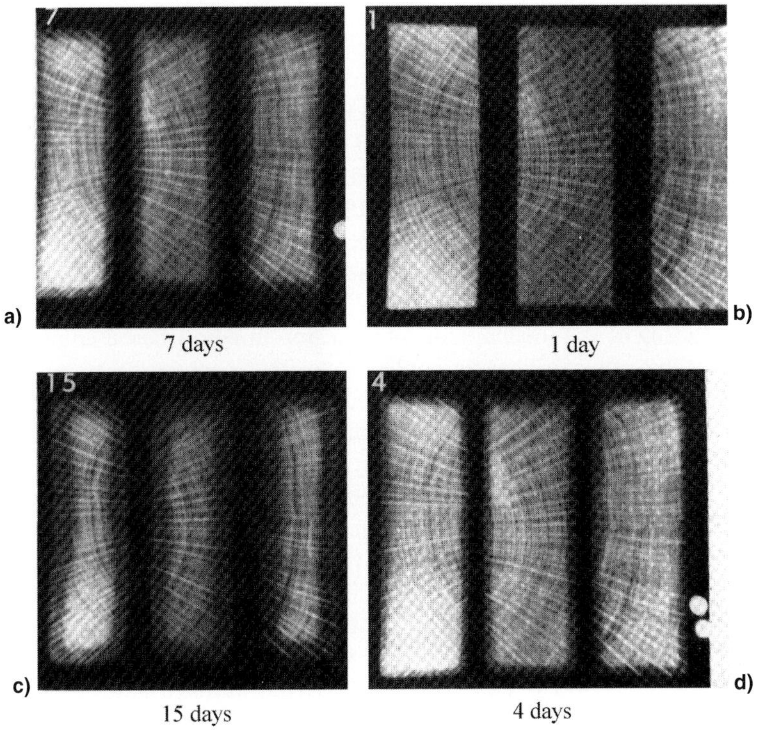

Fig. 6.40a–d. Transverse views of oak lumber (50-mm thickness) during natural drying. **a** One-day drying; **b** 4-day drying; **c** 7-day drying; **d** 15-day after cutting. (Wang et al. 1990; courtesy of Forest Products Society)

higher moisture content than the surrounding annual ring area. It is evident that the drying process takes place from the outside to the side of the board. The kinetics of drying can be observed for all specimens and information about the moisture gradient and distribution can be deduced. A 25-mm-thick board dried at a faster rate than the 50-mm board because it simply takes longer for water to diffuse out from the center of the thicker boards. The dark area is dependent on orientation of the growth ring. The brighter area in the middle region of the cores indicates that diffusion is faster from the central core areas than from the outer surfaces of the board.

We have shown that NMR imaging is a valuable nondestructive and non-contact method for qualitative and quantitative studies of the kinetics of wood drying. Its only drawback is the cost of such equipment, which is similar to the medical device used in hospitals. However, important information can also be obtained by 1-D scanning, with much less expensive spectrometers, using small core samples taken in vivo at various processing stages.

6.4.3.2 Quality Control of Wood-Based Composites

The application of solid-state NMR techniques to the quality control of wood-based composites is related to two main aspects: the nondestructive measurement of the moisture content and the analysis of the adhesive bond lines. For measurement of the moisture content of wood-based composites, the literature is very scarce. Wolter and Netzlmann (1996) reported the utilization of single-side access NMR facilities for measurement of moisture content and density distributions in wood-composite boards. The goal of this preliminary research was to develop an in situ device

An understanding of the mechanisms of adhesion in wood-based composites is important for new engineering applications. Physicochemical phenomena that govern adhesion are related to molecular interactions between the adherent and the adhesive. In wood particles or flakes, as in all solid polymers, molecular ordering and motion can be investigated by measuring relaxation times. Changes in molecular structure of wood composites can be observed if the number of molecular associations between the adhesive molecules and the wood molecules is large enough. Gil and Neto (1999) and Marcinko et al. (1998, 1999) have reported information on the molecular dynamics in wood and adhesives like urea-formaldehyde (UF) and polymeric diphenylmethane diisocyanate (PMDI). ^{13}C spectra were collected using a standard proton longitudinal relaxation time $^{1}HT_{1\rho}$ pulse sequence at 50.77 MHz. Southern pine particles were mixed with resin and cured in a forced-air oven at 140 °C for 1 h. Table 6.9 gives the $^{1}HT_{1\rho}$ data for lignin, hemicelluloses and cellulose. Data obtained for southern pine solid wood and UF-coated southern pine particles are similar. Relaxation times corresponding to PMDI-coated southern pine particles are about 25% less, and this is probably due to the fact that PMDI changes the macromolecular motions of wood molecules. UF has a relatively small effect on the molecular-level dynamics of southern pine wood. In contrast, PMDI reacts strongly with wood molecules and penetrates deeply into the adherent wood, inducing rapid stiffening of the composite.

Table 6.9. Proton longitudinal relaxation time $^{1}HT_{1\rho}$ (ms) for solid pine wood, for PMDI/pine and for UF/pine expressed as a function of the chemical shift (ppm) for lignin, hemicelluloses and cellulose. (Marcinko et al. 1998, courtesy of Forest Products J)

Components	Chemical shift (ppm)	$^{1}HT_{1\rho}$ for pine (ms)	$^{1}HT_{1\rho}$ for UF coated-pine cured (ms)	$^{1}HT_{1\rho}$ for PMDI coated-pine cured (ms)
Lignin	54	8.0	8.5	6.3
Lignin	150	8.5	8.8	6.2
Hemicellulose	21	9.2	9.5	5.5
Hemicellulose	173	9.5	9.2	6.2
Cellulose	73	9.7	8.8	7.9
Cellulose	105	10.1	8.9	8.0

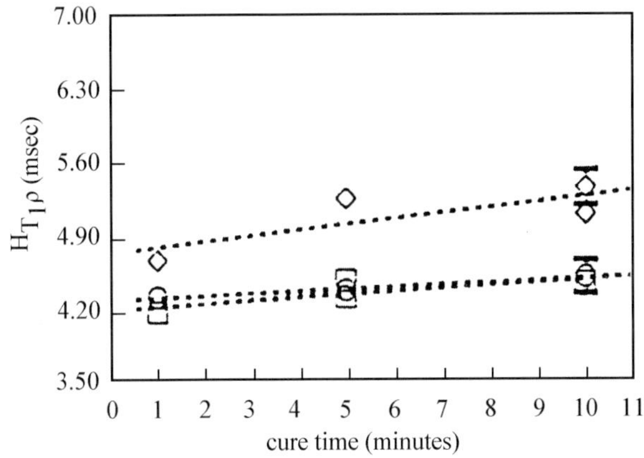

Fig. 6.41. The relaxation time ^{1}H $T_{1\rho}$ (ms) for flakes cured with phenol formaldehyde resin at 110°, 135° and 175°. (Schmidt and Frazier 2000; courtesy of Wood Fiber Science)

The challenge of NMR laboratory testing is to the analyze the adhesive bond lines on pilot and industrial levels. Schmidt and Frazier (2000) demonstrated the viability of cross-polarization magic angle spinning (CP/MAS) NMR measurements for evaluating the degree of adhesive cure (phenol-formaldehyde) in poplar flakeboards. Figure 6.41 shows the relaxation time $^1HT_{1\rho}$ (ms) for flake samples cured with ^{13}C phenol-formaldehyde resin. The curing temperatures investigated were 110, 135 and 175 °C. An increase in the relaxation time $^1HT_{1\rho}$ with curing time and temperature was observed. These results give information on the effects of temperature on the adhesive bond lines. CP/MAS spectra of ^{13}C flake samples of the skin and core of the boards at 13 and 24% moisture contents are shown in Fig. 6.42. Skin samples are different from core samples in the range of 35 and 63 ppm, which correspond to methylene and hydroxymethyl, respectively. Skin samples have an increased methylene peak and a diminished hydroxymethyl peak compared with core samples. The mean ratios of corrected peak integrals of methylene carbon to the hydroxymethyl carbon at 13% moisture content was for the core 0.8 and for skin, 1.6. At 24% moisture content, these ratios were 1.0 and 1.8. Data for mean ratios of corrected peak integrals for the core and skin were significantly different at $K < 0.001$. This difference is explained by the temperature difference between the high temperature surface regions (200 °C) and the cold face and the core (150 °C).

In conclusion, it can be noted that for pilot-scale flakeboards, the relative degrees of resin cure are measurable by proton longitudinal relaxation time ($^1HT_{1\rho}$) and by the ratios of corrected peak resin integrals on CP/MAS spectra of ^{13}C nuclei.

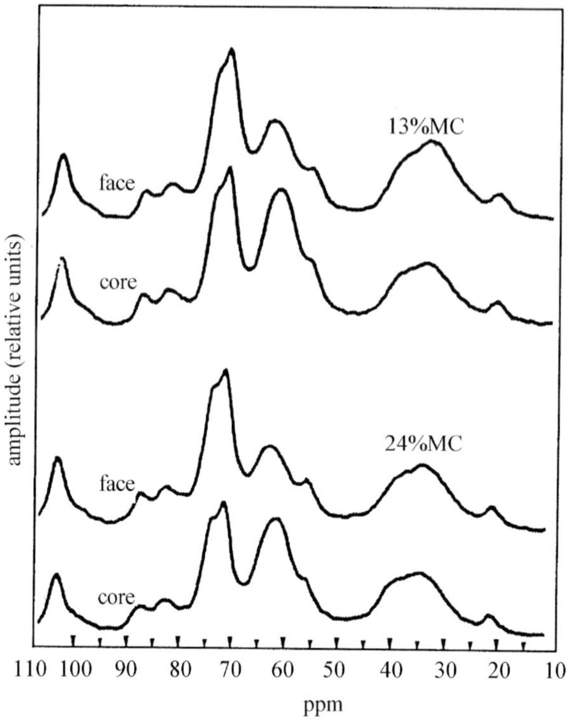

Fig. 6.42. CP/MAS spectra of ^{13}C-labeled flake samples of the core and of the face of the board at 13 and 24% moisture contents. (Schmidt and Frazier 2000; courtesy of Wood Fiber Science)

6.4.3.3 Control of Adhesion

Industry requires reliable methods to evaluate the properties of adhesive-bonded systems. Niemienen and Koenig (1988) used NMR imaging to test the joint between polyvinyl acetate and pine wood. The behavior of the adhesive bond formed between water-based polyvinyl acetate adhesive and pine was followed during curing. Cylindrical specimens of furniture grade wood of 35 mm diameter and 70 mm length were studied. The wood specimens were typical furniture grade pine wood of 35 mm diameter and 70 mm length. The specimens were split into two equal parts. In addition, a trough of 50 × 11 × 3 mm was cut into one of the wood pieces to make it possible to follow the behavior of the adhesive. The specimens were bonded with 50 g/m² of adhesive emulsion, and after 1 min the system was pressed by hand and tightened with paraffin.

All the images taken during curing were obtained on an NMR imaging spectrometer at 300 MHz. The experiment used a standard Carr-Purcell spin-echo pulse sequence. As can be seen in Fig. 6.43, the kinetics of drying as a function

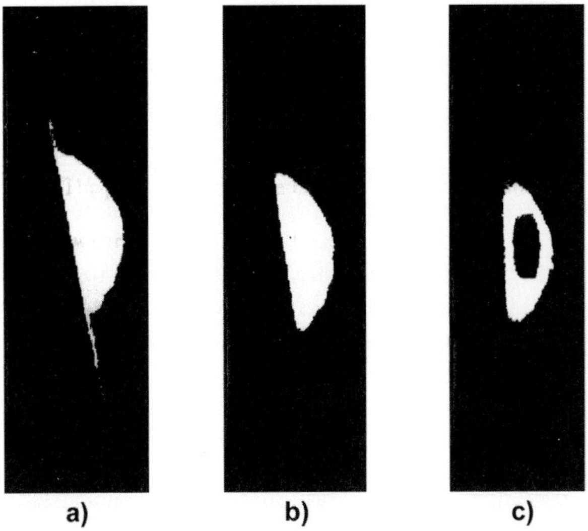

Fig. 6.43a–c. Kinetics of drying a water-based polyvinyl acetate adhesive emulsion between two pine specimens as reported by Niemienen and Koenig (1988) after **a** 10 min; **b** 4 h; **c** 16 h. (VSP International Science Publisher, with permission)

of time shows that after 3 min (Fig. 6.43a) a very thin glue line between the wood specimens had been formed. The 11-mm trough in the specimen is also evident on the right side of the wood sample.

After 4 h of drying (Fig. 6.43b), the adhesive line between the specimens is dry and no longer visible on the slice. Only the still wet adhesive in the trough is imaged. After 16 h, the adhesive is still not completely dry, and an empty space in the trough is visible, probably because the external pressure applied to the adhesive joint was not sufficient. As expected, the adhesive strength of the structure is decreased because of the void presence. The NMR signals are due to water in the adhesive and not water that had penetrated from the adhesive to the wood. The authors noted also that the mobility of thermoplastic polymers increases at elevated temperatures and that NMR images can be recorded even at 180 °C.

In routine industrial situations, voids in the adhesive layer must be detected as early as possible to guarantee the quality of the adhesive bonds in mass production of furniture, veneer, glue-lam, etc. The limitation of the proposed technique is related to the difficulty of detecting solid inclusions in the joints, such as hard resins, because their very short relaxation times are not imaged. The main advantage of NMR imaging for adhesive bonded structures is related to the ease with which the images can be interpreted and to the information provided about the adhesive itself and about the whole structure under inspection.

6.4.3.4 Control of the Impregnation Process

Another attractive idea is to use NMR imaging for the visualization of the impregnation process of wood with preservatives or other liquids, that is, the process of fluid absorption.

First, we will discuss the state of the art of the subject related to other materials, such as fiber-reinforced composites and rocks, and then consider measurements on wood.

Rothwell and Holecek (1984) demonstrated the feasibility of using NMR imaging to follow the dynamics of water diffusion in inorganic materials. The mechanism of water absorption depends on the physical and chemical parameters of the composite studied (glass fiber-reinforced epoxy-resin composite). Rapid characterization of the distribution of absorbed fluid in a polymeric material is possible.

Kleinberg (1999) and many other authors before him, referred to the earlier contributions that NMR had made to studies of the mesoscopic properties of rocks, such as pore sizes and their distribution. Relaxation time measurements can be used to determine a volumetrically weighted distribution of pore sizes spanning several orders of magnitude. The quantitative measurement of the amplitude of the NMR signal and the calibration of proton spin density measurements provide a very accurate method to determine the porosity with a noncontact and noninvasive technique. Similar methods can be used to investigate the porosity and penetrability of wood during preservation.

Hall and Rajanayagam (1986), using three-dimensional proton NMR volume imaging, studied the impregnation of cylindrical maple specimens of 1.2 cm diameter with preservative containing oxides of Cr, Cu and As and with 1 M manganese chloride solution ($MnCl_2$). The specimens were soaked overnight with these substances. With a very coarse spatial resolution, Fig. 6.44 shows the relative intensity of the NMR signals from the "spin warp" technique, measured on a sample compared with the spin lattice relaxation time T_1 of the water protons. These images depend on the relationship between the pulse intervals chosen for the image measurements and the characteristic relaxation time T_1 of the water present in the wood. If the time between pulse sequences is too short, the water protons in wood have not enough time to relax during the measurement, and the image can have a low intensity. Moreover, a part of the signal can disappear before it can be measured, if the signal from the water has a short spin–spin relaxation time T_2 The manganese chloride solution has $T_2 = 10^{-5}$ s, much shorter than that of the water in the wood that is $T_2 = 10^{-3}$ s. The Mn ions are magnetic and significantly reduce the T_1 and T_2 relaxation times of the absorbed fluids, which is an important technical aspect of these measurements. Consequently, with a relatively short time between pulse sequences, only a signal from the manganese chloride solution is observed. This enables the diffusion of the $MgCl_2$ into the bath to be measured. The influence of impregnation time is shown in Fig. 6.45, where one can see the effect after 3.5 and 0.5 h compared with the control measurements on a fresh wood specimen.

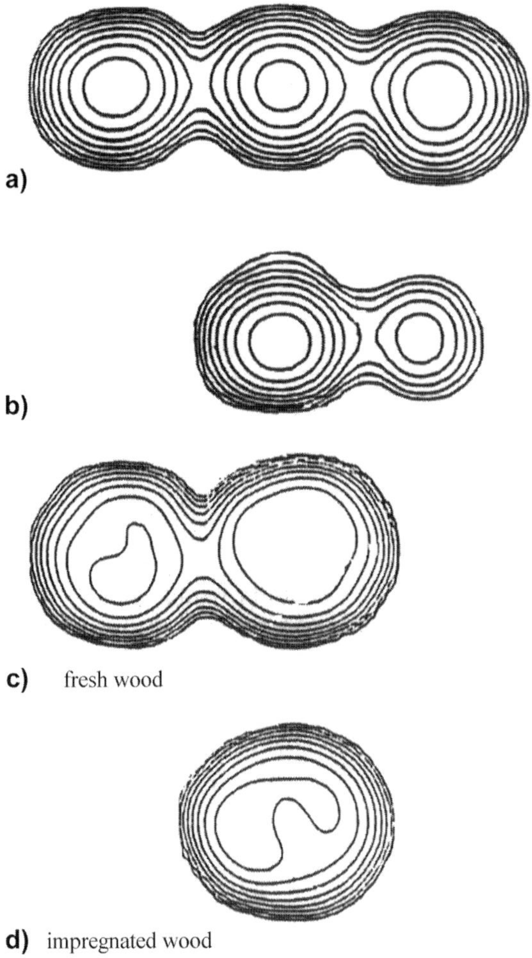

Fig. 6.44a–d. Imaging with the "spin warp" technique of maple specimens impregnated with different preservatives, compared with a control specimen of fresh-cut wood, at 80 MHz frequency for protons and 15 min of total experimental time. The contour levels are logarithmically spaced. The top contour level is at 0.7 of the tallest peak. The following level is at 0.7 of the previous level height. The relative intensity of the NMR signal has been compared with the spin-lattice relaxation time of water protons. **a** Planar image of freshly cut maple. **b** Effect of heating one of them for 3 h at 110 °C. **c** Planar image of three pieces of freshly cut wood. The piece in the middle has been used as reference. **d** Piece soaked in a preservative that shows a substantial change. (Hall and Rajanayagam 1986; courtesy of Wood Science Technology)

6.4.3.5 Examination of Archaeological Wood

In this section, we describe NMR applications to the study of wooden archaeological objects. High resolution techniques using ^{13}C-NMR have been used successfully for this purpose (Lambert et al. 2000). The CP/MAS spectra of archeological wood and fresh wood are similar, as can be seen by comparing

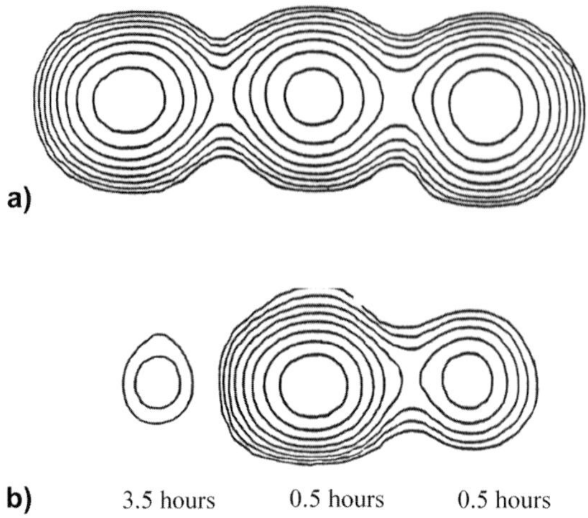

Fig. 6.45a,b. The influence of impregnation time with a 1-M solution of $MnCl_2$ on the distribution of the preservative in wood. **a** Fresh wood. **b** Sample after it had been soaked 3.5 and 0.5 h. (Hall and Rajanayagam 1986; courtesy of Wood Science Technology)

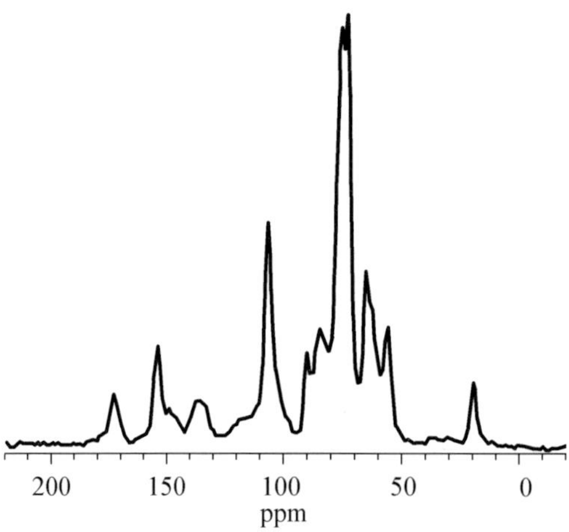

Fig. 6.46. The CP/MAS spectrum of wood. (Lambert et al. 1991, with permission)

Fig. 6.46 with Fig. 6.2. Carbohydrate resonances are found with chemical shifts, δ, between 50 and 90 ppm. The lignin peaks are located between $\delta = 100$ ppm and $\delta = 160$ ppm. Archeological wooden artifacts, like elements of old shipwrecks, are of particular interest for the development of NMR imaging. These objects survive only because they are water-logged and proton NMR imaging is the most appropriate technique for studying their internal structure. The very long-term action of water changes the physical and chemical properties of wood. Different parts of excavated ships show different degrees of degradation.

Cole-Hamilton et al. (1990) examined samples from a Spanish Armada ship "La Trinidad Valencera", which was excavated in 1970 on the northern Irish coast, after more than 380 years of submersion. The wood species was clearly identified as ash. The grain structure and the growth rings were visible with structural details such as rays. The image resolution was 10 μm. Macroscopically, the structure of the old ash was not very different from that of fresh ash. Hatcher et al. (1981) compared the ^{13}C spectra of contemporary spruce with spruce samples buried 450 years ago in Rotterdam and a 10,000-year-old spruce from northern Michigan. The spectra were not very different. Small differences were observed in the distribution of the carbohydrate peaks. In the Michigan sample, the $\delta = 75$ ppm peak was relatively strong while the $\delta = 105$ ppm peak was much weaker. These changes may be due to the beginning of the coalification of the wood.

Wilson et al. (1993) used solid state ^{13}C-NMR spectroscopy to analyze the structure of wooden elements of five old ships submersed in the Indian Ocean. The structural elements submersed in an anaerobic site appeared to have undergone little or no degradation. However, samples in the proximity of a marine environment lost carbohydrates, whereas the inner wood remained relatively intact. Carbohydrates are subject to hydrolysis and leaching. Fengel (1991) noted that for most waterlogged woods in a moist and strongly salty environment, part of the hemicelluloses are solubilized.

Studies on the permeability of waterlogged archaeological wood with a view to conservation were reported by Robertson and Packer (1999). One-dimensional NMR profile imaging in oak timber from the *Mary Rose* ship sunk in 1545 in Portsmouth, UK, were obtained. Increment cores of 4 mm diameter were extracted at different depths, ranging from 3 to 26 mm from a 56-mm-thick, water-saturated wood piece. After cleaning the wood of all soluble paramagnetic salts, the samples were introduced into a tube filled with D_2O, and the subsequent uptake of D_2O was monitored at specific time intervals in a one-dimensional ^1H-NMR profile experiment. The profiles were obtained with a spectrometer operating at 20 MHz and equipped with a pulsed field gradient capacity. The images obtained with a spin-echo sequence were produced with a 90° pulse length of 2.4 μs for radiofrequency excitation and with a 180° pulse applied after 2.5 ms to form a spin echo in the presence of field gradient of 0.26t/m, which gave a spatial resolution of 80 μm. The record of the complete echo was obtained with 1024 points in both real and imaginary channels at a 1-μs dwell time. The one-dimensional profile of the diffusion of D_2O in the longitudinal anisotropic direction of the wood (Fig. 6.47) was obtained from the spatially encoded frequency domain signal, determined by Fourier transformation. The ^1H signal outside the sample does not change with time. Inside the sample, the concentration of D_2O increases from the surface until it reaches equilibrium. The diffusion coefficients for D_2O were calculated as a function of the depth into the sample and the anisotropic axis, as can be seen in

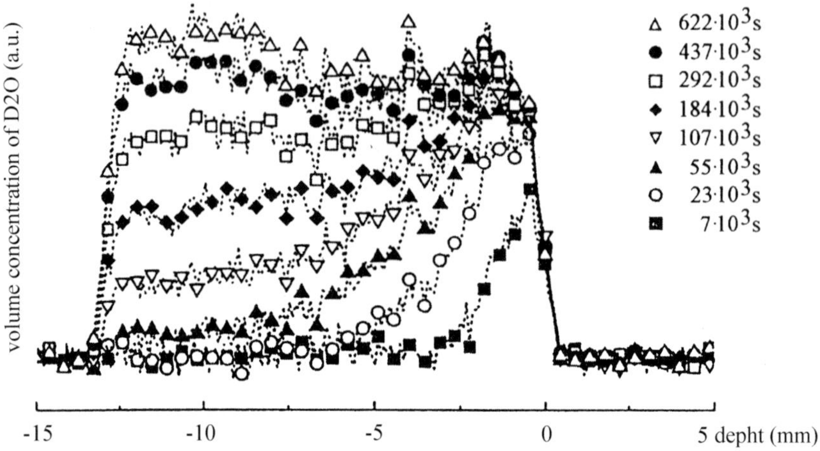

Fig. 6.47. One-dimensional NMR profile of the ingress of D_2O in archeological oak (from the *Mary Rose* ship) in longitudinal direction as a function of the depth of the sample. (Robertson and Packer 1999, with permission)

Fig. 6.48. The higher diffusion coefficient, as expected, was measured in the longitudinal direction. In this direction the diffusion pathway is along the fibers and vessels and is similar to the pathways of fresh wood, but in an archeological sample the diffusion can be blocked by deposits of calcareous material, which can also block the diffusion through the medullary rays. In addition, the permeability in the radial direction can be enhanced by the deterioration of pit membranes. The permeability of archeological wood is considerably modified by the degradation process. The solid core is less degraded and has a high content of cellulosic components, but it is surrounded by a heavily degraded surface layer. The variation of the diffusion coefficients in all anisotropic directions, expressed as a function of the depth of the specimen corresponds to a typical pattern of deterioration of buried archaeological wood, with more degradation at the surface than in the internal part of the specimen.

The potential areas of use to NMR imaging and spectroscopic techniques are related to the identification of the tree species of archeological objects, estimation of the age of samples from the ratio of unsaturated (lignin) to saturated (carbohydrate) resonances, to assessment of the level of decomposition of archeological wood prior to conservation and to the measurement of the ingress of preservatives into the wood in order to stabilize it

6.4.4 Further Applications for the Measurements of Elastic Constants

NMR imaging for the measurement of the elastic properties of bulk solids was first proposed and realized by Lewa (1991, 1992, 1996). This technique uses the

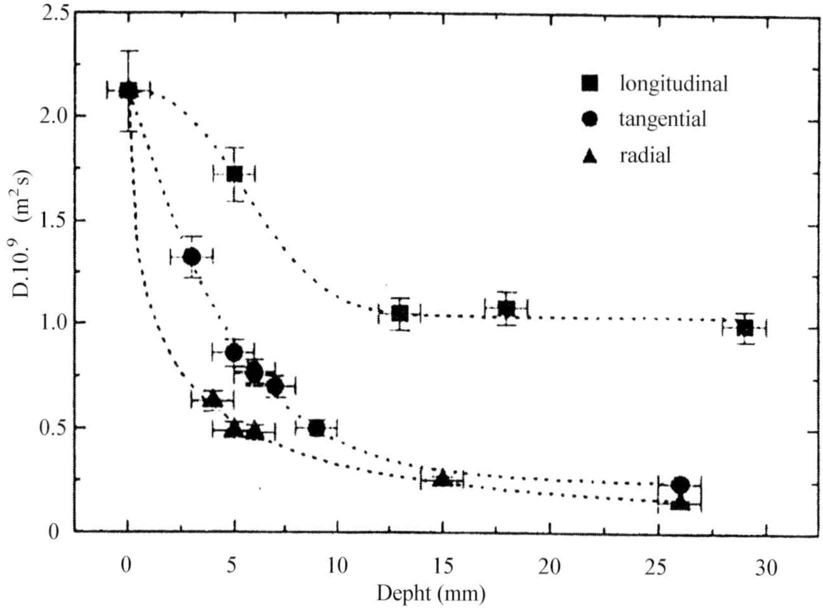

Fig. 6.48. Diffusion coefficient of D_2O as a function of depth and of anisotropic direction in archeological oak from the *Mary Rose* ship. (Robertson and Parker 1999, with permission)

magnetic field gradient located in time and space and relates the spatial coordinates with frequency. Such transformations are possible when the position of an element or the value of the magnetic field intensity are time-dependent. For a time-independent magnetic field with a constant gradient, the Larmor frequency can be modulated by varying the position of the sample. This can be accomplished by using a longitudinal mechanical wave. The modulation of the magnetic field intensity affects the nuclei present in the displacement element and depends on the argument of Bessel's functions. This argument is expressed as a function of the angle between the displacement and the field gradient vectors and the coefficient of the adiabatic compressibility of the medium. The latter depends on the elastic properties of the solid and on instrumental factors such as: the type of paramagnetic nuclei, the adiabatic compressibility and the density of the solid, and wavelength and velocity of the mechanical wave.

The presence of the mechanical wave in the MRI field enables the display of the bulk object by modulation of the Larmor frequency with a mechanical wave applied to the sample in the presence of the magnetic field gradient. The possibility of generating mechanical waves with wave-lengths comparable to the MRI resolution and the cellular size, enables observations at a cell level. Until today, the main application of this elastomagnetic resonance spectroscopy has been in the biomedical field, but extensions of such a technique to characterization of materials, acoustic emission and photo-acoustics appear to be possible and should be investigated.

6.5 Summary

NMR imaging is one of the most powerful and versatile techniques for the characterization of materials. The development of high speed computers and the introduction of high field superconducting magnets have increased the speed, sensitivity and flexibility of such characterizations. Only the cost of such equipment limits the utilization of NMR imaging in the fields of wood science and technology. Standard NMR parameters are the resonance frequency, the magnitude of the signal proportional to the density of the nuclei, the spin-lattice relaxation time T_1, the spin–spin relaxation time T_2, the diffusion coefficient, the flow velocity, the spin–spin coupling and the spectrum of correlation times. The values of these parameters depend on species, moisture content, physiological parameters of wood and on several instrumental and measurement factors such as the Larmor precession frequency, temperature, etc.

The NMR tomographic image depends on the scanning technique used, pulse sequence, and magnetic field intensity. The NMR imaging technique relies on the interaction of nuclear magnetic moment (nuclei) in only a small, controlled zone of the sample under inspection, and is achieved by placing the measured body in a spatially inhomogeneous field. Its nuclear resonance frequency is matched to the RF signal only in the corresponding zone of the object.

The NMR imaging technique can be used to investigate the spatial distribution of all parameters that can be determined by NMR, such as densities, T_1, T_2, diffusion terms, etc. Usually, the nuclear spin density and relaxation time are mapped as a function of their spatial position. Using basic spatial encoding and slice selection principles, different techniques are available to form 1-D, 2-D or 3-D images, using various spin-echo, stimulated-echo and gradient-echo pulse sequences. NMR signals inherently depend on nuclear relaxation time constant, which, in turn reflect the structural environment of the emitting nuclei. There are several modalities of spatially encoding the signals. One of these is to apply a linear magnetic field to the original static field. In this way, nuclei on one side of the sample will feel a weaker total magnetic field than those on the other side. From such a set of data, the image of the sample is reconstructed with an appropriate algorithm. Conventional NMR spectroscopy can be coupled with the corresponding imaging technique, and chemical structures of the specimen can be determined.

The main advantages of the NMR imaging technique are: the method is nondestructive, noncontact, relatively rapid, can be used in situ and in vivo, and does not induce any structural damage. In addition to providing a relative mapping of solid structural inhomogeneities, fluid (water, preservative solutions, etc.) distributions can be observed. The technique has the potential to provide an absolute measure of fluid absorption. A T_2 relaxation map can distinguish between free and bound water. The bound water is strongly bonded to the cellulose and has a much shorter T_2 than the free water. Because the NMR relaxation rates depend on the freedom of molecules to move, they are sensitive to indicators of the chemical and physical characteristics of the sample. For measurements in the presence of fluid–solid interfaces, the decay curve is a probe of the length scale of the structure. Applications such as measurement of the moisture content distribution in wood and in wood-based composites, continuous monitoring of lumber drying, adhesive curing and impregnation of wood with preservatives can be implemented in industry. NMR imaging is one of the most powerful new techniques for wood science and technology, and can be used to monitor industrial processes on a continuous basis in a production line. Future availability of low-cost and easy to use devices will contribute to the more widespread use of this technique.

7 Neutron Imaging

7.1 Introduction

Neutron radiography is a nuclear method for wood structure imaging. Relevant aspects about the physics of neutron scattering and about materials science-oriented neutron scattering studies have been published in reference books and articles such as: Guinn and Lukens (1965), Bacon (1975), Garret and Berger (1977), Kostorz and Lovesey (1979), Nicklow (1979), Domanus (1992), and Dreele (1994).

Neutrons are subatomic particles with zero charge and are the major constituent of the nuclei. Their characteristics are: mass $m_n = 1.675 \times 10^{-27}$ kg, spin 1/2, magnetic moment $\mu_n = -1.913$ nuclear magneton. They have a half-life of 624 s. They carry no charge and their properties can be expressed in terms of momentum, velocity, wave vector, wavelength, and kinetic energy. Neutrons in thermal equilibrium with a moderator material near room temperature are called thermal neutrons and are relatively slow. The Maxwellian spectrum of thermal neutrons, which peaks at 25 meV at room temperature, corresponds to a neutron wavelength of about 0.18 nm. The velocity of neutrons is 2200 m/s at 290 K (Price and Sköld 1986). In the case of wood, which is a hygroscopic material, a specific hydrogen image can be obtained with neutron radiography. Today, the neutron radiography provides images with the highest resolution (15 μm) for water in wood, either in small, clear specimens or in samples of large size.

Neutron imaging is one of the most recently developed testing techniques in wood science and forestry and, at the same time, it is one of the most promising. The main research field of neutron imaging is related to water flow in solid wood and trees. Because of its high specificity to water, the neutron beam technique can image water movement in living cells during meristem development with a high resolution. So far, only few laboratories have been involved in the development of wood structure imaging by the neutron technique, namely, in Japan at Tokyo University and in Switzerland at ETH Zürich. We note the pioneering work of Nakanishi (Nakanishi et al. 1998a; Furukuwa et al. 2001) on the development of the specific neutron imaging technique to visualize the moisture content in plants and in fresh wood specimens cut from living trees.

The applications of neutron scattering to specific problems in wood science and technology are still very scarce. The interest of wood scientists in neutron activation analysis was originally related to studies on ion migration in wood

(Loos 1965a,b; Langwig and Meyer 1973), to diffusion of the preservatives in solid wood (Siau and Meyer 1973), to nondestructive moisture content measurements (Chang 1975) and resin penetration (White et al. 1977), and, 20 years later, to plants and tree physiology (Nakanishi 1994; Nakanishi and Matsubayashi 1997), forestry ecology (Schultz and Ward Whicker 1982), to timber (Nakanishi et al. 1998b) and structural elements (Niemz et al. 2000a,b; Lehmann et al. 2001). This chapter is concerned with a survey of several applications of neutron radiography to wood science.

7.2 Basic Aspects

The development of neutron imaging techniques (Kostorz and Lovesey 1979) is based on the following:

- the interaction of a neutron beam with a solid is weak;
- the absorption is very small compared with X-rays, and large samples can be inspected;
- the wavelength can reveal the atomic arrangement in the sample;
- the nonmagnetic scattering amplitude for neutrons varies in a nonmonotonic way from one element to another and, in contrast to electromagnetic radiation, it does not depend on the number of electrons in an atom;
- being a nuclear property, the scattering amplitudes may differ considerably for different isotopes of a given chemical species.

The principles of neutron radiography are largely presented in various books. Detailed comments on this subject are beyond the purview of this section. A neutron radiographic image of an object is obtained by the attenuation coefficient of a neutron beam that interacts with the matter. The attenuation coefficient can be calculated from the equation:

$$\mu = N\sigma_t = N(\sigma_a + \sigma_s) \tag{7.1}$$

where μ is the attenuation coefficient, N is the number of nuclei per cm^3 of attenuating material, σ_t is total cross section, which is the probability that a nuclear reaction will occur and corresponds to the sum of the absorption σ_a and scattering cross sections σ_s.

This equation is valuable for assessing the relative change in transmitted radiation intensity that might be obtained through the inspected object. In a radiographic situation, this phenomenon is recorded. The object is placed in a neutron beam in front of an image detector. Neutron radiography is an effective method that can provide a specific high resolution hydrogen image because the attenuation of H is 100 to 1000 times higher than that of the other elements. Images of moisture content distribution will be presented in this section. The H amount of different components of wood (cellulose, hemicelluloses, lignin, resins, etc.) can be neglected when this amount is compared with the H present in the moisture of wood (Nakanishi et al. 1998a).

Development of neutron imaging with computer tomography, in progress in different laboratories (Furukawa et al. 2001), will give new opportunities in the future to studies in wood science and tree physiology.

7.3 Equipment

The neutron imaging of wood reported by Nakanishi and coworkers (Nakanishi et al. 1998a,b) was obtained with a neutron beam produced by an atomic reactor (JJR-3 M) installed at the Japan Atomic Energy Institute in Ibaraki prefecture. The wood specimens were sealed in vacuum together with a gadolinium n/γ converter and a film. The specimens were irradiated for 19 s with a neutron flux of 8.7×10^8 n/cm^3. The image obtained on the film was recorded on a computer with a CCD camera. The neutron images obtained of the wood structure were three-dimensional, with *z*-axis indicating the amount of moisture present.

The equipment for neutron radiography used by Niemz et al. (2000a,b,c) has the characteristics given in Table 7.1. This equipment is located in the Paul Scherrer Institute at Villigen, Switzerland, and has been described by Bauer (1998). The neutrons are produced by spallation. The neutron beam line is shown in Fig. 7.1. The corresponding detector setup is seen in Fig. 7.2. The moisture content in structural members was determined by image analysis by subtracting the images obtained for samples under wet and dry conditions.

Table 7.1. Technical characteristics of the thermal neutron beam line NEUTRA used by Niemz et al. (2000c) for neutron imaging of wood structural members (with permission)

Technical parameters	Units	Values
Proton energy	MeV	590
Proton beam intensity	mA	2
Thermal neutron flux at sample position	cm^{-2} s^{-1}	3.4
Mean neutron energy	meV	25
Beam parallelism (l/d)	–	550
Fraction of thermal neutrons	–	100
Field of view at sample position	mm	350
Spatial resolution of imaging plate	µm	50
Number of pixels per line	–	1500
Recording area	cm^2	400
Dynamic range	bit	16

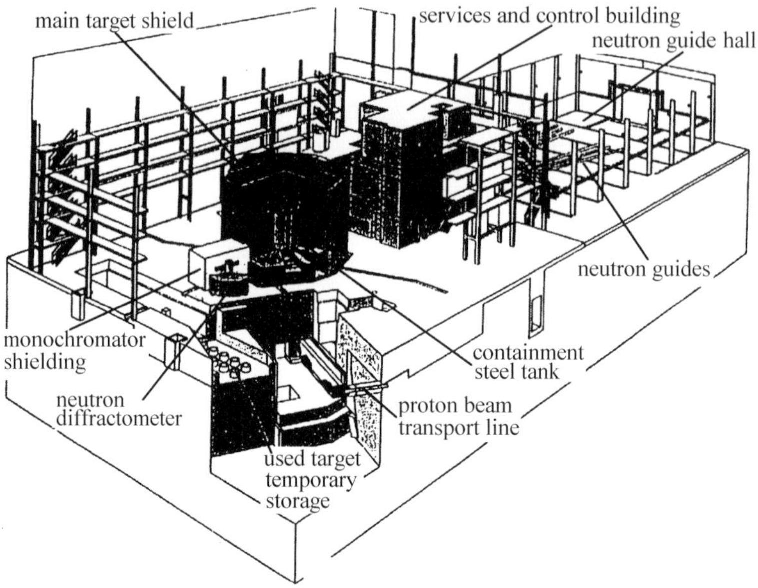

Fig. 7.1. Thermal neutron beam line (*SINQ*) at the Paul Scherrer Institute in Villigen, Switzerland. (Bauer 1998, with permission)

Fig. 7.2. Sample detector setup for neutron radiography. (Niemz et al. 2000c, with permission)

7.4 Applications

In this section, the imaging of the moisture in wood will be studied first on disks taken from living trees during drying and then on small, clear specimens during absorption of water For the study of drying phenomena, two cases will be analyzed, corresponding to short and relatively long drying times. Imaging of water distribution in structural elements submitted to weather conditioning will also be presented.

7.4.1 Imaging of Water Distribution in Trees

The storage of water in a tree is influenced by different factors, such as: the capillarity of the wood structure and the elasticity of the cell tissue. The water content in trees is subjected to seasonal changes and is different in sapwood and in heartwood. Development of new methods for determination of the water distribution in trees should be emphasized. Among the nondestructive methods, neutron imaging seems to be promising. Water distribution was studied with neutron radiography by Nakanishi and Watanabe (1995) who used trees of different species and by Nakanishi et al. (1998b) with trees belonging to the same species.

7.4.1.1 Water Distribution in Trees of Different Species

Nakanishi and Watanabe (1995) investigated the distribution of water in young trees of five species (*Pinus thunbergii, Metasequoia glyptostroboides, Chamecyparis obtusa, Quercus serata* and *Robinia pseudoacacia)* as can be seen from Fig. 7.3. Table 7.2 gives the characteristics of the disk specimens selected from those trees. The specimens were irradiated in the atomic reactor of the Japan Atomic Energy Research Institute with thermal neutrons during 19 s. The total

Table 7.2. Characteristics of the specimens irradiated for 2.3 min. (Data from Nakanishi and Watanabe 1995, with permission)

Species	Age of the tree (years)	Specimen diameter (cm)	Specimen thickness (cm)	Moisture content
Pinus thunbergii	18	13.5	1–3	Green
Metasequoia glyptostroboides	10	13.2	1–3	Green
Chamecyparis obtusa	19	12.6	1–3	Green
Quercus serata	18	16.5	1–3	Green
Robinia pseudoacacia	8	12.3	1–3	Green

a b c

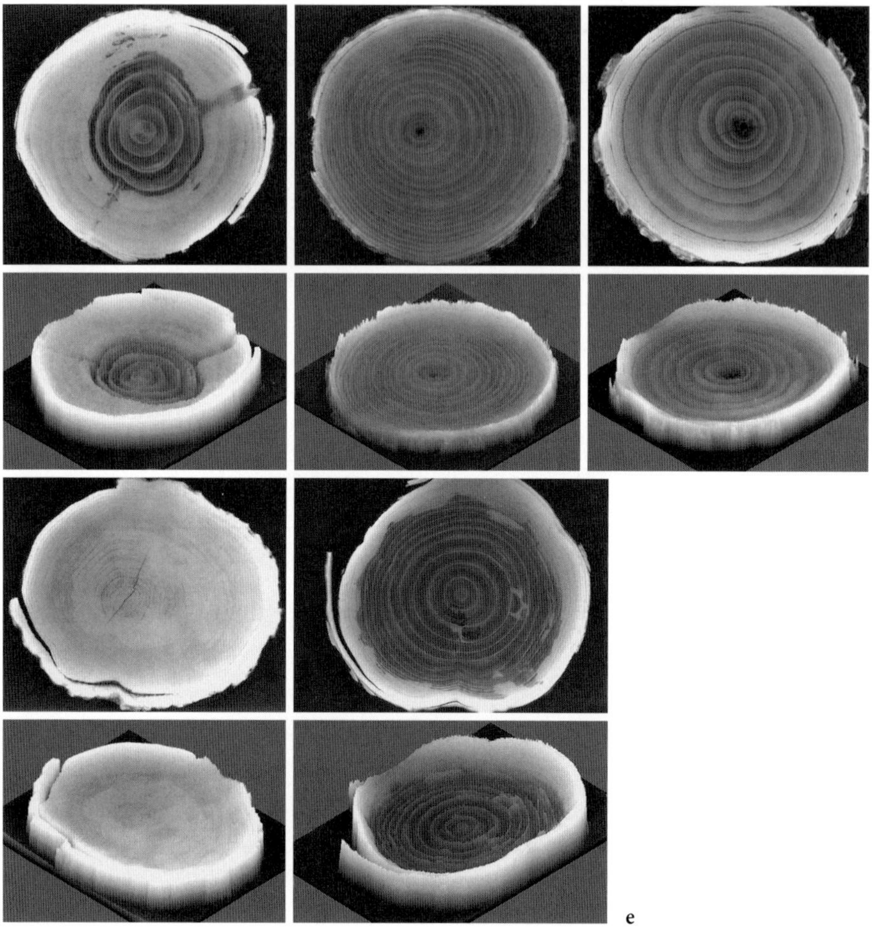

d e

Fig. 7.3a–e. Photographic images and neutron radiography of different species. **a** *Pinus thunbergii*. **b** *Metasequoia glyptostroboides*. **c** *Chamecyparis obtusa*. **d** *Quercus serata*. **e** *Robinia pseudoacacia*. (Nakanishi and Watanabe 1995, with permission)

flux was obtained with (n, γ) reaction and was $8.7 \times 10^7\,n/cm^2$. The water distribution image was obtained on a film. The resolution of the image was $15\,\mu m$. The degree of brightness in the image was proportional to the water content of the sample.

Having these images as a background, it can be noted that in *Pinus thunbergii* within the same ring, a heterogeneous water distribution has been observed. The black zone in the pith indicates the decrease of water content, suggesting the beginning of the necrosis of the tissue. In the image for *Metasequoia glyptostroboides,* we can see that only a small amount of water is observed in the central zone. The maximum amount of water seems to be located in the last annual ring. The knots are not precisely defined on neutron radiography probably because of their lower moisture content. In *Chamecyparis obtusa* the distribution of moisture is very nonuniform, as can be seen from the image of white irregular rings.

The exact pattern of moisture distribution in *Quercus serata* is difficult to identify. The annual ring with water deficiency is represented by a dark zone that is easy to recognize. Parallel zones with different flow properties occur in oaks, which have very large earlywood vessels, resulting in a very high longitudinal permeability. The majority of other tissues in the annual ring consist of thick-walled fibers with extremely low permeability making it difficult for liquids to penetrate except at very high pressure. This extreme nonhomogeneity of wood structure is one of the principal reasons why it is difficult to observe the moisture content pattern in oak wood.

The photographic image for *Robinia pseudoaccacia* clearly shows the presence of the heartwood. In neutron radiography, the heartwood zone was not revealed, and we can suppose that there was no difference in moisture content between heartwood and sapwood. The vessels with tyloses in which no water is present are observed as dark points. The climatic conditions during the growth of the second, third and fourth annual rings are well illustrated by the presence of the false rings, which are represented by different gray levels.

7.4.1.2 Water Distribution in Trees of the Same Species

Nakanishi et al. (1998b) studied the moisture distribution in the inner part of the sapwood of different cultivars of sugi *(Cryptomeria japonica)* with special emphasis on the identification of the presence of a so called white zone in the sapwood, characterized by a high moisture content. This zone is probably caused by genetic factors and environmental conditions. It is present in all specimens, as shown in Fig. 7.4.

In three-dimensional representation of water distribution, the aspect of the "white zone" is very well defined for all specimens (Fig. 7.5). In addition, near the bark, this distribution of moisture is not superimposed on the pattern of the annual rings. Nakanishi and Matsubayashi (1997) suggested that this bright zone observed by neutron radiography could also indicate the presence of different chemical components rich in H, which are different from those in water from the cell wall.

In conclusion, it can be said that with neutron radiography, the visualization of a patterns of water distribution across the transverse section of a tree, mainly near the bark, was possible for all species studied, Neutron imaging allows studies on water distribution in zones near the bark, which are difficult to perform with other techniques.

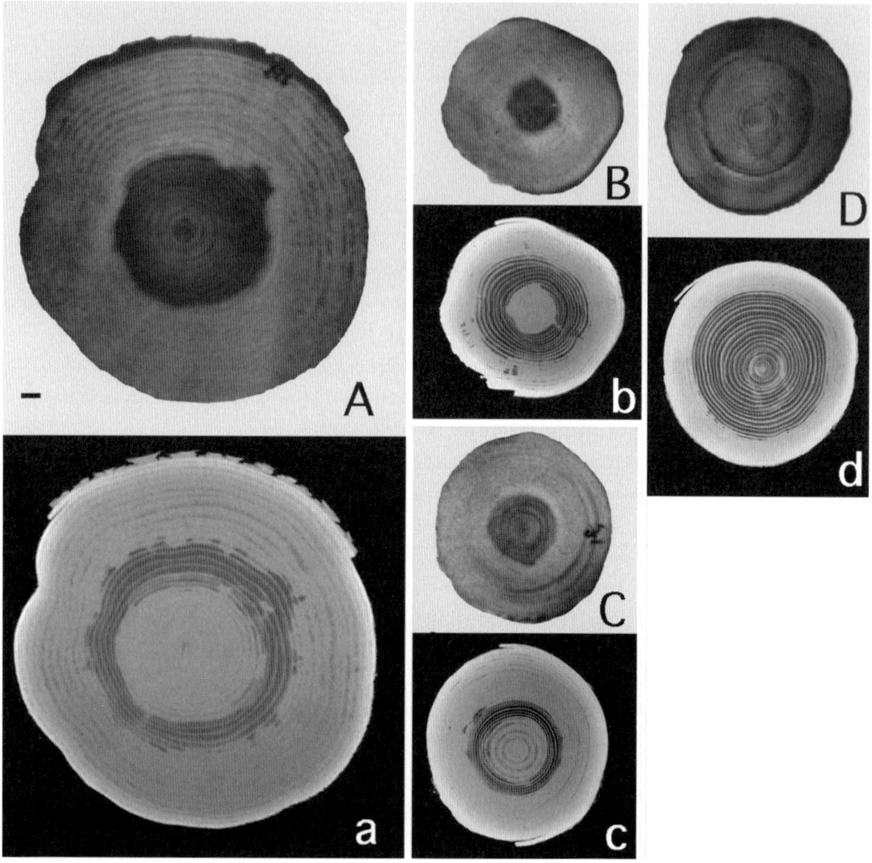

Fig. 7.4. Photographic images (**A**, **B**, **C**, **D**) and corresponding neutron images (**a**, **b**, **c**, **d**) of different cultivars of sugi (*Cryptomeria japonica*). **A** 24-year-old cultivar, 25 Gou. **B** 25-year-old cultivar Honjiro. **C** 29-year-old cultivar 1-Gou. **D** 30-year-old cultivar, Sanbusugi. (Nakanishi et al. 1998b, with permission)

7.4.2 Imaging of Moisture Content in Lumber During Drying

Established wood drying techniques are based on the removal of the large amount of moisture present in green wood and have the purpose of protecting wood materials from excessive drying stresses that can cause defects and degradation. For wood quality control during drying, several nondestructive methods have been developed. In this section, we will discuss some laboratory measurements undertaken with the neutron technique for imaging the kinetics of drying.

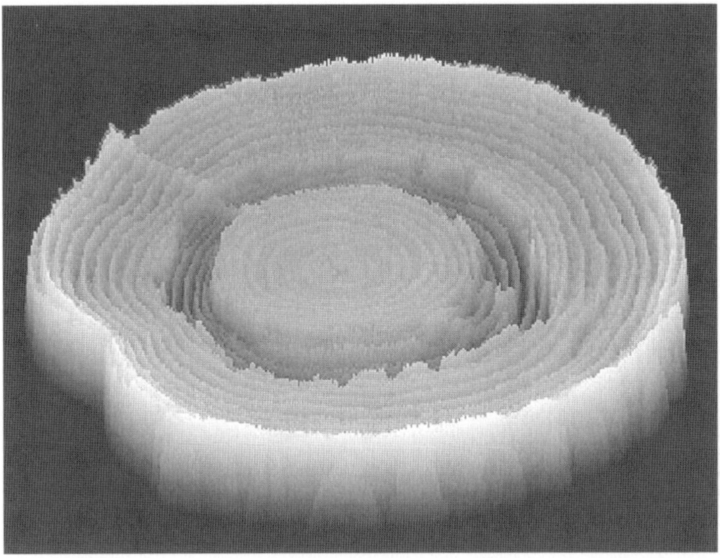

Fig. 7.5. Three-dimensional water distribution in a disk of 1-cm thickness of green *Cryptomeria japonica*. (Nakanishi and Matsubayashi 1997, with permission)

7.4.2.1 Short Drying Time

Nakanishi et al. (1998b) studied the kinetics of drying of *Cryptomeria japonica* wood (Fig. 7.6) during 6 h at 60 °C and at 90% relative humidity, using the same specimens described in Section 7.4.1. At the initial stage, which corresponds to green saturated conditions, all disks give a very bright image. An important difference between the sapwood and the heartwood is well visible with disks A, B and C. The moisture distribution seems to be uniform within the annual rings. Disk D shows a similar and very high moisture content in both sapwood and heartwood. During the drying process of 2, 4 and 6 h, the moisture content still remains high in the sapwood. The heartwood becomes darker with increasing drying time. Figure 7.7 is relevant for the profile of moisture content in wood. The value of the moisture content in sapwood and in heartwood was calculated precisely as a function of the corresponding area. For all disks, the moisture content near the bark was higher than that in the sapwood and in the heartwood. For disk A, the effect of drying time on the profile of the moisture content in the heartwood seems to be less important than on sapwood. After 6 h of drying the moisture content in heartwood was very near that in the sapwood.

From the profile of moisture content distribution, it can be noted that at the beginning of the drying process, disks B and C have a similar behavior as disk

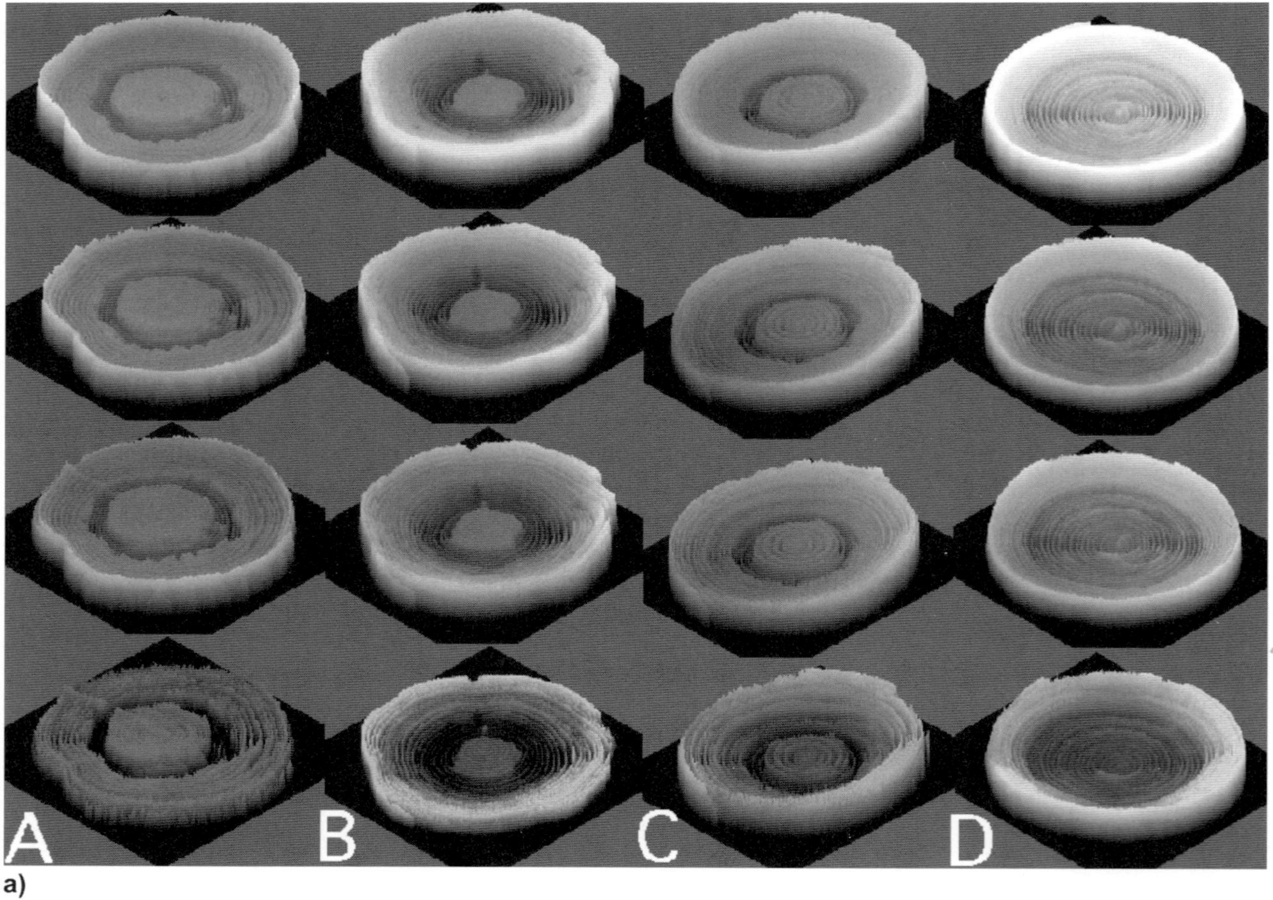

Fig. 7.6a. Imaging of the kinetics of moisture content distribution. **a** For 6 h drying at 60 °C at 90% relative humidity of four disks (*A*, *B*, *C*, *D* as in Fig. 7.4) of 1-cm thickness of *Cryptomeria japonica*.

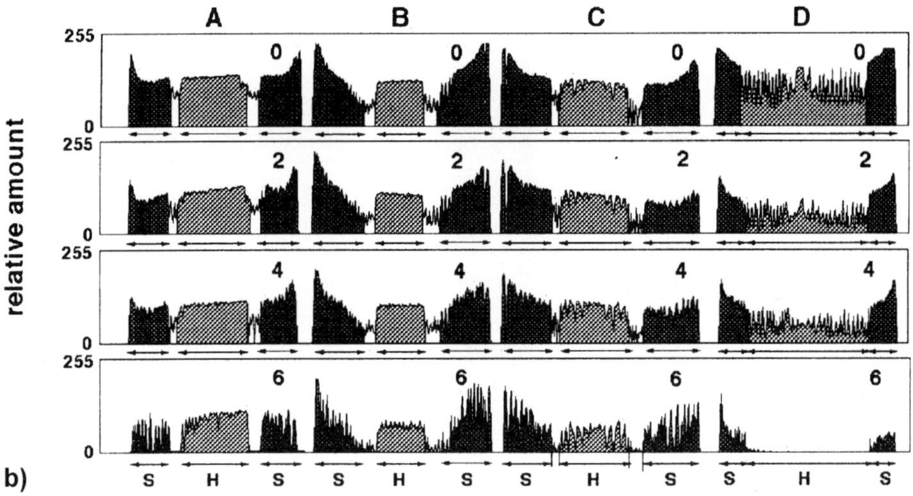

Fig. 7.6b. Variation of the moisture content profile of four disks *A*, *B*, *C*, *D* for which the numbers *0*, *2*, *4*, *6* are the drying hours. (Nakanishi et al. 1998b, with permission)

2h 6h 22h 30h 45h

Fig. 7.7. Neutron radiography of spruce drying as a function of time. (Lehmann et al. 2001, with permission)

A, but after 6 h of drying, the moisture content is higher in the sapwood than in the heartwood. For all these three disks the heartwood represents about 33% of the transverse section. By contrast, for disk D, the zone occupied by the heartwood seems to amount to about 70% of the transverse section. After 6 h of drying, the heartwood lost a considerable amount of water compared with the sapwood. The causes of the different moisture distributions are unknown, but the effects of such differences on the wood during drying can be disastrous because of the development of a nonuniform drying stress.

7.4.2.2 Long Drying Time

A longer drying process (45 h) was studied with spruce specimens (Fig. 7.7) by Lehmann et al. (2001). During the drying process at 20 °C and 65% relative

Fig. 7.8. Kinetics of spruce drying. (Lehmann et al. 2001, with permission)

humidity, the wood moisture content decreased from a saturated state to air dry at about 18%. After 22 h of drying, some anatomic details at the transverse section of specimens were observed. After 45 h, the disposition of the annual rings was clearly visible, together with shrinkage of the specimen. The regular square cross section of the specimen in the water-saturated state had been transformed during drying into a rhomboid-shaped section. The minimum shrinkage was along the radial direction of the wood (at 45° in the figure) which corresponds to the direction of the rays. The maximum shrinkage occurred in the tangential direction to the annual rings. The drying kinetics (Fig. 7.8) have been investigated by advanced image processing, by subtracting image characteristics in two different drying stages. The image of the saturated sample has been considered as a reference. During drying, a crack observed in the upper part of the image (after 8 h of drying) increased in size and propagated in the tangential direction to the annual ring. After 70 h of drying, the crack is well visible on the external, dried area of the specimen.

7.4.2.3 Imaging of Water Absorption

Wood can be protected from the attack of biological agents by applying chemical preservatives. The uniform distribution and sufficient penetration of these substances can greatly increase the life of wood structures. For an understanding of the complex phenomena during the impregnation and for the determination of preservative effectiveness, it is necessary to utilize nondestructive techniques able to visualize in real time the absorption of liquids by the materials. This section deals with the determination of water flow in solid wood and wood-based composites with the neutron imaging technique.

Different wood species (beech, pine, spruce, sweet chestnut), particleboard and MDF specimens (Lehmann et al. 2001) were partially immersed in water, as shown in Fig. 7.9. Quantitative water absorption as a function of time is shown in Fig. 7.10. After 5 h of immersion, the beech samples absorbed 19 g of water, while all other specimens absorbed about 4 g of water. After 22 h of immersion, the maximum water absorption was 25 g for beech. During the same period, water absorption was at a minimum, (about 1 g) for the MDF

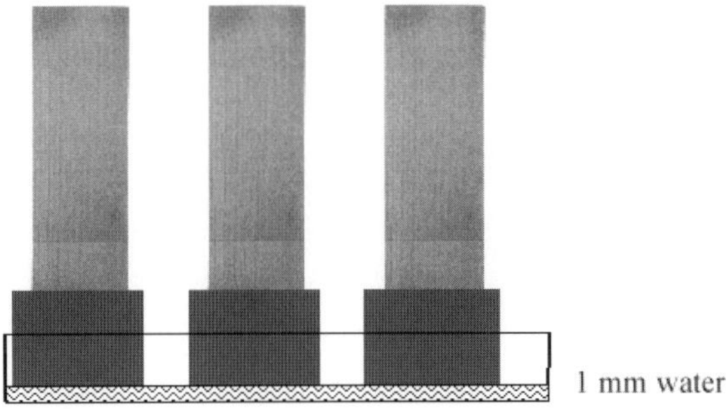

Fig. 7.9. Experimental device for water absorption by capillarity in wood samples 200 mm long × 10 mm wide × 16 mm thick. (Lehmann et al. 2001, with permission)

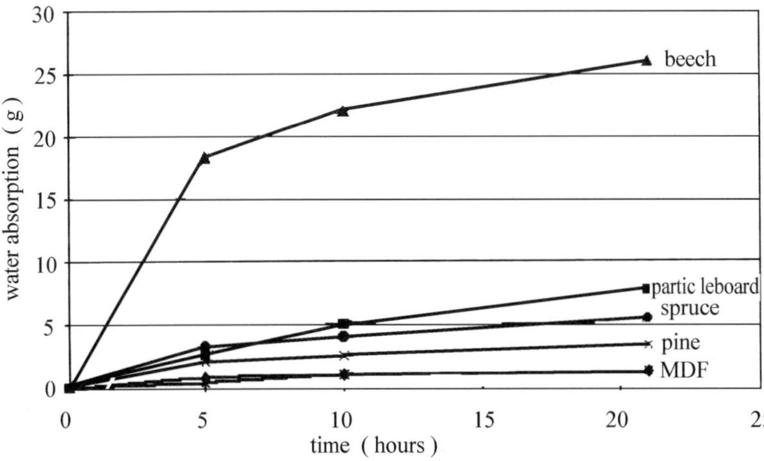

Fig. 7.10. Water absorption as a function of time for beech, spruce, pine, chestnut, particleboard and MDF samples. (Lehmann et al. 2001, with permission)

sample. The kinetics of water absorption for beech, pine and particleboard samples, with neutron radiography as a function of time, is shown in Fig. 7.11. On the image, the darker zone, which corresponds to the maximum amount of absorbed water, is observed for the beech samples.

7.4.3 Imaging of Moisture Content Distribution in Structural Elements

The widespread use of wood in the construction of buildings, windows, doors, etc. has both an economic and an esthetic basis. The beauty of wood is diffi-

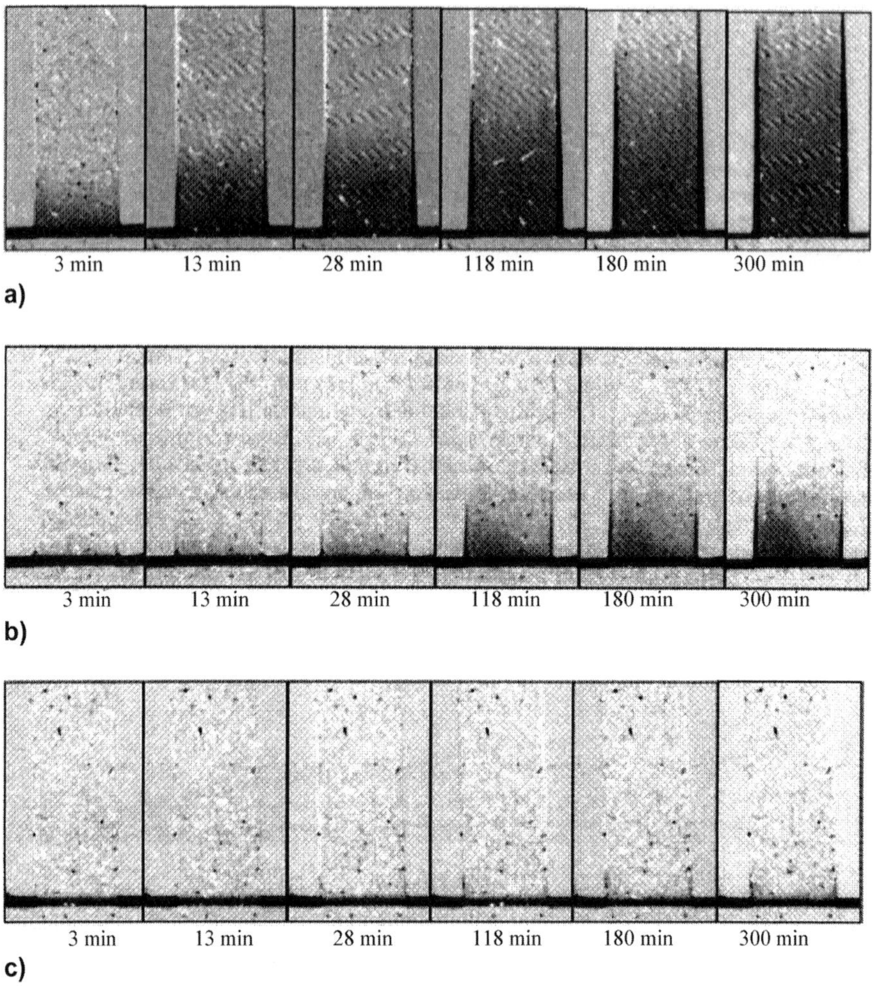

Fig. 7.11a–c. Neutron imaging of water absorption in different samples as a function of time. **a** Beech; **b** pine; **c** particleboard. (Lehmann et al. 2001, with permission)

cult to match with other materials, when architectural considerations are taken into account. The importance of a proper connection design cannot be overstated (Natterer et al. 2000).

On the connections, the effect of weathering may be undesirable, depending on the requirements for a particular wood product. Natural weathering can produce on wood connections a nonuniform moisture content distribution that induces shrinkage and swelling. The accompanying stress thus induced by shrinkage and swelling causes excessive dimensional changes, such as warping, surface roughening and checking, loss of some surface fibers, etc. The physi-

Table 7.3. Specimen characteristics for neutron radiography. (Data from Niemz et al. 2000c, with permission)

Sample	Type of joint	Material	Connection	Joint opening gap size (mm)
1	Obtuse joggler	Spruce/spruce	Obtuse meeting, doweled	0.1–0.4
2	Joint, plug	Spruce/spruce	Slot-mortise joint	0.1–1.0
3	Joint, doweled	Spruce/spruce	Bevel, doweled	0–0.4
4	Obtuse insert, glued	Spruce/spruce	Obtuse insertion, glued	0.30–5
5	Obtuse insert, glued	Spruce/beech	Obtuse insertion, glued	1–1.5

cal deterioration of wood during weathering depends on wood species, density, growth annual rate, ring orientation, proportion of latewood, etc. The weathering process is usually accompanied by fungal attack.

During artificial weathering, qualitative and quantitative nondestructive estimation of moisture content distribution in real size window connections is possible by using neutron radiography, as demonstrated by Niemz et al. (2000c). For this purpose, different types of joints have been used, as can be seen from Table 7.3. For producing structural elements, a single species (spruce) or two different species (spruce and beech) were used. The dimensions of the samples were 200 mm length, 50 mm width and 30 mm thickness. The specimens were exposed to different cyclic conditions of relative humidity and temperature as for the first cycle, 120 h at 25% relative humidity and 30 °C followed by the second cycle of 120 h at 99% relative humidity and 30 °C. After the second cycle, the samples were conditioned for 24 h at 50 °C and 2% relative humidity. To induce maximum water absorption in the joints, and to simulate rain, the specimens were submerged in water for 2 h. Radiographic images were taken of each sample in dry and moisturized stages, as can be seen from Fig. 7.12a for the spruce/spruce doweled corner connection and in Fig. 7.12b for an obtuse insertion (spruce/beech, glued). For all samples, a visible gap was observed, ranging from 0.1 to 1.5 mm. The profile of the moisture content as a function of the geometry of the sample was determined (Fig. 7.13). Water absorption is higher in the transverse section of the spruce specimen. Moisture content is increasing continuously in the spruce up to the joint. In beech, the moisture content seems to be constant and no absorption was observed in the transverse section.

Performance of window connections can be optimized with neutron imaging and can be evaluated at three levels – short-term, long-term and product quality assurance. In the short term, the designer must take into consideration the properties of wooden connections (moisture level, temperature, strength and rigidity of the members). The long-term performance of the connections is equated with the ability of the joint and windows to withstand the

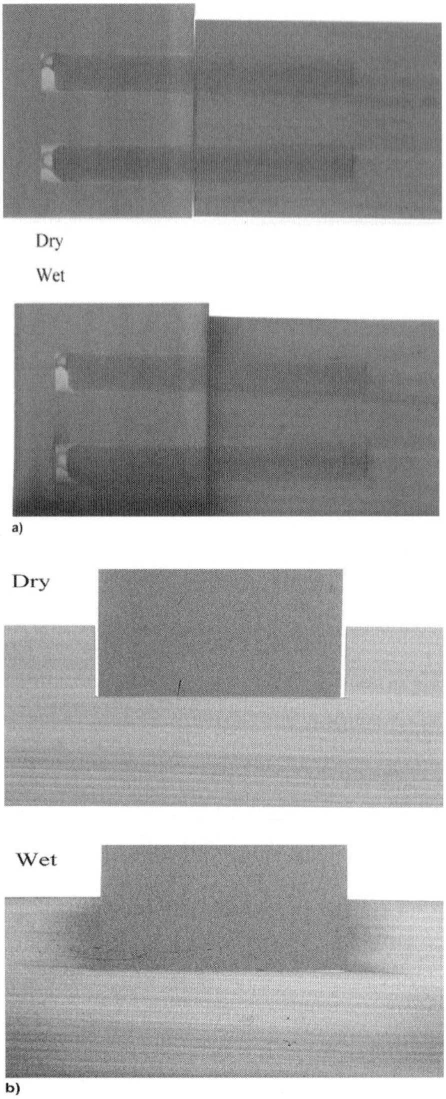

Fig. 7.12a,b. Moisture content distribution in corner connections for dry and moisturized samples. **a** Spruce/spruce doweled. **b** Obtuse insertion spruce/beech, glued. (Niemz et al. 2000c, with permission)

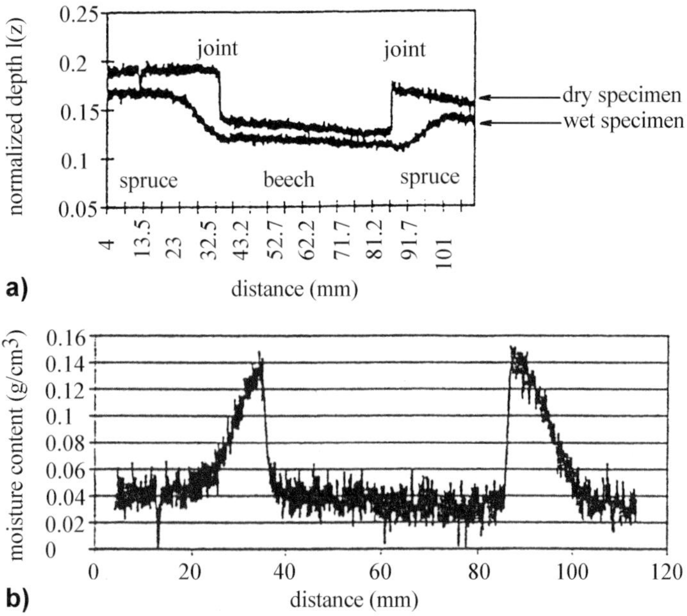

Fig. 7.13. Moisture distribution profile as a function of the geometry of the joint for obtuse insertion (spruce/beech). **a** Normalized depth as a function of distance (length of the specimen). **b** The corresponding moisture content as a function of distance. (Niemz et al. 2000c, with permission)

degradation induced by an increasing rate of moisture level, temperature, the presence of microorganisms, the mechanical level, etc. The maintenance of the manufacturing process to assure that the product will perform to a certain quality level is the major concern of the product quality assurance, in detecting unacceptable connectors and products, and in detecting the cause and correcting it.

7.5 Summary

Neutron beam imaging of wood structure is one of the most recent nondestructive techniques developed for this material. The neutrons are subatomic particles with zero charge. They are the major constituents of nuclei and are produced in fission reactors. Neutrons in thermal equilibrium with a moderator material near room temperature are called thermal neutrons. Neutron radiography of wood provides a specific hydrogen image of high resolution (15 μm) and allows the imaging of water distribution in wood and in living wood cells. The interaction of neutrons with wood is weak, and the absorption is small and, consequently, large samples can be inspected. Today, the applications of neutron imaging to specific problems of wood science and technology are still very scarce because of cost limitations.

This chapter is concerned with several applications of neutron radiography for the imaging of water distribution in wood at microscopic and macroscopic levels. Water distribution imaging in disk-type specimens recovered from living trees was obtained by irradiation with a thermal neutron flux of 8.7×10^7 n/cm^2 for 19 s. The degree of brightness of the image is proportional to the water content of the sample. The kinetics of the drying process in lumber can also be observed. The method was successful for the imaging of moisture content distribution in corner connections of wood elements of real size.

References

Abragam A (1961) Principles of nuclear magnetism. Clarendon Press, Oxford, 599 pp

American Society for Testing and Materials (1986) Standard test for loss characteristics and dielectric constant of solid electrical insulating materials. ASTM D, Philadelphia, pp 150–174

Andersen AH, Kak AC (1984) Simultaneous algebraic reconstruction technique (SART): a superior implementation of the art algorithm. Ultrasonic Imaging 6:81–94

Anfodillo T, Sigalotti GB, Tomasi M, Semenzato P, Valentini R (1993) Application of a thermal imaging technique in the study of the ascent of sap in woody species. Plant Cell Environ 16:997–1001

Araujo CG, MacKay AL, Hailey JRT, Whittall KP (1992) Proton magnetic resonance techniques for characterization of water in wood; application to white spruce. Wood Sci Technol 26: 101–113

Arfvidsson J, Lindgren O, Wiberg P (1997) The development of a moisture sorption model to predict moisture distribution within solid wood based on nondestructive moisture gradient measurements using CT-scanning and digital image processing. Int Conf COST Action E 8 Theme: Wood – Water Relations Copenhagen Denmark, pp 107–120

Arita K, Kuratanik L (1984) Wooden pole tester for determining the strength of decayed wooden poles. Jpn Telecommun Rev 26 3:167–173

Asplund T, Johansson LG (1984) Feasibility study of X-ray computerized tomography in research and development for wood-mechanical industry and forestry. (Forstudie-datortomograf for trateknisk och skoglig forskning och utveckling). Trateknik Rapport, Sweden, no 53, Svenska Traforskningsinstitutet, A no 904, Stockholm, 35 pp

Bacon GE (1975) Neutron diffraction. Oxford Univ Press, London, 426 pp

Badel E (1999) Determination of elastic and shrinkage properties of an annual ring of oak in transversal plan: description of the morphology, measurements of microscopic properties and homogenisation calculations. (Determination des propriétés elastiques et du retrait d'un cerne annuel de chêne dans le plan transverse: description de la morphologie, mesure des propriétés microscopiques et calcul d'homogeneisation). Thèse de Doctorat à l'ENGREF Nancy, France, 176 pp

Bahr AJ (1982) Microwave nondestructive testing. Gordon and Breach Science Publishers, New York, 86 pp

Bahyl V (1998) Tomography of the internal structure of wood. In: Kurjatko S, Kudela J (eds) Wood structure and properties. Arbora Publishers, Zvolen, pp 55–58

Baker JR, Mitchell PL, Cordey RA, Settle JJ (1994) Relationship between physical characteristics and polarimetric radar backscatter for Corsican pine stands in Thetford Forest UK. Int J Remote Sensing 15:2827–2849

Balageas DL, Levesque P, Nacitas M, Krapez JC, Gardette G, Lemistre M (1996) Microwave holography revealed by photothermal films and lock-in IR thermography: application to electromagnetic materials NDE. Proc SPIE 2944, pp 55–66

Barducci I, Pasqualini G (1948) Misura dell' attrito interno delle costanti elastiche del legno. (Measurement of the internal friction and of the elastic constants of wood). Nuovo Cimento 5: 416–502

Barnes D, Admirall L, Pike RL, Mathur VNP (1976) Continuous system for the drying of lumber with microwave energy. For Prod J 26 5:31–42

Bauer G (1998) Operation and development of the new spallation neutron source SINQ at Paul Scherrer Institut. Nuclear Instrum MethodsPhys Res B 139:65–71

Bauer G, Dai Y, Wagner W (2001) SINQ, Layout, operation applications and R&D to high power. Proc Symposium Structural materials for hybrid systems: a challenge in metallurgy, Paris, 29–31 October, 6 pp

Bauer M, Guntrum C, Ota M, Rippel W, Busse G (1992) Thermographic characterization of defects and failure in polymer composites. In: Balageas D, Busse G, Carlomagno GM (eds) Proc Quantitative Infrared Thermography (QIRT92). Editions Européennes Thermiques et Industrie, Paris, pp 141–144

Beall F (1996) Future of nondestructive evaluation of wood and wood-based materials. Holzforsch Holzverwert 5:73–75

Beatty R, Choffel D, Charpentier P (1999) Towards products tracking into wood industries. Proc IWMS:325–334

Beaudoin JL, Merienne R, Danjoux R, Egee M (1985) Numerical system for infrared scanners and application to the subsurface control of materials by photothermal radiometry. Infrared Techn Appl SPIE 590:287

Bennet CA, Patty RR (1982) Thermal wave interferometry: a potential application of the photoacoustic effect. Appl Opt 21:49–54

Benson-Cooper DM, Knowles RL, Thomson FJ, Cown DJ (1982) Computed tomographic scanning for the detection of defects within logs. FRI Bull For Res Inst N Z 8:9

Beravs K, Oven P, Sersa I, Torelli N, Demsar F (1998) Electric current density imaging of pedunculate oak (*Quercus robur* L.) twigs by magnetic resonance imaging. Holzforschung 52: 541–545

Beres C, Fenyvesi A (1994) The role of computer tomographical methods in investigation of water transport in oak trees. Proc 1st Eur Symp on Nondestructive Evaluation of Wood. University of Sopron, Sopron, Hungary, 525

Berglind H, Dillenz A (2000) Detection of glue deficiency in laminated wood with thermography. 12th Symp Nondestructive Testing of Wood, University of Western Hungary, Sopron, pp 413–420

Bernard P, Chalebois A, Pearson N, Samson M (1996) Application of infrared ellipsometry to detection of local grain organization and spiral grain in wood. 10th Symp NDT of Wood. Press Univ Romandes, Lausanne, pp 165–174

Bernatowicz G (1994) Determination of moisture distribution and check formation by means of thermography. (Bestimmung des Feucheverteilung und Rissbildung mittels Thermographie). Ann Warsaw Agric Univ For Wood Technol 45:65–69

Berndt H, Johnson GC (1994) Examination of wave propagation in wood from a microstructural perspective. In: Thompson DO, Chimenti DE (eds) Review of progress in quantitative nondestructive evaluation, vol 14b. Plenum Press, New York, pp 1661–1668

Berndt H, Schniewind AP, Johnson GC (1999) High resolution ultrasonic imaging of wood. Wood Sci Technol 33:185–198

Berndt H, Schniewind AP, Johnson GC (2000) Ultrasonic energy propagation through wood: where, when how much. Proc 12th Symp Nondestructive Testing of Wood. University of Western Hungary, Sopron, pp 57–65

Bhardwaj MC (2000) High transduction piezoelectric transducers and introduction of noncontact analysis. In: Harvey JA (ed) Encyclopedia of smart materials. Wiley, New York, 25 pp

Bhardwaj MC, Neeson I, Langron ME, Vandervalk E (2000) Contact free ultrasound: the final frontier in nondestructive materials characterization. Paper at 24th Conference: an International conf on engineering ceramics and structures. 24–27 Jan 2000, The American Ceramic Society, Cocoa Beach, FL, USA

Biagi E, Gatteschi G, Masotti L, Zanini A, Cerofolini M, Lorenzi A (1994) Tomografia ad ultrasuoni per la caratterizzazione difettologica del legno. (Ultrasonic tomography for defect characterization in wood). Alta Frequ Riv Elettronica 6 2:48–57

Biernacki JM, Beall FC (1993) Development of an acousto-ultrasonic scanning system for nondestructive evaluation of wood and wood laminates. Wood Fiber Sci 25:289–297

Birkeland R (1985) The status of tomographic scanning as a tool for detecting internal log defects. Proc 5th Nondestructive Testing of Wood Symposium, Washington State University. Eng Publications, Pullman, pp 231–237

Birkeland R (1990) NDE facilities and research program in Norway. 7th Symp on Nondestructive Testing of Wood, Washington State University, Madison, pp 57–62

Birsk AS (1972) Particleboard blow detector. For Prod J 22 6:23–26

Bodig J (1994) NDE of wood in North America. Concept and applications. Proc First Eur Symp on Nondestructive Evaluation of Wood. University of Sopron, Sopron, Hungary, pp 1–13

Bodig J, Jayne BA (1982) Mechanics of wood and wood composites. Van Nostrand Reinhold, New York, 712 pp

Bolomey JC, Pichot C (1990) Microwave tomography: from theory to practical imaging systems. Int J Imag Syst Technol 2:144–156

Bourbié T, Coussy O, Zinszner B (1987) Acoustics of porous media. Gulf Publ Co, Houston, TX, 334 pp

Bowyer JL (2000) Wood science in a changing world. Where are we headed? Wood Sci Technol 34:175–181

Brown DP, Pratum TK, Bledsoe C, Ford ED, Cothern JS, Perry D (1991) Nonivasive studies of conifer roots: nuclear magnetic resonance imaging of Douglas fir seedlings. Can J For Res 21: 1559–1566

Brown JH, Davidson RW, Skaar C (1963) Mechanism of electrical conduction of wood. For Prod J Oct:455–459

Brownstein KR (1980) Diffusion as an explanation of observed NMR behavior of water absorbed on wood. J Magn Reson 40:505–510

Buckley J (2000) Air-coupled ultrasound – a millennial review. 15th World Conf On NDT, Roma http://www.ndt.net/article/wcndt00/paper/idn507/idn507.htm

Bucur V (1980) Anatomical structure and some acoustical properties of resonance wood. Catgut Acoust Soc Newsl, series I, no 33:24–29

Bucur V (1986) Les termes non diagonaux de la matrice des rigidités du bois. (Off-diagonal terms of stiffness matrix of wood) Holzforschung 40:315–324

Bucur V (1990) NMR imaging of green wood. INRA Document a distribution limité, Institut National de la Recherche Agronomique, Champenoux, France, 10 pp

Bucur V (1992) Anisotropy characterization of structural flakeboards with ultrasonic methods. Wood Fiber Sci 24:337–346

Bucur V (1995) Acoustics of wood. CRC Press, Boca Raton, 284 pp

Bucur V, Archer RR (1984) Elastic constants of wood by an ultrasonic method . Wood Sci Technol 18:255–265

Bucur V, Rasolofosaon PNJ (1998) Dynamic elastic anisotropy and nonlinearity in wood and rock. Ultrasonics 36:813–824

Bucur V, Rocaboy F (1987) Surface wave propagation in wood: prospective method for the determination of wood off-diagonal terms of stiffness matrix. Ultrasonics 26:344–347

Bucur V, Lanceleur P, Rogé B (2001) Acoustic properties of wood in tridimensional representation of slowness surfaces. Ultrasonic International Symp Delft, p 6

Burov VA, Rumyantseva OD (1996) Linearized inverse problem of scattering in monochromatic and pulse modes. Acoust Phys 40:34–42

Busse G (1979) Optoacoustic phase angle measurement for probing a metal. Appl Phys Lett 35: 759–760

Busse G (1988) Imaging with optically generated thermal waves. Physical acoustics, vol XXII. Academic Press, New York, pp 403–478

Busse G (1994) Nondestructive evaluation of polymer materials. NDT Eval Int 27 5:253–262

Busse G (2001) Lockin thermography. In: Maldague X (ed) Nondestructive testing handbook, vol 3. Infrared and thermal testing. American Society for Nondestructive Testing Inc, Columbus, OH, pp 318–327

Busse G, Bauer M, Rippel W, Wu D (1992a) Lockin vibrothermal inspection of polymer composites. In: Balageas D, Busse G, Carlomagno GM (eds) Proc Quantitative Infrared Thermography (QIRT92). Editions Europpé1ennes Thermiques et Industrie, Paris, pp 154–159

Busse G, Wu D, Karpen W (1992b) Thermal wave imaging with phase sensitive modulated thermography. J Appl Phys 71 8:3962–3965

Byrne GF, Fenn MD, Burgar MI (1986) Nuclear magnetic resonance studies of water in tree sections. Agric For Meteorol 38:307–317

Callaghan PT (1991) Principles of nuclear magnetic microscopy. Oxford University Press, Oxford, 492 pp

Callaghan PT (1995) Pulsed-gradient spin-echo NMR of planar cylindrical and spherical pores under conditions of wall relaxation. J Magn Reson A 113:53–59

Carlomagno GM, Berardi PG (1976) Unsteady thermographyy in nondestructive testing. Proc 3rd Biennal Exchange, St Louis, pp 33–39

Carr HY, Purcell EM (1954) Effects of diffusion on free precession in nuclear magnetic resonance experiments. Phys Rev 94:630–638

Castagnoli A, Marchetti M, Bottai L, Chirici G, Oradini A (1997) Using a hyperspectral MIVIS (Multispectral Infrared/Visible imaging Spectrometer) to map forest cover at medium and large scales. Genio-Rurale 60 (9):54–61

Castera P (1998) How far can we control the heterogeneity of wood and wood based materials through design? In: Frantziskonis GN (ed) Proc PROBAMAT – 21st century: probability and materials. Test, models and applications for the 21st century. Kluwer, Dordrecht, pp 65–72

Catalano M, Catena G, Palla L (1986) Infrared technique for some pathological phenomena in trees. (Impiego dell' infrarosso per l'individuazione di alcuni fenomeni morbosi negli alberi) Agric Ricerca 59:7–16

Catena G (1989) Thermography for the estimation of trees stability (Impiego della termografia nel processo di valutazione della stabilita delle piante). Atti de la Fondazione Giorgio Ronchi LIII:755–766

Catena G (1991) Remote sensing in forestry (Telerilevamento nel settore forestale). Monti Boschi 1:17–24

Catena G (1992) Application of thermography in phytopathology. Diagnosis of sanitary state of ornamental trees. (Une application de la thermographie en phytopathologie. Diagnostic de l'é1tat sanitaire d'arbrres d'ornement). Phytoma – La dé1fense des vé1gé1taux 439:46–48

Catena G (1993a) Use of the thermal scanner to diagnose cavities in trees. (Utilizzazione di uno scanner termico per la diagnosi di cavita negli alberi). Riv Ing Agraria 24 2:67–75

Catena G (1993b) La termografia nelle indagini fitosanitarie di insiemi arborei. Economia Montana. Linea Ecol XXV 3:51–56

Catena G (1997) A tree seems to be very damaged it must be harvested or conserved? The thermography helps for a right decision. (Un albero appare molto dannegiato: lo si deve abbattere o conservare? La tomografia aiuta a prendere to decisione giusta). Economia Montana. Linea Ecol XXIX 1:9–14

Catena G, Catena A (2000) Termography for the evaluation of cavities and pathological tissues in trees. (Evidenziazione mediante la termografia di cavita e tessuti degradati negli arberi). Agric Ricerca 185:47–64

Catena G, Palla L, Catalano M (1990) Thermal infrared detection of cavities in trees. Eur J For Pathol 20:201–210

Catena G, Palla L, Catalano M, Mazzola S (1995) Indagine campione sullo stato del verde urbano di Roma. Seconda nota. Agric Ricerca 105:11–20

Chang BY (1975) Investigation of the nondestructive determination of wood moisture content by neutron – capture gamma radiation. Diss Abstr Int B 36 (4):1533

Chang CTP, Watson AT, Edwards CM (1999) NMR imaging of fluids and flow in porous media. In: Wang P-Z (ed) Methods in the physics of porous media, vol 35. Academic Press, San Diego, pp 387–423

Chang SJ, Cohen M, Wang PC (1989a) Ultrafast scanning of hardwood logs with a NMR scanner. Proc 3rd Int Conf On Scanning Tech, Sawmilling
Chang SJ, Olson JR, Wang PC (1989b) NMR imaging of internal features in wood. For Prod J 39 6:43–49
Charles JA, Wilson DW (1981) A model of passive thermal nondestructive evaluation of composite laminates. Polym Composites 2 3:105–111
Chazelas JL (1991) Caractéristiques physiques et mécaniques locales du bois dans la zone des noeuds. Thèse de Doctorat Université Blaise Pascal, Clermont Ferrand, 156 pp
Chazelas JL, Vergne A, Bucur V (1988) Analyse de la variation des propriétes physiques et mécaniques locales du bois autour des nœuds. (Wood local properties variation around knots). Actes du Colloque "Comportement Mécanique du Bois". GS Rhéologie du Bois, Bordeaux, France, pp 344–347
Chelidze T, Gueguen Y, LeRavalec M (1998) From classic fractal mechanics of disordered media: self-consistency versus self-similarity. In: Frantziskonis GN (ed) Proc PROBAMAT - 21st century: probability and materials. Test, models and applications for the 21st century. Kluwer, Dordrecht, pp 197–231
Chiorescu S, Grönlund A (2000) Validation of a CT-based simulator against a sawmill yield. For Prod J 50 6:69–76
Choffel D (1999) Automation of wood mechanical grading. Coupling of vision and microwave devices. SPIE 3836:114–121
Choffel D, Martin P (1996) Microwaves and vision device for mechanical grading. Proc 10th Symp Nondestructive Testing of Wood, Press Polytechniques et Universitaires Romandes, Lausanne, pp 331–339
Choffel D, Goy B, Gapp D (1992) Interaction between wood and microwaves automatic grading application. Workshop on scanning technology and image processing on wood. Lulea University of Technology, Lulea, Sweden, pp 1–8
Choffel D, Martin P, Gross P (1995) Nondestructutive testing methods on solid timber: microwaves, ultrasound and spectrometry. Proc Int Conf Progress in Forest Products Research, Göttingen, pp 86–89
Choong ET (1963) Movement of water through a softwood in the hygroscopic range. For Prod J 13:489–498
Cielo P, Krapez JC, Lamontagne M (1988) Lumber moisture evaluation by a reflective cavity photothermal technique. Rev Phys Appl 23:1565–1576
Coates ER, Chang SJ, Liao TW (1998) A quick defect detection algorithm for magnetic resonance images of hardwood logs. For Prod J 48 10:68–74
Cohen MH, Mendelson KS (1982) Nuclear magnetic resonance and the internal geometry of sedimentary rocks. J Appl Phys 53 2:1127–1135
Cole-Hamilton DJ, Chudek JA, Hunter G, Martin CJM (1990) NMR imaging of water in wood including water-logged archaeological artefacts. J Inst Wood Sci 12:111–113
Coles ME (1999) X-ray imaging. In: Wong P-Z (ed) Experimental methods in the physical sciences, vol 35. Academic Press, San Diego, pp 301–336
Comino E, Socco V, Martinis R, Nicolotti G, Sambuelli L (2000) Ultrasonic tomography for wood decay diagnosis. In: Backhauss GF, Bader H, Idezak E (eds) Int Symp Plant Health in Urban Horticulture. Mitt. Bundesanst Land-Forstwirtschaft, Braunschweig, p 279
Cormack AM (1963) Representation of a function by its line integrals with some radiological applications. Part I: J Appl Phys 34:2722–2727; Part II: J Appl Phys 35:2908–2913
Corral CZ, Lindquist CS (1998) On implementing Kasa's circle fit procedure. IEEE Trans Instrum Meas 47:789–795
Cown DJ, Clement BC (1983) A wood densitometer using direct scanning with X-rays. Wood Sci Technol 17:91–99
Crawford CR, Kak AC (1982) Multipath artifacts in ultrasonic transmission tomography. Ultrasonic Imaging 4:234–266
Davis WC, Moschler Jr (1986) A direct scanning densitometer to measure density profiles in wood composite products. For Prod J 36, 11/12:82–86

Davis JR, Weels P, Morgan M, Shadbolt P (1989) A field portable X-ray CT pole scanner and CT log scanning. Proc of 7th Nondestructive Testing of Wood Symposium, Washington State University. Eng Publications, Pullman, pp 251–262

Davis JR, Lerdin A, Wells P, Ilic J (1991) X-ray microtomography of wood. J Int Wood Sci, 12,4: 259–261

Dean SR (1983) The Radon transform and some of its applications. Wiley, New York, 289 pp

Delsanto PP, Romano A, Scalerani M, Moldoveanu F (1998) Application of genetic algorithms to ultrasonic tomography. J Acoust Soc Am 104:1374–1381

Dench EC (1973) Advantages of microwave processing. Trans Int Microwave Power Inst 1:13–16

Denis F, Jossinet J, Gimenez G (1986) Tomographie acoustique: états, applicatiations et perspectives. Rev Acoust 79:21–25

Devaney AJ (1982) A filtered back propagation algorithm for diffraction tomography. Ultrasonic Imaging 4:336–350

Devaney AJ (1986) Reconstructive tomography with diffractive wave fields. Inv Prob 2:161–183

Dillenz A, Wu D, Breitrück K, Busse G (2000a) Lock-in thermography for depth resolved defect characterization. 15th World Congress on NDT, Roma, 6 pp, http://www.ndt.net/article/wcndt00/papers/idn311/idn311.htm

Dillenz A, Zweschper T, Busse G (2000b) Elastic wave burts thermography for NDE of subsurface features. Insight 42 12:815–817

Dines KA, Lytle RJ (1979) Computerized geophysical tomography. Proc IEEE 67 7:1065–1073

Divos F, Szegedi S, Raics P (1994) Local densitometry of wood by gamma back-scattering. Holz Roh Werkst 54:279–281

Domanus JC (1992) Practical neutron radiography. Kluwer, Boston

Douglas Mast T (1999) Wideband quantitative ultrasonic imaging by time-domain diffraction tomography. J Acoust Soc Am 106 6:3061–3071

Dreele von RB (1994) Neutron diffraction. In: Cahn RW, Hassen P, Kramer EJ (eds) Material sciences and technology. A comprehensive treatment, vol 2B. VCH, Weinheim, pp 562–609

Drobakhin O (2000) Homographic approach to multifrequency microwave methods of dielectric material NDE. 15th World Conf on Nondestructive Testing, Roma, 5 pp

Dubey YM, Deorani SC (1997) Dielectric properties of coniferous timbers at microwave frequencies. J Indian Acad Wood Sci 1:77–82

Dumoulin JP, de Belleval JF (1987) Flaw classification in composites using ultrasonic echography. In: Alippi A, Mayer W (eds) Ultrasonic methods in evaluation of inhomogeneous materials. Martinus Nijhoff, Dordrecht, pp 207–217

Elbez G (1990) Medical X-ray scanner applied to wood gluing. Centre Technique du Bois et de l'Ameublement, Paris, 5 pp

El-Rayes MA, Ulaby FT (1987) Microwave dielectric spectrum of vegetation. Part I: experimental observations. IEEE Trans Geosci Remote Sensing GE 25 5:541–549

Falls S, Chow T, Yong R, Hutchins D (1989) Acoustic emission analysis and ultrasonic velocity imaging in the study of rock failure. J Acoust Emission 8:S166–S169

Faust TD, Tang M, Bhandarkar S, Smith JW, Toliner EW (1996) Evaluation of growth related features in selected hardwoods logs using X-ray computed tomography and image analysis. Proc 10th Symp Nondestructive Testing of Wood, Presses Polytechniques et Universitares Romandes, Lausanne, pp 201–208

Fengel D (1991) Aging and fossilization of wood and its components. Wood Sci Technol 25: 153–177

Feraz ES de B (1976) Wood density determination by attenuation of low-energy gamma radiation. IPEF Piracicaba Bresil 12:61–68

Filippini B, Lambert P, Martin P, Mallick G, Leban JM (1990) Research on wood properties and defects with microwave probe (Investigation des propriélréls du bois et des singularitéls par capteur micro-ondes). Actes du 3-é2me Colloque. Sciences et Industries du Bois. ARBORA, Bordeaux Tome I, pp 113–121

Fink M (1983) Ultrasonic imaging (L'imagerie ultrasonore). Rev Phys Appl 18:527–556

Fioravanti M, Ricci R (1991) L'impiego della tomografia computerizzata per misure densitometriche sul legno: indagine sperimentale e risultati metodologici. (Use of computed tomography for densitometric measures on wood: experimental researches and methodological results). Ann Acad Ital Sci For XL:1–35

Flibotte S, Menon RS, MacKay AL, Hailey JRT (1990) Proton magnetic resonance of western red cedar. Wood Fiber Sci 22:362–376

Forrer JB, Funck JW (1998) Dielectric properties of defects on wood surfaces. Holz Roh Werkst 56:25–29

Fortunko C, Renken M, Murray A (1991) Examination of objects made of wood. Proc IEEE Ultrasonic Symp, pp 1099–1103

Frantziskonis GN, Blodgett MP (1998) Multiscale material characterization and applications. In: Frantziskonis GN (ed) Proc PROBAMAT – 21st century: probability and materials. Test, models and applications for the 21st century. Kluwer, Dordrecht, pp 367–378

Frick J, Sawade G, Schuh H (1997) Defect localization in pillars of marble by ultrasonic tomography. Otto Graf J 8:174–188

Fridel MJ, Tweeton DR, Jackson Jessop JA, Billington S (1992) Mining application of seismic tomography. 62nd Annu Int Meet Soc Exploration Geophys BG1 7, pp 58–62

Fukada E (1965) Piezoelectric effect in wood and other crystalline polymers. 2nd Symp NDT of wood, Washington State University, Pullmann, pp 143–170

Fukada E, Yasuda S, Kohara J, Okamoto H (1956) Dynamic Young's modulus and piezoelectric constants of old timber. Bull Kabayasi Inst Phys Res 6:104–107

Funt BV, Bryant EC (1985) A computer vision system that analyses CT-scans of sawlogs. Proc of IEEE Computer Soc Conf on Computer vision and pattern recognition, San Francisco, CA, 9–13 June, pp 175–177

Funt BV, Bryant EC (1987) Detection of internal log defects by automatic interpretation of computer tomography images. For Prod J 37 1:56–62

Furukawa J, Nakanishi TM, Matsubayashi M (2001) Water uptake activity in soybean root revealed by neutron beam imaging. Nondestr Test Eval 16:335–334

Gadian DG (1982) Nuclear magnetic resonance and its applications to living systems. Clarendon Press, Oxford, 197 pp

Garrett DA (1977) The technological development of neutron radiography. Atomic Energy Rev 15:125–142

Garrett DA, Berger H (1977) The technological development of neutron radiography. Atomic Energy Rev 15, 2:125–142

Gierlik E, Dzbenski W (1996) Wood densitometry based on the radiometric methods. Proc 10th Symp Nondestructive Testing of Wood, Presses Polytechniques et Universitaires Romandes, Lausanne, pp 217–225

Gil AM, Neto CP (1999) Solid state NMR studies of wood and other lignocellulosic materials. Annu Rep NMR Spectrosc 37:75–117

Gilardi G, Abis L, Cass AEG (1995) Carbon 13 CP/MAS solid-state NMR and FT-IR spectroscopy of wood cell wall biodegradation. Enzyme Microbial Technol 17:268–275

Gilbert P (1972) Iterative method for the tridimensional reconstruction of an object from projections. J Theor Biol 36:105–117

Gilboy WB (1984) X- and γ-ray tomography in NDE applications. Nuclear Instrum Meth Phys Res 221:193–200

Gilboy WB, Foster J (1982) Industrial applications of computerized tomography with X- and gamma-radiation. In: Sharpe RS (ed) Research techniques in nondestructive testing, vol 6. Academic Press, London, pp 255–287

Giordano G (1971) Tecnologia del legno. Vol I. Materia prima. Unione Tipografico – Editrice Torinese, Milano, 1086 pp

Glos P (1982) Machine grading of sawn timber: current technology and comparison of methods. Holz Zentralbl 108:153–155

Golosovsky M, Lann AF, Davidov D, Frenkel A (2000) Microwave near-field imaging of conducting objects of a simple geometric shape. Rev Sci Instrum 71 10:3927–3932

Goy B, Martin P, Leban JM (1992) The measurements of wood density by microwave sensor. Holz Roh Werkst 50:163–166

Granier A, Anfodillo T, Sabatti M, Cochard H, Dreyer E, Tomasi M, Valentini R, Breda N (1994) Axial and radial water flows in trunks of oak trees: a qualitative and quantitative analysis. Tree Physiol 14:1383–1396

Greaves H (1998) Trends in wood research and utilization – a view from the institute of wood science, Australian branch. J Inst Wood Sci 14 6:293–300

Green RE Jr (1987) Ultrasonic nondestructive materials characterization. In: Vary A (ed) Materials analysis by ultrasonics. Noyes Data Corp, Park Ridge, USA, pp 1–29

Greenleaf JF (1981) Computerized transmission tomography. In: Marton L, Marton C (eds) Methods of experimental physics, vol 19. Ultrasonics. Academic Press, New York, pp 563–589

Grimberg R, Savin A, Lupu A, Rotundu C, Iancu L (2000) A method to determine the debonding zones in multilayer wood materials. 15th World Conference on Nondestructive Testing, Roma, 2000, 6 pp

Grönlund A, Grundberg S, Grönlund U (1994) The Swedish stem bank – an unique database for different silvicultural and wood properties. IUFRO S5.01–04 Workshop Proc, Hook, Sweden, pp 71–77

Grundberg S, Grönlund A (1997) Simulated grading of logs with an X-ray log scanner. Grading accuracy compared with manual grading. Scand J For Res 12:70–76

Grundberg S, Grönlung A, Grönlund U (1995) The Swedish Stem Bank – a data base for different silvicultural and wood properties. Res Rep TULEA [BR3] 1995:31. Lulea Unic Technol, Skelleftea, Sweeden, 14 pp

Grundberg S, Lindgren O, Grönlung A (1996) The development of an X-ray based scanner for log grading at full production speed. Proc 10th Symp Nondestructive Testing of Wood Swiss Federal Institute of Technology, Press Polytechniques et Universitaires Romandes, Lausanne, p 407

Guinn VP, Lukens HR (1965) Nuclear methods. In: Morrison GH (ed) Trace analysis: physical methods. Interscience, New York

Habermehl A(1982a) A new nondestructive method for determining internal wood condition and decay in living trees. I. Principles, method and apparatus. Arboric J 6:1–8

Habermehl A (1982b) A new nondestructive method for determining internal wood condition and decay in living trees. II. Results and further developments. Arboric J 6:121–130

Habermehl A, Ridder HW (1992a) Metodik der Computer – Tomographie zur zerstörungsfreien Untersuchung des Holzkörpers von stehenden Bäumen. (Computerized tomography applied to studies on wood of living trees). Holz Roh Werkst 50:465–474

Habermehl A, Ridder HW (1992b) Computer Tomographie am Baum (Computed tomograhy for tree). Materialprüfung 34:325–329; 357–360

Habermehl A, Ridder HW (1993) Anwendungen der mobilen Computer-Tomographie zur zerstörungsfreien Untersuchung des Holzkörpers von stehenden Bäumen (Application of mobile computed tomography for nondestructive investigations on wood of living trees) Holz Roh Werkst 51:1–6

Habermehl A, Ridder HW (1994) Applications of computerized tomography in forest and tree sciences. Int Symp on Computerized Tomography for Industrial Applications, 8–10 June, Berlin, pp 71–81

Habermehl A, Ridder HW (1995) Computerised tomographic investigations of street and park trees. Arboric J 19:419–437

Habermehl A, Ridder HW (1996) Computer Tomographie in der Forstwirtschaft und Baumpflege. (Teil 1 und 2) DGZfP/DACH – Zeitung no 55:48–55 and no 56:47–55

Habermehl A, Ridder HW (1998) γ-ray tomography in forest and tree sciences. Proc SPIE (Int Soc Optical Eng) 3149:234–244

Haegglund G, Lindgren O (1985) Maetning av densitets-och fuktvariationer i traematerial med hjaelp av datortomografi. (Measuring of density and moisture variations in wood with the help of computed tomography). Traeteknikrapport no 67. Stockholm Traeteknik Centrum, Stockholm, 21 pp

Hagman POG, Grundberg SA (1995) Classification of scots pine (*Pinus sylvestris*) knots in density images from CT scanned logs. Holz Roh Werkst 53:75–81

Hailey JRT, Menon RS, Mackay A, Burgess AE, Swanson JS (1985) Nuclear resonance scanning for log characterization. 5th Symp Nondestructive Testing of Wood, Washington State University, Pullman

Halabe UB, Ganga Rao H, Rao Hota V (1995) Nondestructive evaluation of wood using frequency domain analysis. In: Thompson DO, Chimenti DE (eds) Review of progress in quantitative nondestructive evaluation, vol 14. Plenum Press, New York, pp 1653–1660

Hall LD, Rajanayagam V (1986) Evaluation of the distribution of water in wood by use of three dimensional proton NMR volume imaging. Wood Sci Technol 20:329–333

Hall LD, Rajanayagam V, Stewart WA, Steiner PR (1986a) Magnetic resonance imaging of wood. Can J For Res 16:423–426

Hall LD, Rajanayagam V, Stewart WA, Steiner PR, Chow S (1986b) Detection of hidden morphology of wood by magnetic resonance imaging. Can J For Res 16:684–687

Halloin JM, Cooper TG, Potchen EJ, Vozzo JA (1994) Magnetic resonance imaging, a technology for noninvasive plant analysis. In: Research and applications of chemical sciences in forestry. USDA, GTR SO-104, New Orleans, Louisiana, 9 pp

Harley J, Morris PI (1988) Application of scanning and imaging techniques to assess decay and wood quality in logs and standing trees. Forintek Canada Corp no FO 42-91-42-(1988)E, 10 pp

Harpole GB, McDonald KA (1981) Investment opportunity: a scanning – ultrasonics cut stock manufacturing system. USDA, FPL, Research Paper 390, 7 pp

Hatcher PG, Berger IA, Earl WL (1981) NMR studies of ancient buried wood. I. Observations on the origin of coal to the brown coal stage. Org Geochem 3:49

Hattori Y, Kanagawa Y (1985) Nondestructive measurement of moisture distribution in wood with a medical X-ray CT scanner. Part I. Mokuzai Gakkaishi 31:974–982

Hauffe P, Mahler G (2000) Evaluation of internal log quality using X-ray and ultrasound. 12th Symp Nondestructive Testing of Wood, Sopron, pp 259–263

Hearmon RFS (1948) The elasticity of wood and plywood. Dept Sci Ind Res For Prod Res Spec Report no 7. HMSO, London, 87 p

Hearmon RFS (1965) The assessment of wood properties by vibrations and high frequency acoustic waves. 2nd Symp NDT, Washington State University, pp 49–66

Hearmon RFS, Burcham JN (1954) The dielectric properties of wood. For Prod Res Special Report no 8, Dept of Sci and Industry, London, 25 p

Henneke EG, Jones TS (1979) Detection of damage in composite materials by vibrothermography. ASTM Spec Tech Publ 696:83–95

Hennecke EG, Reifsnider KL, Stinchcomb WW (1979) Thermography – an NDE method for damage detection. J Metals Sept:11–15

Herman GT (1980) Image reconstruction from projections. The fundamentals of computerized tomography. Academic Press, London, 316 pp

Herman GT, Lent A, Rowland S (1973) Mathematics and applications: a report on the mathematical foundations and on applicability to real data of the algebraic reconstruction technique. J Theor Biol 43:1–32

Hippel von A (1961) Dielectric materials and applications. Wiley, New York, 302 pp

Hoag M, McKimmy MD (1988) Direct scanning X-ray densitometry of thin wood sections. For Prod J 38 1:23–26

Hodges DG, Anderson WC, McMillin CW (1990) The economic potential of CT scanners for hardwood sawmills. For Prod J 40 3:65–69

Hörig H (1935) Theory of elasticity of anisotropic solids applied to wood (Anwendung der Elastizitätstheorie anisotroper Körper auf Messungen an Holz). Ing Arch VI:8–14

Hounsfield GN (1972) A method and apparatus for examination of a body by radiation such as X-ray or gamma radiation. Patent Specification 1283915, The Patent Office, London

Hoyle RJ (1961) A nondestructive test for stiffness of structural lumber. For Prod J 11 6: 251–254

Hsi E, Hossfeld R, Bryant RG (1977) NMR relaxation study of water absorbed on milled North white cedar. J Colloid Interface Sci 62:389–395

Hutchins DA, Jansen DP, Edwards P (1993) Lamb wave tomography using noncontact transducers. Ultrasonics 31:97–103

Hutchins MT, Windsor CG (1987) Industrial applications. In: Sköld K, Price DL (eds) Neutron scattering, vol 23, part C. Academic Press, Orlando, pp 405–407

Iancu L, Iancu G, Grimberg R, Lupu A (2000) Quantification of defects in wood by use of ultrasonics in association with imagistic methods. 15th World Conf On NDT, Roma, http://www.ndt.net/article/wcndt00/paper

Ida N (1992) Microwave nondestructive testing. Kluwer, Amsterdam, 300 pp

Ida N, Wang JS (1996) Models for microwave nondestructive testing of materials. Mater Sci Forum 210–213:93–100

Jacquiot C, Trenard Y, Dirol D (1973) Atlas d'anatomie des bois des angiospermes. Tome 1. Centre Technique du Bois, Paris, 150 pp

Jain VK, Sanyal SN (1996) A preliminary note on the attenuation of microwaves by few wood species. Indian For 122, 5:428–431

James WL (1959) A method for rapid measurement of the rate of decay of free vibrations. USDA, FPL Bull no 2154, USDA, Madison, WI

James WL (1975) Dielectric properties of wood and hardboard: Variation with frequency, moisture content and grain orientation USDA For Serv Res Pap FPL-245, Madison, WI

James WL (1989) Electrical properties. In: Schniewind AP (ed) Concise encyclopedia of wood and wood-based materials. Pergamon Press, Oxford, pp 99–102

James WL, Hamill DW (1965) Dielectric properties of Douglas fir measured at microwave frequencies. For Prod J 15 2:51–56

James WL, Yen YS, King, RJ (1985) A microwave method for measuring moisture content, density and grain angle of wood. USDA For Serv Res Notes FPL-250, Madison, WI, 9 pp

Jartti P, Luukkala M (1977) Ultrasonic method for web speed measurements. Tappi J 60 11:167

Javadpour Z, Hughes D, Keating JG, Feeney FE, Evertsen JA (1996) Assessment of timber quality using neural networks on CT images. Proc 10th Symp Nondestructive Testing of Wood Swiss Federal Institute of Technology, Press Polytechniques et Universitaires Romandes, Lausanne, pp 303–311

Jayne BA (1955) A nondestructive test of glue bond quality. For Prod J 5 5:294–301

Jayne BA (1959) Vibrational properties of wood as indices of quality. For Prod J 9 11:413–416

Jayne BA (1972) Theory and design of wood and fiber composite materials. Syracuse University Press, Syracuse, NY, 418 pp

Johansson LG (1985) Experiences from using X-ray tomography, isotope-based nontomographic measuring and NMR for testing logs before sawing 5th Symp Non destructive testing of Wood, Washington State University, Pullman, pp 205–229

Kabir MF, Daud WM, Khalid K, Sidek HHA (1998) Dielectric and ultrasonic properties of rubber wood. Effect of moisture content, grain direction and frequency. Holz Roh Werkst 56:223–227

Kabir MF, Schmoldt DL, Schafer ME (2000a) Ultrasonic detection of knots, cross grain and bark pockets in wooden pallet parts. Proc World Conf Timber Eng 2000, Whistler, British Columbia, Canada, pp 7521–7528

Kabir MF, Schmoldt DL, Schafer ME (2000b) Detection of defects in red oak deckboards by ultrasonic scanning. Proc 4th Int Conf Image Proccesing and Scanning of Wood, Mountain Lake Resort, Virginia, pp 89–96

Kabir MF, Schmoldt DL, Schafer ME (2001) Roller transducer scanning of wooden pallet parts for defect detection. In: Thompson DO, Chimenti DE (eds) Review of progress in quantitative nondestructive evaluation, vol 20. Plenum Press, New York, pp 1218–1225

Kaestner AP, Baath L (2000) Microwave polarimetry based wood scanning. 12th Symposium on Nondestructive Testing of Wood, Sopron, pp 349–356

Kätzel R, Ridder HW, Habermehl A (1997) Investigations of ammonia stressed Scots pine trees (*Pinus sylvestris*) with computed tomography in an immission area around a cattle farm. Phyton 37 1:141–149

Kak AC (1979) Computerized tomography with X-ray emission and ultrasound sources. Proc IEEE 67:1245–1272

Kak, AC Slaney M (1988) Principles of computerized tomographic imaging. IEEE Press, New York, 329 pp

Kanagawa Y, Hattori Y (1985) Nondestructive measurement of moisture distribution in wood with a medical X-ray CT scanner. Part II. Changes in moisture distribution with drying. Mokuzai Gakkaishi 31:983–989

Kanagawa Y, Furuyama Y, Hattori Y (1992) Nondestructive measurement of moisture diffusion coefficient in wood drying. Drying Technol 10:1231–1248

Karsulovic JT, Leon LA, Dinator Mi (1999) The use of linear attenuation coefficients of gamma radiation for detecting knots in *Pinus radiata*. For Prod J 49 2:73–76

Kasischke ES, Melack JM (1997) The use of imaging radars for ecological applications. Remote Sensing Environ 59 2:141–156

Keller E, Thiercelin F (1975) The influence of big medulary rays on some properties of beech. (Influence de gros rayons ligneux sur quelques propriéltéls du bois de hé3tre). Ann Sci For 32 2:113–129

Kessler R (1987) Applicability of imaging radar for classification of forest vegetation. Photogrammetria 41 4:221–232

Khan MA, Rizvi TZ, Naheed A, Zaheer MY, Iqbal MZ, Shoaib S (2000) Dielectric observed consequences of microbial treatment. Holzforschung 54:335–339

Kharadly M (1985) Microwave diagnostic for stress rating of dimension lumber. Proc 5th Nondestructive Testing of Wood, Washington State University, Pullman, pp 445–464

Kim YS, Singh AP (2000) Michromorphological characteristics of wood biodegradation in wet environments: a review. IAWA J 21:135–155

King RJ (1978) Microwave electromagnetic nondestructive testing of wood. Proc 4th Nondestructive Testing of Wood, Washington State University, Pullman, pp 121–134

King RJ (2000) On-line industrial applications of microwave moisture sensor. In: Baltes H, Göpel W, Hesse J (eds) Sensor update, vol 7. VCH Verlag, Weinheim, pp 109–170

King RJ, Basuel JC (1993) Measurement of basis weight and moisture content of composite boards using microwaves. For Prod J 43 9:15–22

King RY, Yen YH (1981) Probing amplitude, phase and polarization of microwave field distributions in real time. IEEE Trans Microwave Theor Tech 29:1225–1231

Kinney JH, Johnson QC, Bonse U, Nichols MC, Saroyan RA, Nusshardt R, Pahl R, Brase JM (1988) Three dimensional X-ray computed tomography in material science. MRS Bull 31 1:13–17

Kittel C (1968) Introduction to solid state physics. Wiley, New York, 399 pp

Klein P, Vogel H (1993) Computertomography – a help for dendrochronology and wood identification? Proc 4th Int Conf on Nondestructive Testing of Works of Art, Berlin, pp 85–91

Kleinberg RL (1999) Nuclear magnetic resonance. In: Wang P-Z (ed) Methods in the physics of porous media, vol 35. Academic Press, San Diego, pp 337–385

Kliger IR, Perstorper M, Johansson G, Pellicane PJ (1995) Quality of timber products from Norway spruce. Part 3. Influence of spatial position and growth characteristics on bending stiffness and strength. Wood Sci Technol 29:397–410

Kline R, Sullivan C, Mignona RB, Delsanto PP (1996) A parallel processing algorithm for acoustic tomography. In: Tortoli P, Masotti L (eds) Acoustical imaging. Plenum Press, New York, pp 561–566

Kline RA, Wang YQ (1992) A technique for ultrasonic tomography in anisotropic media. J Acoust Soc Am 91:878–884

Kline RA, Wang YA, Mignogna RB, Delsanto PP (1994) A finite difference approach to acoustic tomography in anisotropic media. J Nondestr Eval 13:75–83

Koch M, Hunsche S, Schuacher P, Nuss MC, Feldmann J, Fromm J (1998) THz-imaging: a new method for density mapping of wood. Wood Sci Technol 32:421–427

Kollmann F (1951) Technology of wood and wood based composites (Technologie des Holzes und der Holzwerkstoffe). Springer, Berlin Heidelberg New York, 1048 pp

Kollmann FFP, Côté WA (1968) Principles of wood science and technology. Springer, Berlin Heidelberg New York, 592 pp

Kollmann FFP, Krech M (1960) Dynamic measurements of wood elastic properties and damping. (Dynamische Messung der elastischen Holzeigenschaften und der Dämpfung). Holz Roh Werkst 18 2:41–45

Kostorz G, Lovesey SW (1979) Neutron scattering. General introduction. Treatise Mater Sci Technol 15:1–67

Kucera LJ (1990) Current use of the NMR on wood at the Swiss Federal Institute of Technology: overview and outlook. 7th Int Symp on Nondestructive Testing of Wood, Washington State University, Pullman, pp 71–72

Kuo PK, Feng ZJ, Ahmed T, Favro LD, Thomas RL, Hartikainen J (1987) Parallel thermal wave imaging using a vector lock-in video technique. In: Hess P, Pelzl J (Eds) Photoacoustic and photothermal phenomena. Springer, Berlin Heidelberg New York, pp 415–418

Kuroda N, Suzuki Y (1996) Effects of morphological factors on the electrical behavior of wood with the presence of free water. Current behavior from voltage reversals and tracheary elements. Mokuzai Gakkaishi 42:817–824

Lambert JB, Frye JS, Carriveau GW (1991) The structure of oriental lacquer by solid state NMR spectroscopy. Archaeometry 33 87:93

Lambert JB, Shawl CE, Stearns JA (2000) Nuclear magnetic resonance in archeology. Chem Soc Rev 29 3:175–182

Lanceleur P, de Belleval JM, Mercier N (1998) Synthetic tridimensional representation of anisotropic materials. Acta Acust 84:1047–1054

Langwig JE, Meyer JA (1973) Ion migration in wood verified by neutron activation analysis. Wood Sci 6 1:39–50

Lau SK, Almond DD, Milne JM (1991) A quantitative analysis of pulsed video thermography. NDT Eval Int 24 4:195–202

Laufengerg TL (1986) Using gamma radiation to measure density gradient in reconstructed wood products. For Prod J 36 (2):59–62

Laurence GR, Banner A (1980) The application of thermography for locating potential frost pockets in forest cutovers. In: Proc of the 6th Canadian symposium on remote sensing, Halifax, Nova Scotia. Canadian Aeronautics and Space Institute, Halifax, Canada, pp 369–374

Lebowitz C, Zoughi R, Spicer JWM, Osinder R (1996) Comparison of ultrasonic, microwave and photothermal imaging of defective graphite-epoxy composite panels. Proc SPIE 2944:67–74

Lehmann E, Vontobel P, Niemz P (2001) Application of neutron radiography as method in the analysis of wood. (Anwendung der Methods der Neutronenradiographie zur Analyse von Holzeigenschaften). Holz Roh Werkst 59:463–471

Lehto A, Jaarinen J, Tiusanen M, Jokinen M, Luukkala M (1981) Amplitude and phase in thermal wave imaging. Electr Lett 17:364–365

Lewa CJ (1991) Magnetic resonance imaging in the presence of mechanical waves. Spectrosc Lett 24 1:56–67

Lewa CJ (1992) NMR detection of elastic wave effects. In: Alippi A (ed) Acoustic sensing and probing. World Science, Singapore, pp 435–438

Lewa CJ (1996) Elasto-magneitc resonance spectroscopy. Europhys Lett 35 1:73–76

Liezers M, Miller RM, Spillane DEM, Ryan TH (1985) Developments in thermal wave imaging. Proc Int Conf Acoustic Emission and Photoacoustic spectroscopy and applications, London, pp 65–69

Lin RT (1967) Review of the dielectric properties of wood and cellulose. For Prod J 17 7:61–66

Lin RT (1973) Wood as an orthotropic dielectric material. Wood Fiber 5 3:226–236

Lindberg H, Grahn O, Lindgren A, Heligren C (1996) Watter diffusion through acrylate latex paint films measured by computed tomography. Proc of 10th Nondestructive Testing of Wood Symposium. Swiss Federal Institute of Technology, Press Polytechniques et Universitaires Romandes, Lausanne, 408 p

Lindegaard-Andersen A, Notea A, Bushlin Y, Rheinlander J (1990) Image-centred frame of reference in rotational film-based X-ray tomography. NDT Int 23 6:332–334

Lindgren LO (1985) On the relationship between density / moisture content in wood and X-ray attenuation in computer tomography. Proc 5th Symp Nondestructive Testing of Wood; Washington State University, Eng Publications, Pullman, pp193–203

Lindgren O (1991a) Medical CAT-scanning: X-ray absorption coefficients, CT-numbers and their relation to wood density. Wood Sci Technol 25:341–349

Lindgren O (1991b) The accuracy of medical CT-scan images for nondestructive density measurements in small volume elements within solid wood. Wood Sci Technol 25:425–432

Lindgren O (1994) NMR for nondestructive wood moisture content measurement. First European Symp on NDE of Wood, University of Sopron, Sopron, pp 124–129

Lindgren O, Davis J, Wellq P, Shadbolt P (1992) Nondestructive wood density distribution measurements using computed tomography. Holz Roh Werkst 50:295–299

Liu CJ, Olson JR, Tian Y, Shen Q (1988) Theoretical wood densitometry: I. Mass attenuation equations and wood density models. Wood Fiber Sci 20:22–33

Loos WE (1965) Determining moisture content and density of wood by nuclear radiation techniques. For Prod J 15 3:102–106

Loos WE (1965) Neutron activation of Engelmann spruce. For Prod J 15 4:178

Luong PM (1995) Infrared scanning of failure process in wood. SPIE 2473:298–309

Luong PM (1996) Infrared thermography of damage in wood. 10th Symp NDT of Wood, Press Univ Romandes, Lausanne, pp 175–185

Luong PM (1998) Fatigue limit evaluation of metals using an infrared thermographic technique. Mech Mater 28:155–163

Luukkala M, Heikilla P, Surakka Y (1971) Plate-wave resonance. A contactless test method. Ultrasonics 9:201–208

Lyamshev ML, Stanullo J, Busse G (1995) Thermo acoustic vibrometry. Remote in situ monitoring of ceramic sintering. Materialprüfung 37 1/2:22–24

Lynnworth LC (1989) Ultrasonic measurements for process control. Theory, techniques, applications. Academic Press, Boston, 694 pp

Lynnworth LC, Magory V (1999) Industrial process control sensors and systems. In: Papadakis EP (ed) Ultrasonic instruments and devices. Academic Press, San Diego, pp 276–470

Lynnworth LC, Nguyen TH, Smart CS, Khrakovsky OK (1997) Acoustically isolated paired air transducers for 50, 100, 200 or 500 kHz applications. IEEE Trans UFFC 44 5:1087–1100

Lytle RJ, Dines K (1980) Iterative ray tracing between boreholes for underground image reconstruction IEEE Trans Geosci Remote Sens GE 18:234–240

MacFall JS, Johnson GA, Kremer PJ (1990) Observation of water-depletion region surrounding loblolly pine roots by magnetic resonance imaging. Proc Natl Acad Sci USA 87:1203–1207

MacGregor RP, Peemoeller H, Schneider MH, Sharp AR (1983) Anisotropic diffusion of water in wood. J Appl Polymer Sci Polymer Symp 37:901–909

Madsen B (1994) Radiological density scanning – a portable gamma camera based on backscatter tomography. Proc 9th Symp Nondestructive Testing of Wood, Washington State University, Madison, pp 131–137

Makoviny I (1988) The anisotropy of wood dielectric constants as a function of anatomical structure. (Zur Anisotropie der Dielektrizitätskonstanten des Holzes in grundlegenden anatomischen Richtungen). Holztechnologie 29 4:210–213

Malan FS, Marais PG (1992) Some notes on the direct gamma ray densitometry of wood. Holzforschung 46:91–97

Maloney TM (1992) Composites for the future. Invited paper. Proc IUFRO Conference, Division 5, Nancy, 605p p

Maloney TM (1996) The family of wood composite materials. For Prod J 46 2:19–26

Mandelis A (2000) Diffusion waves and their uses. Physics Today 53 No 8 Part I:29–34

Mansfield P, Morris PG (1982) NMR imaging in bio-medicine. In: Waugh J (ed) Advanced magnetic. Resonance Suppl no 2. Academic Press, New York

Marcinko JJ, Devathala S, Rinaldi PL, Bao S (1998) Investigating the molecular and bulk dynamics of PMDI/wood and UF/wood composites. For Prod J 48 6:81–84

Marcinko JJ, Rinaldi PL, Bao S (1999) Exploring the physicochemical nature of PMDI/wood structural composite adhesion. For Prod J 49 5:75–78

Marok M, Kundela J, Cunderlik I (1996) Identification of reaction beech wood by X-ray computed tomography. Holz Roh Werkst 54:97–98

Martin P, Collet R, Barthelemy P, Roussy G (1987) Evaluation of wood characteristics: internal scanning of the material by microwave. Wood Sci Technol 21:361–371

Martinis R (2002) Nondestructive techniques for decay diagnosis on standing trees. (Analisi di tecniche non destruttive per la diagnosi di carie su alberi in piedi). Thesis University of Florence, Florence, Italy, 300 pp

Masuch G, Franz JT, Marsmann H, Gross D, Kettrup A (1991) ^{1}H NMR micro-imaging and correlated SEM studies of spruce needles from healthy and declined forest sites. Int J Environ Anal Chem 45:179–191

Masuda M, Takahashi S (1998) Change of thermal images of lumber including knots, finger joints and metal connectors. Bull Kyoto Univ For (Kyoto Daigaku Nogaku Bu Enshurin Hokoku) 69:114–128

Masuda M, Takahashi S (1999) Thermographical detection of knots and other defects. Proc INCEUPT '99 For Prod Assoc ROC Bull 16:181–187

Masuda M, Takahashi S (2000) Thermographical detection of defects of lumber and glued joints. 12th Symp NDT of Wood, University of Western Hungary, Sopron, pp 421–430

Masuda M, Fujimoto K, Takino S, Sadoh T (1995) Change of thermal images of timber including knots under repeated bending. Bull Kyoto Univ For 67:167–173

McDonald KA (1978) Lumber defect detection by ultrasonics. Forest Prod Lab Res Paper FPL 311, USDA, Madison, WI, 25 pp

McGaughey WJ, Young RP (1990) Comparation of ART, SIRT, Least-Squares, and SVD two-dimensional tomographic inversions of field data. 60th Annual Meeting of Soc Exploration Geophysicists SEG, pp 74–77

McKeon JCP, Hinders MK (1999) Parallel projection and crosshole Lamb contact scanning tomography. JASA 106 5:2568–2577

McLauchlan TA, Norton JA, Kusec DJ (1973) Slope of grain indicator. For Prod J 23 5:50–55

McMillin CW (1982) Application of automatic image analysis to wood science. Wood Sci 14: 97–105

Meiboom S, Gill D (1958) Modified spin-echo method for measuring nuclear relaxation times. Rev Sci Instrum 29:688–691

Menon RS, MacKay AL, Flibotte S, Hailey JRT, Blook M, Burgess AC, Swanson JS (1987) A NMR determination of the physiological water distribution in wood during drying. J Appl Polymer Sci 33:1146–1155

Menon RS, MacKay AL, Flibotte S, Hailey JRT (1989) Quantitative separation of NMR images of water in wood on the basis of T(2). J Magn Reson 82 205–210

Millard P, Chudek JA (1993) Imaging the vascular continuity of *Prunus avium* petioles during leaf senescence using NMR spectroscopy. J Exp Bot 44:599–603

Miller WH (1988) Design and implementation of a portable computerized axial tomography system for field. Nucl Instrum Meth Phys Res 270:590–597

Moksin MM (1993) Improvement of an opto-thermal technique for wood. Wood Sci Technol 28: 53–58

Moschler WW Jr, Dougal EF (1988) Calibration procedure for a direct scanning densitometer using gamma radiation. Wood Fiber Sci 20:297–303

Moschler WW Jr, Winistorfer PM (1990) Direct scanning densitometry: an effect of sampling heterogeneity and aperture area. Wood Fiber Sci 22:31–38

Mothe F, Duchanois G, Zannier B, Leban JM (1998) Analyse microdensitométrique appliquée au bois. Ann Sci For 55:301–313

Murata K, Sadoh T (1994) Heat absorption and transfer in softwoods and their knot surfaces. Mokuzai Gakkaishi 40:1180–1184

Musial MW (1988) A nondestructive method for determining the degree of flake orientation in OSB. Wood Sci Technol 22:371–378

Naito S, Sawada Y, Fujii Y, Okumura S (1998) Thermography of wood specimens in static bending test. Kyoto Daigaku Nogaku Bu Enshurin Hokoku 69:104–113

Naito S, Fujii Y, Sawada Y, Okumura S (2000) Thermographic measurement of slope of grain using the thermal anisotropy of wood. Mokuzai Gakkaishi 46:320–325

Nakanishi TM (1994) The water image of plants using neutron radiography. Chem Regul Plants 29 2:182–190

Nakanishi TM, Matsubayashi M (1997) Nondestructive water imaging by neutron beam analysis in living plants. J Plant Physiol 151:442–445

Nakanishi TM, Watanabe S (1995) Water distribution inside the woods by neutron radiography. Bull Tokyo Univ For 93:13–19

Nakanishi TM, Karakama I, Sakura T, Matsubayashi M (1998a) Moisture imaging of a camphor tree by neutron beam. Radioisotopes 47:387–391

Nakanishi TM, Okano T, Karakama I, Ishihara T, Matsubayashi M (1998b) Three dimensional imaging of moisture in wood disk by neutron beam during drying process. Holzforschung 52:673–676

Nanassy AJ (1970) Overlapping of dielectric relaxation spectra in oven-dry birch at temperatures from 20 to 100 degrees C. Wood Sci Technol 4:104–121

Nanassy AJ (1972) Dielectric measurements of moist wood in a sealed system. Wood Sci Technol 6:67–77

Nanassy AJ (1973) Use of wide line NMR for measurement of moisture content in wood. Wood Sci 5:187–193

Nanassy AJ (1974) Water sorption in green and remoistured wood studied by the broad-line component of the wide line NMR spectra. Wood Sci 7:61–68

Natterer J, Sandoz JL, Rey M (2000) Bois matélriau et struture bois. Press Polytheqnique et Universitaire Romande, Lausanne, 488 pp

Neuenschwander J, Niemz P, Kucera LJ (1997) Studies for visualizing wood defects using ultrasonic in reflection and transmission mode. Holz Roh Werkst 55:339–340

Newmann RH (1992) NMR study of spatial relationship between components in wood cell walls. Holzforschung 46:207–210

Nicklow RM (1979) Phonons and defects. Treatise Mater Sci Technol 15:191–225

Nicolotti G, Miglietta P (1998) Using high-technology instruments to assess defects in trees. J Arbor 24 6:297–302

Niemienen AOK, Koenig JL (1988) NMR imaging – a promising new adhesive evaluation technique. J Adhesion Sci Technol 2:407–414

Niemz P (1989) Use of online moisture-measuring instruments in the woodworking industry: literature review (Einsatz von on-line-Feuchtemessgeräten in der holzverarbeitenden Industrie – eine Literaturübersicht). Holztechnologie 30 1:9–13

Niemz P, Sander D (1989) Mesuring techniques in wood industrie. Prozessmesstechnick in der Holzindustrie. VEB Fachbuchverlag, Leipzig, 288 pp

Niemz P, Kucera LJ, Flisch A, Blaser E (1997) Anwendung der Computertomographie an Holz. (Computerized tomography of wood). Holz Roh Werkst 55:279–280

Niemz P, Bodmer HC, Kucera LJ, Ridder HW, Habermehl A, Wyss P, Zürcher E, Holdenrieder O (1998a) Eignung verschiedener Diagnosemethoden zur Erkennung von Stammfäulen bei Fichte (Suitability of different methods for decay detection in Norway spruce). Schweiz.Z Forstwes 149:615–630

Niemz P, Kucera LJ, Ridder HW, Habermehl A, Flisch A (1998b) Durchblick mit Computertomographie (Transversal view with computer tomography). Wald Holz 9:7–10

Niemz P, Kucera LJ, Schob M, Scheffer M (1999) Experiments for defect detection in wood with ultrasound (Experimentelle Untersuchungen zur Erkennung in Holz mit Ultraschal). Holz Roh Werkst 57:96–102

Niemz P, Kucera LJ, Lehmann E, Vontobel P, Hanschke S (2000a) Investigation of wooden corner connection by neutron radiography methods in order to evaluate the ingress behavior of water. 12th Symp Nondestructive Testing of Wood, Sopron, p 463

Niemz P, Lehmann E, Vontobel P (2000b) Application of neutron radiography for investigation of wood. 12 th Symp Nondestructive Testing of Wood, Sopron, p 462

Niemz P, Lehmann E, Vontobel P, Haller P, Hanschke S (2000c) Investigation using neutron radiography for evaluations of moisture ingress into corner connections of wood. 12th Symp on Wood, University of Western Hungary, Sopron, pp 489–497

Nishino Y, Norimoto M (1990) Structure and anisotropy of dielectric constant in hardwood. Wood Resh Tech Notes Kyoto Univ 26:78–90

Norimoto M (1976) Dielectric properties of wood. Wood Res Kyoto 59/60:106–152

Norimoto M, Yamada (1972) The dielectric properties of wood. VI. On the dielectric properties of chemical constituents of wood and the dielectric anisotropy of wood. Wood Res Bull Wood Res Inst Kyoto Univ:31–43

Norimoto M, Zhao G (1993) Dielectric relaxation of water adsorbed on wood II. Mokuzai Gakkaishi 39:249–257

Norimoto M, Hayashi S, Yamada T (1978) Anisotropy of dielectric constant in coniferous wood. Holzforschung 32:167–172

Norton SJ, Linzer M (1980) Ultrasonic reflectivity imaging in three dimensions: exact inverse scattering solutions for plane, cylindrical and circular apertures. IEEE Trans Biomed Eng 28:202–220

Nyström J, Earl Kline D (2000) Automatic classification of compression wood in green Southern yellow pine. Wood Fiber Sci 32:301–310

Obataya E, Yokoyama M, Norimoto M (1996) Mechanical and dielectric relaxations of wood in a low temperature range. I. Relaxations due to methylol groups and adsorged water. Mokuzai Gakkaishi 42:243–249

Oja J, Temnerud E (1999) The appearance of resin pockets in CT-images of Norway spruce (Picea Abies (L) Karst). Holz Roh Werkst 57:400–406

Oja J, Grundberg S, Grönlund A (2000) Predicting the strength of sawn products by X-ray scanning of logs: a preliminary study. Wood Fiber Sci 32:203–208

Okumura S, Suzuki R, Fujii Y (1996) Preliminary study on thermography of wood under compression load. Kyoto Daigaku Nogaku Bu Enshurin Hokoku 68:161–169

Olek W, Guzenda R, Baranowska HM, Olszewski J (1994) On possible application of nuclear magnetic relaxation measurements to investigation of water penetration into wood. First European Symp on NDE of Wood, University of Sopron, Sopron, pp 130–137

Olek W, Guzenda R, Weres J (2000) Computed-aided estimation of thermal conductivity of Scots pine wood. 12th Symp NDT of Wood, University of Western Hungary, Sopron, pp 431–438

Olson JR, Arganbright DG (1981) Prediction of mass attenuation coefficients of wood. Wood Sci 14:86–90

Olson JR, Liu CJ, Tian Y, Shen Q (1988) Theoretical wood densitometry: II. optimal X-ray energy for wood density measurement. Wood Fiber Sci 20:187–196

Olson JR, Chang SJ, Wang PC (1990) Nuclear magnetic resonance imaging: a noninvasive analysis of moisture distributions in white oak lumber. Can J For Res 20:586–591

Onoe M, Tsao JW, Tamada H, Nakamura H, Kogure J, Kawamura H, Yoshimatsu M (1984) Computed tomography for measuring the annual rings on a live tree. Nuclear Instruments and Methods in Physics Research 221:213–220

Pang S (1996) Moisture content gradient in a softwood board during drying: simulation from 2-D model and measurement. Wood Sci Technol 30:165–178

Pang S, Wiberg P (1998) Model predicted and CT scanned moisture distributed in a *Pinus radiata* board during drying. Holz Roh Werkst 56:9–14

Pang S, Keey RB, Laugrish TAG, Walker JCF (1994) Air flow reversals in high temperature kiln drying of *Pinus radiata* boards:1: drying of a single board. N Z J For Sci 24:83–103

Papadakis EP (1999) Ultrasonic instruments and devices. Reference for modern instrumentation, techniques and technology. Academic Press, San Diego, 809 pp

Passialis C (1997) Physico-chemical characteristics of waterlogged archeological wood. Holzforschung 51:11–113

Pearce RB, Sumer S, Doran SJ, Carpenter TA, Hall LD (1994) Noninvasive imaging of fungal colonization and host response in the living sapwood of sycamore (*Acer pseudoplaranus*) using nuclear magnetic resonance. Physiol Plant Pathol 45:359–384

Pearce RB, Fisher BJ, Carpanter TA, Hall LD (1997) Water distribution in fungal lesions in the wood of sycamore (*Acer pseudoplatanus*), determined gravimetrically and using nuclear magnetic resonance imaging. New Phytol 135:675–688

Pei J, Yousuf MI, Degertrkin FL, Honein BV, Kuri-Yakub BT (1995) Lamb wave tomography and its application in pipe erosion/corrosion monitoring. Proc IEEE Ultrasonics Symp 1995, Seattle, Washington, pp 795–798

Pellerin RF (1965) A vibrational approach to nondestructive testing of structural lumber. For Prod J 15 3:93–101

Peters RA (1995) A new algorithm for image noise reduction using mathematical morphology. IEEE Trans Med Imag 4:554–568

Peterson KR (1993) The role of nondestructive evaluation in assuring the wise use of our timber resource. 9th Symp NDT of Wood, Washington State University Press, Madison, pp 7–9

Peyskens E, Pourq M de, Stevens M, Schalck J (1984) Dielectric properties of softwood species at microwave frequencies. Wood Sci Technol 18:267–280

Pfeffer PE, Gerasimowicz WV (1989) Nuclear magnetic resonance in agriculture. CRC Press, Boca Raton, 441 pp

Polge H (1966) L'analyse densitomé1trique de cliches radiographiques: Une nouvelle methods de determination de la structure du bois. Ann Sci For 20 4:530–581

Polge H (1978) Fifteen years of wood radiation densitometry. Wood Sci Technol 12:187–196

Poliszko S, Hoffmann G (1985) Dielectric behavior of wood-polystyrene composite. J Appl Polymer Sci 30:799–804

Price DL, Sköld K (1986) Introduction to neutron scattering. In: Sköld K, Price DL (eds) Neutron scattering, vol 23, part A. Academic Press, Orlando, pp 1–98

Purslow DF (1971) The use of microwave moisture meter for studying moisture changes in timber. J Inst Wood Sci 5 4:40–46

Puttick KE (1987) Thermal NDT methods. In: Summerscales J (ed) Nondestructive testing of fibre-reinforced plastics composites. Elsevier, London, pp 65–103

Quick JJ, Hailey JRT, Mac Kay AL (1990) Radial moisture profiles of cedar sapwood during drying: a proton magnetic resonance study. Wood Fiber Sci. 22:404–412

Quin F Jr, Steele PH, Shmulsky R (1998) Locating knots in wood with an infrared detector system. For Prod J 48 10:80–84

Radon J (1917) Über die Bestimmung von Funktionen durch ihre Integralwerte langs gewisser Manningfaltigkeiten (On the determination of functions from their integrals along certain mainfolds). Ber Sach Akad Wiss 29:262–277

Rafalski J (1966) On the dielectric properties of compress red beech (Über die dielekkyrischen Eigenschaften unterschiedich verdiichteten Rotbuchen vollholzes). Holztechnologie 7 2:118–122

Ranta L, May HA (1978) Measuring the density profile of particleboard by gamma radiation. Holz Roh Werkst 36:467–474

Rantala J, Wu D, Busse G (1996) Vibrothermography applied for nondestructive evaluation of polymer materials. Mater Sci Forum 210–213:433–438

Raschi A, Tognetti R, Ridder HW, Beres C (1995) Water in the stems of sesilis oak (*Quercus petraea*) by computer tomography with concurrent measurements of sap velocity and ultrasound emission. Plant Cell Environ 18:545–554

Reimers P, GilboyWB, Goebbels J (1984) Recent development in the industrial application of computerized tomography with ionizing radiation. NDT Int 17 4:197–207

Reynolds WN, Wells GM (1984) Video compatible thermography. Br J NDT 26 1:40–44

Rice RW, Steele PH, Kumar L (1992): Detecting knots and voids in lumber with dielectric sensors. Ind Metrol 2 3/4:309–315

Richards JA (1990) Radar backscatter modeling of forests: a review of current trends. Int J Remote Sens 11:1299–1312

Rinn F (1991) Three dimensional impulse tomograph for examination of trees and timber. Engineering and Distribution, Heidelberg, 4 pp

Robertson MB, Packer KJ (1999) Diffusion of D_2O in archaeological wood measured by 1-D NMR profiles. Appl Magn Reson 17:49–64
Rosencwaig A, Busse G (1980) High resolution photoacoustic thermal wave microscopy Appl.Phys.Lett 36 : 725–727
Rosencwaig A (1983) Photoacoustics in material science. In: Herman H (ed) Treatise on material science and technology, vol 198. Academic Press, NewYork, pp 67–117
Ross RJ, Pellerin RF (1991) Nondestructive evaluation of wood past, present and future. In: Ruud CO, Green RE (eds) Nondestructive characterization of material, vol IV. Plenum Press, New York, pp 59–64
Ross RJ, Pellerin RF (1994) Nondestructive testing for assessing wood members in structures. A review. Gen Tech Rep FPL-GTR-70, USDA, Forest Product Laboratory, Madison, WI, 40 pp
Rothwell WP, Holecek DR (1984) NMR imaging: study of fluid absorption by polymer composites. J Polymer Sci Polymer Lett Ed 22:241–247
Roux S (1998) Scales. In: Frantziskonis GN (ed) Proc PROBAMAT – 21st Century: probability and materials. Test, models and applications for the 21th century. Kluwer, Dordrecht, pp 573–577
Rust S (2000) A new tomographic device for the nondestructive testing of trees. 12th Symp on Nondestructive Testing of Wood, University of Western Hungary, Sopron, pp 233–237
Sadoh T, Murata K (1993) Detection of knots in hinoki and karamatsu lumber by thermography. Mokuzai Gakkaishi 39:13–18
Sahimi M, Arbabi S (1993) Mechanics of disordered solids II. Phys Rev B 47:703–712
Salerno A, Wu D, Busse G, Rantala J (1997) Thermographic inspection with ultrasonic excitation In: Thompson DO, Chimenti DE (eds) Review of progress in quantitative nondestructive evaluation, vol 16. Plenum Press, New York, pp 345–352
Salerno A, Dillenz A, Wu D, Rantala J, Busse G (1998) Progress in ultrasonic lockin thermography. In: Balageas D, Busse G, Carlomagno GM (eds) Quantitative infrared thermography. QIRT 98, Lodz, pp 164–160
Samson M (1988) Traverse scanning for automatic detection of general slope of grain in lumber. For Prod J 38 7/8:33–38
Sasov A, van Dyck D (1998) Desktop X-ray microscopy and microtomography. J Microsc 191 pt 2:151–158
Schau D (1984) Fourier transform NMR. Elsevier, Amsterdam, 344 pp
Schechter RS, Chaskelis HH, Mignogna RB, Delsanto PP (1994) Real time parallel computation and visualization of ultrasonic pulses in solids. Science 265:1188–1196
Schechter RS, Mignona RB, Delsanto PP (1996) Ultrasonic tomography using curved ray paths obtained by wave propagation simulations on a massively parallel computer. J Acoust Soc Am 104:2103–2111
Schienwind AP (1990) Physical and mechanical properties of archeological wood. In: Rowell RM, Barbour RJ (eds) Archeological wood properties, chemistry and preservation. American Chemical Soc, Washington, DC, pp 87–109
Schirone A, Lo Monaco A (1988) L'uso dei raggi X in xililogia con particolare riferimento alle indagini dendroclimatologiche (The utilization of X-ray for dendroclimatology). Economia Montana (Linea Ecologica) 5:47–54
Schmidt RG, Frazier CE (2000) Solid state NMR analysis of adhesive bondlines in pilot scale flakeboards. Wood Fiber Sci 32:419–425
Schmidt T (1978) Scanning computing methods for measuring knots and other defects in lumber and veneer. Proc 4th Nondestructive Testing of Wood, Washington State University, Pullman, pp 23–25
Schmoldt DL (1996) CT imaging, data reduction, and visualization of hardwood logs. In: Meyer DA (ed) Proc 24th Annual Hardwood Symposium, National Lumber Assoc Cashiers, National Hardwood Lumber Assoc, Cashiers, North Carolina, pp 69–80
Schmoldt DL, Zhu D, Conners RW (1993) Nondestructive evaluation of hardwood logs using automatic interpretation of CT images. In: Thompson DO, Chimenti DE (eds) Review of progress in quantitative nondestructive evaluation, vol 12. Plenum Press, New York, pp 2257–2264

Schmoldt DL, Li P, Araman PA (1996a) Iterative simulation of hardwood log veneer slicing using CT images. For Prod J 46 4:41–47

Schmoldt DL, Nelson RM, Ross RJ (1996b) Ultrasonic defect detection in wooden pallet parts for quality sorting. Proc SPIE 2944:285–295

Schmoldt DL, Pei Li A, Abbot Lyn A (1997) Machine vision using artificial neural networks with local 3-D neighborhoods. Comput Electr Agric 16:255–271

Schmoldt DL, Occena LG, Lynn Abbott A, Gupta NK (1999) Nondestructive evaluation of hardwood logs: CT scanning, machine vision and data utilization. Nondestr Test Eval 15:279–309

Schmoldt DL, He J, Lynn Abbott A (2000) Automated labeling of log features in CT imagery of multiple hardwood species. Wood Fiber Sci 32:287–300

Schneider R, Netzelmann U, Kröning M (1996) Characterization of green ceramics by microwaves and ultrasound. Mater Sci Forum 210–213:77–84

Schniewind AP (1981) Concise encyclopedia of wood and wood-based materials. Pergamon Press, Oxford, 354 pp

Schomberg H (1982) Nonlinear image reconstruction from ultrasonic time to flight projections. Acoust Imaging 10:381–396

Schulte M, Frühwald A, Welling H, Kruse K, Broker FW (1995). Nondestructive in-line quality control in board production. Tappi Proc of 1995 European plastic laminates forum "The leading edge", Heidelberg, pp 71–79

Schultz V, Ward Whicker F (1982) Radioecological techniques. Plenum Press, New York, 298 pp

Senft JF, Suddarth SK, Angleton RD (1962) A new approach to stress grading of lumber. For Prod J 12 4:183–186

Sepulveda P, Gönlund A (2000) Measurement of spiral grain with computed tomography compared with pattern on boards. Proc 12th Symp Nondestructive Testing of Wood, Sopron, pp 239–244

Sharp AR, Riggin MT, Kaiser R (1978) Determination of moisture content of wood by pulsed nuclear magnetic resonance. Wood Fiber 10:74–81

Siau JF (1984) Transport process in wood. Springer, Berlin Heidelberg New York, 245 pp

Siau JF (1995) Wood: influence of moisture on physical properties. Department of Wood Science and Forest Products, Virginia Polytechnic Institute and State University, Blacksburg, Virginia, 227 pp

Siau JF, Meyer JA (1973) Neutron activation analysis of pentachlorophenol in treated wood. Wood Sci 6 1:19–21

Sieber AJ (1985) Forest signatures in imaging and nonimaging microwave scatterometer data. ESA J 9 4:431–448

Skaar C (1948) The dielectric properties of wood at several radio frequencies. Tech Pub no 69. NY State College of Forestry, Syracuse

Skaar C (1988) Wood in water relations. Springer, Berlin Heidelberg New York, 283 pp

Skatter S (1998) Determination of cross-sectional shape of softwood logs from three X-ray projections using an elliptical model. Holz Roh Werkst 56:179–186

Slichter CP (1978) Principles of magnetic resonance. Springer, Berlin Heidelberg New York, 397 pp

Schniewind AP (1990) Physical and mechanical properties of archaeological wood. In: Rowell RM, Barbour RJ (eds) Archaeological wood: properties, chemistry and preservation. Advances in Chemistry Series no 225, pp 87–109

Sobue N (1993) Nondestructive characterization of wood. Mokuzai Gakkaishi 39:973–979

Socco V, Martinis R, Sambuelli L, Comino E, Nicolotti G (2000) Open problems concerning ultrasonic tomography for wood decays diagnosis. 12th Symp NDT of Wood, University of Western Hungary, Sopron, p 468

Socco V, Sambuelli L, Martinis R, Nicolotti G, Comino E (2002) Feasability of ultrasonic tomography for NDT of decay on living trees. Res Nondestr Eval (in press)

Southon TE, Mattsson A, Jones RA (1992) NMR imaging of roots: effects after root freezing of containerized conifer seedlings. Physiol Plant 86:329–334

Steele PH, Cooper JE (2000) Estimating lumber strength with radio frequency scanning. 12th Symp Nondestructive Testing of Wood, Sopron, pp 343–348

Steele PH, Harless TEG, Wagner FG, Taylor FW (1989) Potential dollar increase from internal log information. Proc 7th Symp on Nondestructive Testing of Wood, Washington State University, Pulman, pp 241–250

Steele PH, Neal SC, McDonald KA, Cramer SM (1992) The slope of grain indicator for defect detection in unplaned hardwood lumber. For Prod J 41 1:15–20

Steele PH, Quin F Jr, Cooper E (1998) Infrared detection of knots in southern yellow pine lumber heated with radiant heat. 11th Symp NDT of Wood, Washington State University, Pullman, pp 57–62

Steele PH, Patton MD, Cooper E (2000a) Factor influencing differential thermal response of knots and clear wood. 12th Symp NDT of Wood, University of Western Hungary, Sopron, pp 403–412

Steele PH, Patton MD, Cooper E (2000b) A thermal recognition system to identify knots in wood. Proc wood technology. Clinic and Shows, Portland, Oregon, pp 255–266

Steele PH, Lalit Kumar, Shmulsky R (2000c) Differentiation of knots, distorted grain and clear wood by radio frequency scanning. For Prod J 50 3:59–62

Sugimori M, Lam F (1999) Macro-void distribution analysis in strand-based wood composites using an X-ray computer tomography technique. J Wood Sci Jpn Wood Res Soc 45:245–257

Sulivan C, Kline R, Mignogna RB, Delsanto PP (1996) A parallel processing approach to acoustic tomograph. J Acoust Soc Am 99:2142–2147

Suryoatmono B, Cramer S, Shi Y, McDonald KA (1994) Within-board lumber density variations from digital X-ray images. 9th Symp Nondestructive Testing of Wood, Washington State University. Eng Publications, Madison, pp 168–175

Swanson JS, Hailey JRT (1987) Scanning and imaging techniques for assessing decay and wood quality in logs and standing trees. Proc 6th Symp Nondestructive Testing of Wood, Washington State University. Eng Publications, Pullman, pp 83–94

Szendrödi L, Habermehl A, Ridder HW (1994) Computer tomographic investigation of standing trees in Hungary. 1st European Symp on Nondestructive Evaluation of Wood, Sopron, pp 503–511

Szymani R, McDonald KA (1981) Defect detection in lumber: state of the art. For Prod J 31 11:34–44

Talpe J (1971) Theory of experiments in paramagnetic resonance. Pergamon Press, Oxford

Tan HS (1981) Microwave measurements and modeling of the permittivity of tropical vegetation samples. Appl Phys 25:351–355

Tanaka T (1994) Thermographic detection of deteriorated locations and nondestructive evaluation of strength of biodeteriorated wood. 9th Symp NDT of Wood, Washington State University, Pullman, pp 78–83

Tanaka T, Divos F (2000) Thermographic inspection of wood. 12th Symp NDT of Wood, University of Western Hungary, Sopron, pp 439–447

Tanaka T, Norimoto M, Yamada T (1975a) Anisotropy of dielectric constant of wood. J Soc Mater Sci 24 264:867–872

Tanaka T, Norimoto M, Yamada T (1975b) Dielectric properties and structure of wood I. Mokuzai Gakkaishi 21:129–134

Taylor FW, Wagner FG Jr, McMillin CW, Lon Morgan I, Hopkins FF (1984) Locating knots by industrial tomography – a feasibility study. For Prod J 34 5:42–46

Thomas RL, Favro LD, Kuo PK, Ahmed T, Han X, Wang L, Wang X, Shepard SM (1995) Pulse echo thermal wave imaging for nondestructive evaluation. Proc 15 th Int Confrerence Acoustics, Trondheim, pp 433–436

Thompson F (1989) Moisture measurements using microwaves. Meas Control 22:210–215

Thompson F (1996) Non destructive moisture measurement using microwaves. Mater Sci Forum 210–213:85–92

Tice D, Euskirchen J (1976) Faster and more effective forest fire surveillance with live IR imaging superimposed on visible terrain view. Proc XVIth IUFRO World Congress, Oslo, pp 53–59

Tiitta M, Olkkonen H, Lappalainen T, Kanko T (1996) Density measurements of particleboard veneer and wood specimens by narrow beam gamma absorption technique. 10th Symp Nondestructive Testing of Wood Swiss Federal Institute of Technology, Press Univ Romande, Lausanne, pp 187–200

Tinelli A, Catena G (1991) Injuries on ornamental trees in Castelporziano Indagine sulle piante monumentali della tenuta di Castelporziano. Monti Boschi 5:9–13

Tiuri M, Jokela K, Heikkela S (1980) Microwave instrument for accurate moisture and density measurement of timber. J Microwave Power 15:251–254

Tognetti R, Raschi A, Beres C, Fenyvesi A, Ridder HW (1996) Comparison of sap flow, cavitation and water status of *Quercus petraea* and *Quercus cerris* trees with special reference to computer tomography. Plant Cell Environ 19:928–938

Tomikawa Y, Iwase Y, Arita K, Yamada H (1985) Nondestructive inspection of rotted or termite damaged wooden poles by ultrasound. Proc 5th Symposium on Ultrasonic Electronics, Tokyo 1984. Jpn J Appl Phys 24 1985 Suppl 24 1:187–189

Tomikawa Y, Iwase Y, Arita K, Yamada H (1986) Nondestructive inspection of wooden pole using ultrasonic computed tomography. IEEE Trans UFFC 33 4:354–358

Torgovnikov GI (1990) Dielektrische Eigenschaften von absolut trockenem Holz und Holzstoff (Dielectric properties of oven dried wood). Holztechnologie 30 1:9–11

Torgovnikov GI (1993) Dielectric properties of wood and wood based materials. Springer, Berlin Heidelberg New York, 196 pp

Tremblay C (1995) Longitudinal and radial variation of slope of grain in black spruce lumber. For Prod J 45 1:79–83

Troughton GE, Clarke MR (1987) Development of a new method to measure moisture content in unseasoned veneer and lumber. For Prod J 37 1:13–19

Tsai WH (1985) Moment-preserving thresholding: a new approach. Comput Vision Graph 29: 377–393

Tsuchikawa S, Takahashi T, Tsutsumi S (2000) Nondestructive measurement of wood properties by using near-infrared laser radiation. For Prod J 50 1:81–86

Ulaby FT, Jedlicka RP (1984) Microwave dielectric properties of plant materials. IEEE Trans Geosci Remote Sensing GE 22:406–414

Ulaby FT, Razni M, Dobson MC (1983) Effect of vegetation cover on microwave radiometric sensitivity to soil moisture. IEEE Trans Geosci Remote Sensing GE 21:51–61

Ulaby FT, Sarabandi K, McDonald K, Whitt M, Dobson MC (1990) Michigan microwave canopy scattering model. Int J Remote Sensing 11:1223–1253

Unger A, Planitzer J, Morgos A (1988) X-ray computer tomography and magnetic resonance tomography for characterizing wet archaeological wood. Holztechnologie 29:249–250

USDA, Forest Product Laboratory, Forest Service (1972) Thermal properties. In: Wood handbook: wood as an engineering material. USDA, Madison, WI, pp 324–326

Uyemura T (1960) Dielectric properties of wood as the indicator of the moisture. Bull Gov For Exp Stn Tokyo 119:95–172

Vansteenkiste D (2000) Mise au point d'une méthode rapide d'analyse quantitative de l'anatomie du bois de chêne: analyse automatique des clichés radiographiques. In: Bobillier E, Jacob D (eds) 6ème journée de la mesure – Electronique, Informatique, Automatique. INRA, Rennes, 8 pp

Varga A, Matusik A, Kiss E, Divos F (2000) Use of medical CT to qualify defects in wood. 12th Symp Nondestructive Testing of Wood, Sopron, p 466

Vavilov V (1992) Thermal non destructive testing: short history and state of art. In: Balageas D, Busse G, Carlomagno GM (eds) Proc Quantitative Infrared Thermograpfy (QIRT92). Editions Européennes Thermiques et Industrie, Paris, pp 179–194

Veres JS, Cofer GP, Johnson GA (1993) Magnetic resonance imaging of leaves. New Phytol 132: 769–774

Vermass HF (1973) Regression equations for determining the dielectric properties of wood. Holzforschung 27:132–136

Vincent L (1993) Morphological grayscale reconstruction in image analysis. Applications and efficient algorithms. IEEE Trans Image Process 2:176–201

Vinegar HJ (1986) X-ray computed tomography and NMR imaging of rocks. J Petrol Technol March:257–259

Völgyi F (2000) Microwave NDT of particleboards. 12th Symp Nondestructive Testing of Wood, Sopron, pp 357–365

Vozzo JA, Halloin JM, Cooper TG, Potchen EJ (1996) Use of NMR spectroscopy and MRI for discriminating *Juglans nigra* L. seeds. Sci Technol 24:457–463

Wagner FG, Taylor FW (1985) Economic returns from internal log scanning. Fifth Nondestructive Testing of Wood Symp, Washington State University, Pullman, pp 267–280

Wagner FG, Taylor FW, Ladd DS, Mcmillin CW, Roder FL (1989a) Ultrafast CT scanning of an oak log for internal defects. For Prod J 39 11/12:62–64

Wagner FG, Taylor F, Ladd D, McMillin C, Roder F (1989b) Ultrafast CT scanning of logs for internal defects. 7th Symp Nondestructive Testing of Wood Symposium, Washington State University. Eng Publications, Pullman, pp 221–229

Wang PC, Chang SJ (1986) Nuclear magnetic resonance imaging of wood. Wood Fiber Sci 18: 308–314

Wang PC, Chang JS, Olson JR (1990). Scanning logs with an NMR scanner. 7th Symp Nondestructive Testing of Wood, Washington State University, Pullman, pp 209–219

Wang YQ, Kline RA (1994) Ray tracing in isotropic and anisotropic materials: application to tomographic image reconstruction. J Acoust Soc Am 95 2525–2532

Weatherwax RC, Stamm AJ (1947) The coefficients of thermal expansion of wood and wood products. Trans ASME May:421–432

Weibe S (1992) Untersuchungen zur Wundenwicklung und Wundbehandlung an Bäumen unter besonderer Beruchsichtigung der Holzfeuchte. PhD thesis, Univ München, München, Germany

Wert CA, Weller M, Caulfield D (1984) Dynamic loss properties of wood. J Appl Phys 56 9: 2453–2458

White MS, Ifju G, Johnson JA (1977) Methods for measuring resin penetration into wood. For Prod J 27 7:52–54

Whittall KP, MacKay AL (1989) Quantitative interpretation of NMR relaxation data. J Magn Reson 84:134–152

Wiberg P, Moren TJ (1999) Moisture flux determination in wood during drying above fiber saturation point using CT scanning and digital image processing. Holz Roh Werkst 57: 137–144

Wiberg P (1996) CT-scanning during wood drying for nondestructive moisture content gradient evaluation. Proc 10th Nondestructive Testing of Wood Symposium. Swiss Federal Institute of Technology, Press Polytechniques et Universitaires Romandes, Lausanne, p 409

Wickman BE (1985) Comparison of the degree-day computer and a recording thermograph in a forest environment. Reserch Note PNW-427. USDA, Forest Service, Madison, WI, 6 pp

Willis RL, Stone TS, Berthelot YH, Madigorski WM (1997) An experimental numerical technique for evaluating the bulk and shear dynamic moduli of viscoelastic materials. J Acoust Soc Am 102:3549–3555

Willis RL, Wu L, Berthelot YH (2001) Determination of the complex Young and shear moduli of viscoelastic materials. J Acoust Soc Am 109:611–621

Wilson DW, Charles JA (1981) Thermographic detection of adhesive-bond and interlaminar flaws in composites. Exp Mech 21:276–280

Wilson MA, Godfrey IM, Hanna JV, Quezada RA, Finnie KS (1993) The degradation of wood in old Indian Ocean shipwrecks. Org Geochem 20 5:599–610

Winistorfer PM, Moschler WW Jr (1986) A direct scanning densitometer to measure density profiles in wood composite products. For Prod J 36 11/12:82–86

Wisniewski M, Lindow SE, Ashworth EN (1997) Observations of ice nucleation and propagation in plants using infrared video thermography. Plant Physiol 113 2:327

Wissing ST, Wellinng J (1995) Nondestructive measurement of the density profile in wood based panels by means of radio frequency. Proc Int Conf Progress in Forest Products Research, Göttingen, p 165

Wolter B, Netzelmann U (1996) Nondestructive testing of wood using a single sided nuclear magnetic resonance. 10th Int Symp on Nondestructive Testing of Wood. Press Polytechniques et Universitaires Romandes, Lausanne, p 409

Wu D (1992) Lockin thermography for multiplex photothermal nondestructive evaluation. In: Balageas D, Busse G, Carlomagno GM (eds) Proc Quantitative Infrared Thermograpfy (QIRT92). Editions Européennes Thermiques et Industrie, Paris, France, pp 371–376

Wu D (1994) Lock in thermography for defect characterization in veneers. In: Balageas D, Busse G, Carlomagno GM (eds) Proc Quantitative Infrared Thermograpfy (QIRT94). Editions Européennes Thermiques et Industrie, Paris, France, pp 298–301

Wu D, Busse G (1995) Remote inspection of wood with lock in thermography. Proc of 1995 European Plastic Laminates Forum "The leading Edge", Heidelberg, pp 27–29

Wu D, Busse G (1996) Remote inspection of wood with lock-in thermography. Tappi J 79 8:119–123

Wu D, Karpen W, Busse G (1992) Lockin thermography for multiplex photothermal nondestructive evaluation Proc Quantitative Infrared Thermography (QIRT 1992). Eurotherm Ser 27:371–376

Wu D, Salerno A, Sembach, J, Maldague X, Rantala J, Busse G (1996) Wood-based products inspection by lockin thermography. Proc 3rd Eurowood Symp and 4th FESYP Technical Conf Braunschweig, Germany, 5 pp

Wu D, Salerno A, Sembach J, Maldague X, Rantala J, Busse G (1997a) Lockin thermographic inspection of wood particleboard. SPIE 3056:230–234

Wu D, Sembach J, Salerno A, Hora G, Busse G (1997b) Ensuring the quality of coated wood-based products by means of lock-in thermography. (Qualitatssicherungg beschichteter Holzwerkstoffe mittels lockin-Thermographie). Holz -Zentralbl 123:50–775

Wycoff W, Pickup S, Cutter B, Miller W, Wong TC (2000) The determination of the cell size in wood by nuclear magnetic resonance diffusion techniques. Wood Fiber Sci 32:72–80

Xu Y, Okumura S, Noguchi M (1993) Thermographic detection of starved joints of wood. Mokuzai Gakkaishi 39:544–549

Xu Y, Okumura S, Noguchi M (1994) Thermographic detection of starved joints of wood. 9th Symp NDT of wood. Forest Product Soc, Madison, pp 209–217

Yamanaka K, Nagata Y, Koda T (1991) Enhancement of acoustic imaging of polymers by cooling. Ultrasonics 29 March:159–165

Yokoyama M, Norimoto M (1996) Contour diagrams of dielectric loss for absolutely dried spruce wood. Wood Res Kyoto Univ 83:37–39

Yokoyama M, Norimoto M (1997) Cole-Cole plots for dielectric properties of absolutely dried wood. Tech Notes 33:71–82

Yokoyama M, Obataya E, Norimoto M (1999) Mechanical and dielectric relaxations of wood in a low temperature range. II. Relaxations due to adsorbed water. Mokuzai Gakkaishi 45:95–102

Yokoyama M, Kanayama Y, Furuta Y, Norimoto M (2000a) Mechanical and dielectric relaxations of wood in a low temperature range. III. Application of search law to dielectric properties due to adsorbed water. Mokuzai Gakkaishi 46:173–180

Yokoyama M, Ohmae K, Kanayama K, Furuta Y, Norimoto M (2000b) Mechanical and dielectric relaxations of wood in a low temperature range. IV. Dielectric properties of adsorbed water at high moisture contents. Mokuzai Gakkaishi 46:523–530

Youngquist JA, Hamilton TE (1999) The next century of wood products utilization: a call for reflection and innovation. Proc Int Conf on effective utilization of plantation timber, Chi-Tou, Taiwan. For Prod Assoc ROC Bull 16:1–9

Zhu D, Beex AA, Conners R (1991) Stochastic field-based object recognition in computer vision. SPIE 1569:174–181

Zielonka P, Dolowy K (1998) Microwave drying of spruce: moisture content, temperature and heat energy distribution. For Prod J 48 6:77–80

Zimmermann MH (1983) Xylem structure and the ascent of sap. Springer, Berlin Heidelberg New York, 143 pp

Zoughi R (1990) Microwave nondestructive testing: theories and applications. Int Adv Nondestr Test 15:225–288

Zoughi R (1996) Principles of and applications of microwave and millimeter wave NDE methodologies: an overview. SPIE Proc 2944:46–74

Zoughi R, Ganchev S, Carriveau GW (1996) Overview of microwave NDE applied to thick composites. Mater Sci Forum 210–213:69–76

Subject Index

List of Notations

λ	thermal wavelength
ε^*	complex dielectric constant
ω	angular frequency of the intensity modulated laser heat source
δ	chemical shift
ρ	density
τ	recovery time
σ	screening constant
α	thermal diffusivity
$\Delta\Phi$	variation of phase of the radio frequency wave
ε' and ε''	components of the complex dielectric constant (real and imaginary)
ΔA	variation of the amplitude of the radio-frequency wave
ε_c	canopy equivalent dielectric constant
ε_d^*	complex permittivity of dielectric material
ε_o	free space perimittivity
ω_o	Larmor frequency
βo	phase constant of the air
λ_o	wavelength of the electromagnetic wave in air
ε_w^*	complex permittivity of water
μ	attenuation coefficient of the ionizing radiation beam
μ	thermal diffusion length
μ_s	linear attenuation coefficient of X-rays for the x voxel in the cross-sectional slice
μ_w	linear attenuation coefficient of X-rays in water
μ'	mass attenuation coefficient
μ_c	attenuation coefficient due to Compton effect
μ_p	attenuation coefficient due to photo-electric effect
a	cell lumen size (μm) with thermal imaging
Bo	static magnetic field
C	specific heat of the solid wood
C	image contrast
c	speed of propagation of the radiation
$C^*(\omega)$	complex capacitance
C_{11}, C_{22}, C_{33}	ultrasonic stiffnesses on axes 1, 2 and 3
d	sample thickness (mm) with thermal imaging
D and *L*	defect depth and length with thermal imaging
e	thermal effusivity

EPV	energy/pulse length of ultrasonic wave
EV	pulse length energy value
f	frequency
FID	free induction decay
G	wood-specific gravity
G	conductance of the sample
H	Hounsfield, the unit used for X-ray imaging
h	Plank constant
I	intensity of transmitted ionizing radiation beam through the sample
Io	intensity of transmitted ionizing radiation beam through the air
k	thermal conductivity
k_1, k_2, k_3	coefficients of dielectric anisotropy
k_a	thermal conductivity of the dead air
k_L	wood thermal conductivity in L direction
k_L^w	thermal conductivity of cellular wall in longitudinal direction
k_T	wood thermal conductivity in transverse direction
L, R, T	anisotropic axes of wood related to longitudinal, radial and tangential directions versus the annual ring
M	total number of projections
MC	wood moisture content (%)
$MC_{\mu w}$	absolute moisture content deduced from microwave technique
m_d	weight of wood on the basis of absolute dry fraction
M_o	magnetization momentum
m_w	weight of water
M_z and M_{xy}	components of magnetization momentum
N	pixel size
p	1/v = slowness
P	power of the electric field dissipated in the specimen as heat
P(T)	temperature-dependent heat losses
Q	power flux absorbed at the front surface
R_F	Fresnel zone
T	absolute temperature (K)
t	time
T_1	spin-lattice relaxation time or longitudinal relaxation time
T_2	transverse relaxation time
TOF	time of flight of ultrasonic wave
V	phase velocity of thermal wave with which the temperature modulation moves along z direction
v	ultrasonic velocity
v_a	porosity of wood
v_m	fraction volume of water
w	width of the pixel

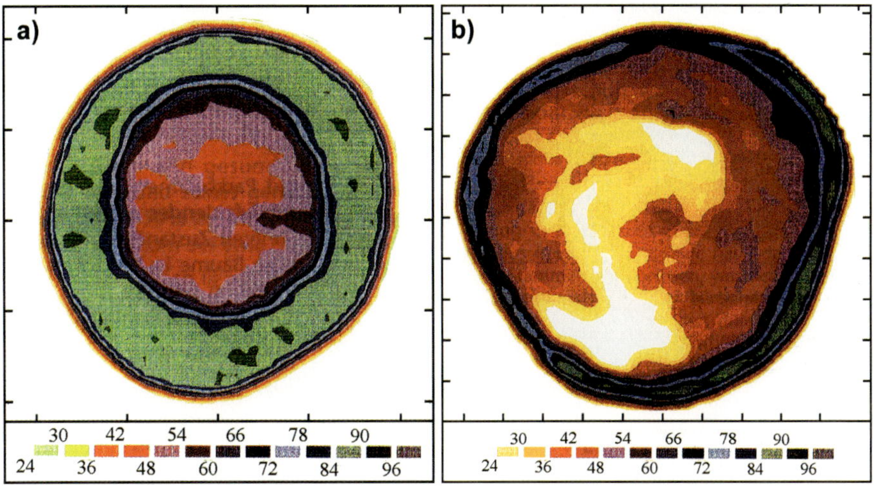

Fig. 2.17a,b. Tomogram on a color scale of a transverse section of a healthy (**a**) and of a decayed (**b**) Norway spruce tree. (Habermehl and Ridder 1996, with permission)

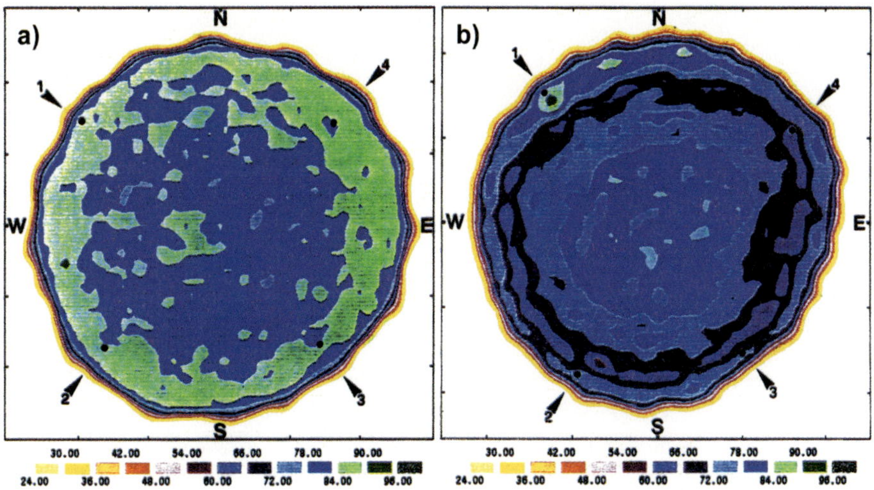

Fig. 2.23a,b. Tomograms of *Quercus petraea* and *Quercus cerris* (Tognetti et al. 1996). **a** *Quercus cerris*; **b** *Quercus* petraea. *N*, *E*, *S*, *W* are the cardinal points; *1*, *2*, *3*, *4* are the position of measured points The density scale in CT numbers is represented by different colors (i.e., CT = 30 corresponds to dry condition and CT = 96 corresponds to high moisture content). (with permission)

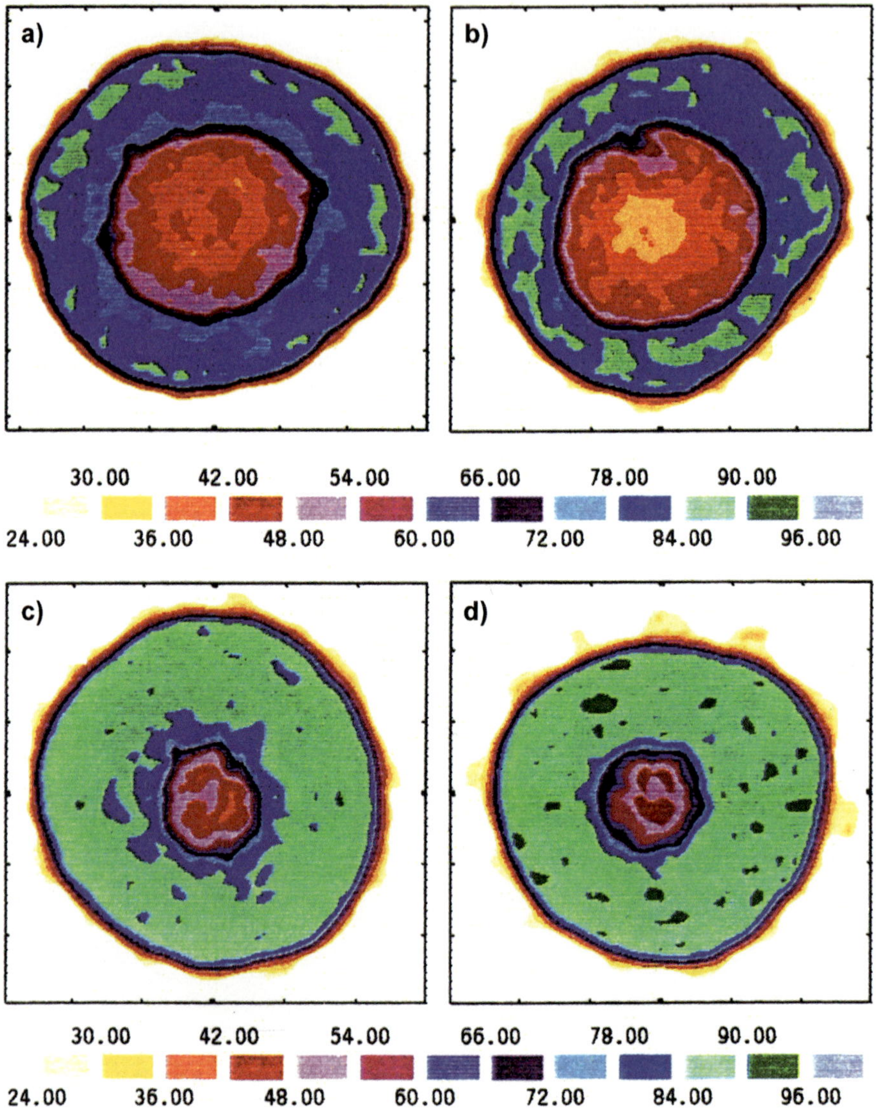

Fig. 2.25a–d. The development of sapwood in Scots pine induced by ammonia pollution, compared with a normal tree at 100 cm stem height. Trees are located in a forest from the district Torgelow, Germany (Kätzel et al. 1997, with permission). **A** Tomogram for tree no. 1, with 624-cm^2 cross section located at site 1, 200 m from the farm (maximum pollution zone). **B** Tree no. 8, with 745-cm^2 cross section located at site 2, 280 m from the farm. **C** Tree no. 2 at site 3, with 513-cm^2 cross section located 2900 m from the farm (minimum pollution zone). **D** Tree no. 8 at site 3 located 2900 m from the farm (minimum pollution zone). The density scale in CT numbers is represented by different colors, CT = 30 corresponds to dry conditions and CT = 96 corresponds to a high moisture content. The tomograms were taken in September 1993 and 1994

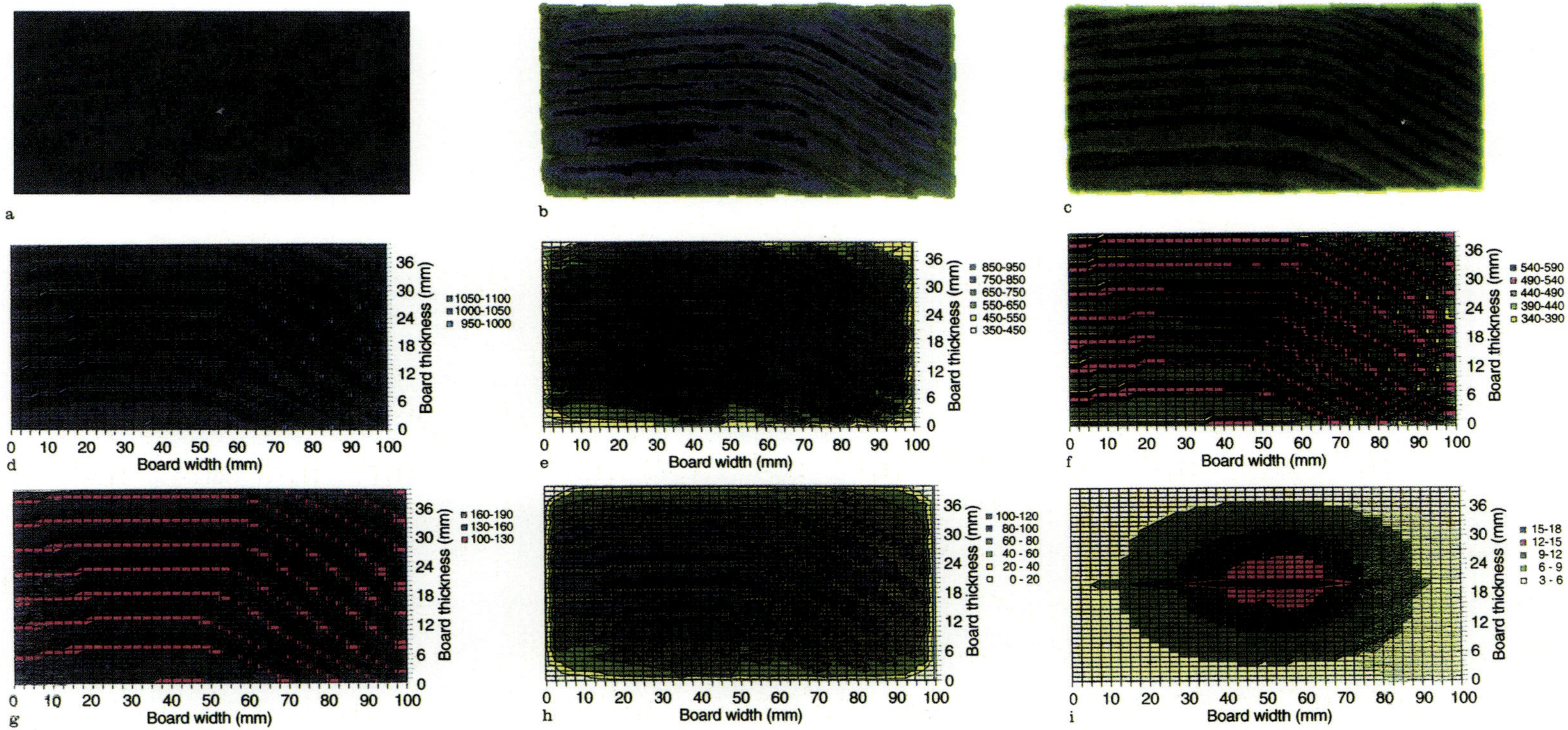

Fig. 2.32a–i. Dynamics of wood drying (Pang and Wiberg 1998). Experimental tomograms (with permission). **a** Scanning of the board before drying – corresponding to wet saturated wood. **b** Scanning of the board before drying – corresponding to wet wood after 9.6 hours of drying, and to the early stage of the drying process. **c** Scanning of the board before drying – corresponding to wet wood after 30.4 h of drying and to the late stage of the drying process. Predicted images for density distribution. **d** Wet wood density before drying. **e** Wood density after 9.6 h of drying. **f** Wood density after 30.4 h of drying. Predicted images for moisture content distribution. **g** Model of predicted wet wood density before drying. **h** Model of predicted wet wood density after 9.6 h of drying. **i** Model of predicted wood density after 30.4 h of drying

Fig. 3.5. Infrared device for the detection of cavities in trees. (Catena and Catena 2000, with permission)

Fig. 3.11a, b. Cavities in standing trees. **a** photographic image. **b** thermographic image of cavity (in blue) in trunk and in branches. (Catena and Catena 2000, with permission)

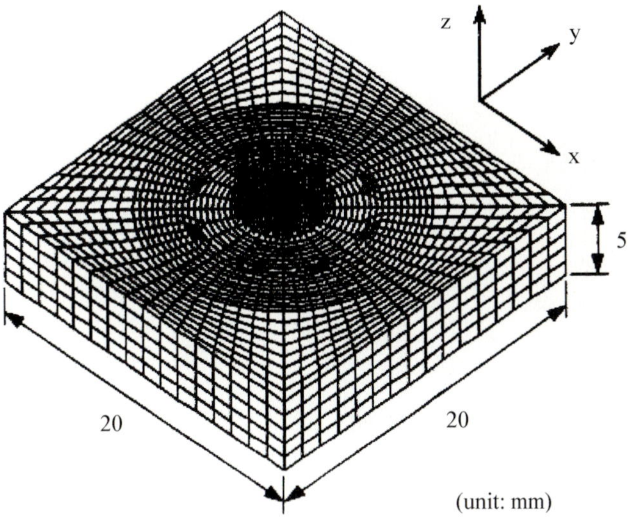

Fig. 3.17. Thermographic measurements of the slope of grain. Finite element model. The heat flux is represented by the *vertical arrows* in the central region. (Naito et al. 2000; courtesy of the Japan Wood Research Society).

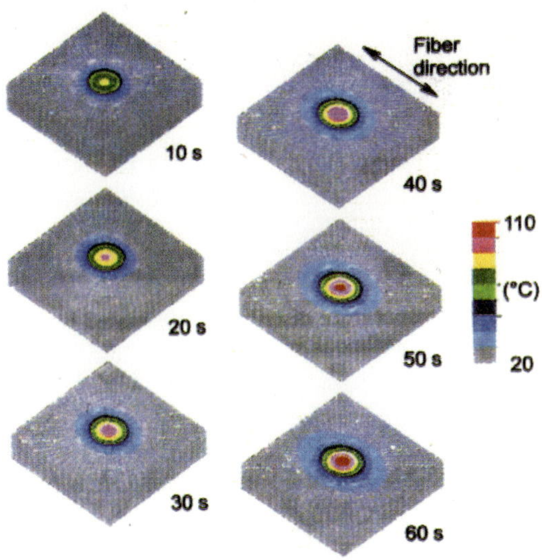

Fig. 3.18. Temperature distribution as a function of heating and slope of grain deduced with the finite element method time obtained with the finite element method. (Naito et al. 2000; courtesy of the Japan Wood Research Society)

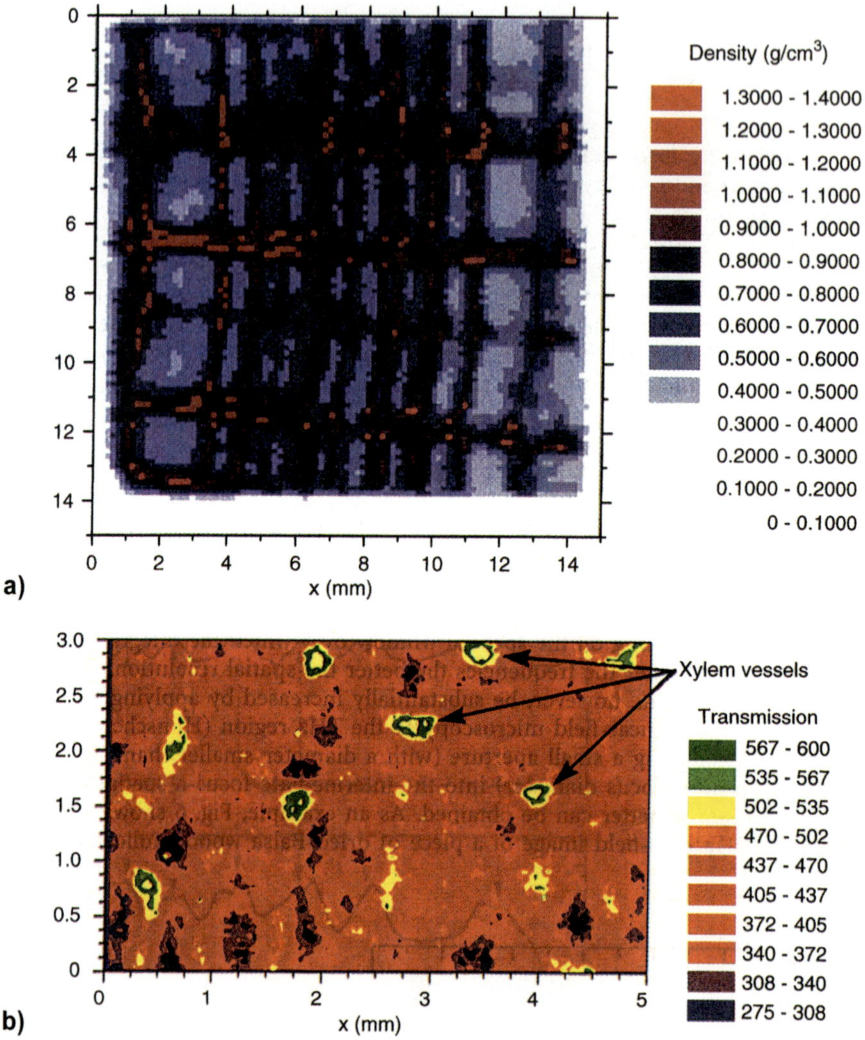

Fig. 3.20a,b. Density mapping of a transverse section of two species: **a** beech and **b** balsa. The size of the specimen was 14 × 14 × 1.7 mm. (Koch et al.1998; courtesy of Wood Science Technology)

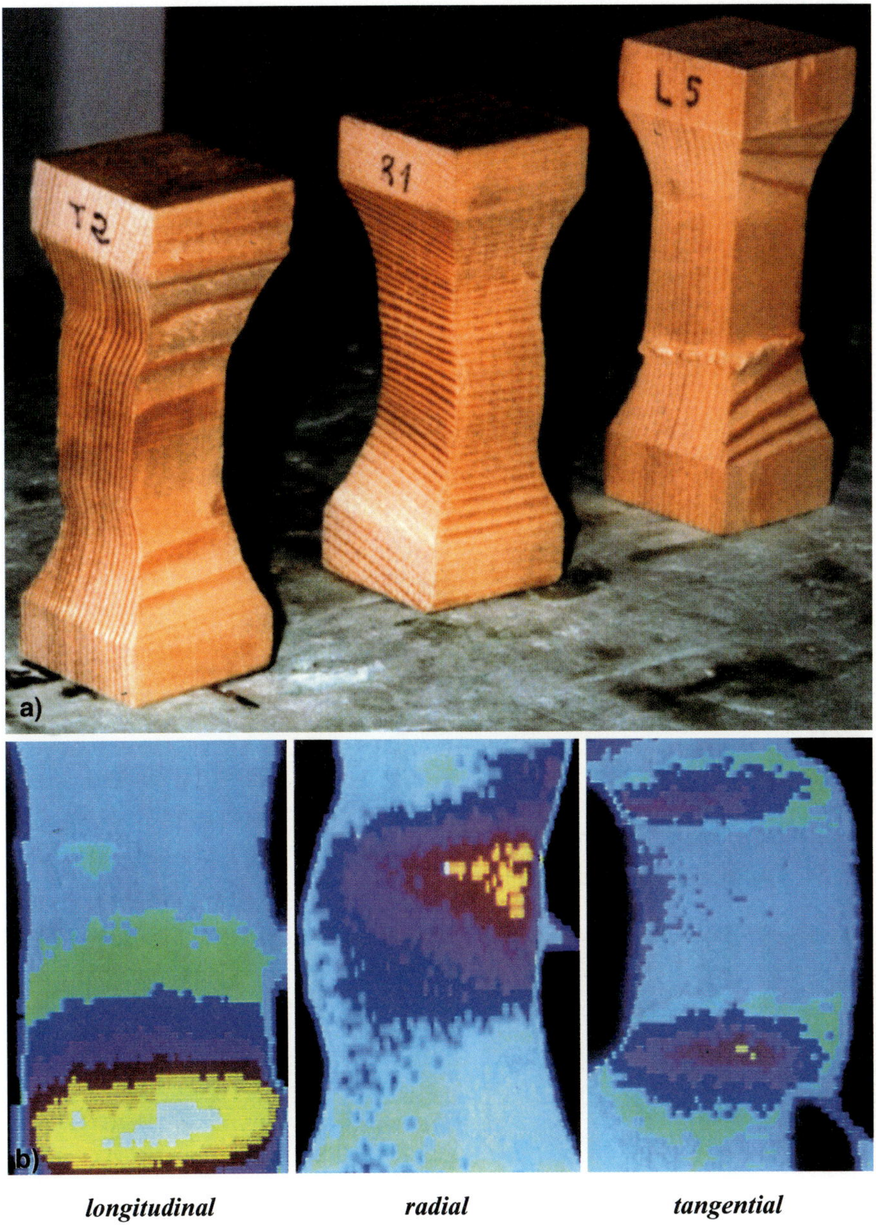

Fig. 3.28a, b. Infrared imaging of pine specimens under compression. **a** Specimens with major axis in longitudinal, radial and tangential directions. **b** Localization of intrinsic dissipation in wood specimens as a function of anisotropy (each color hue corresponds to 0.2 °C) in longitudinal, radial and tangential directions. (Luong 1996, with permission)

Fig. 3.30a–c. Infrared imaging of specimens under compression loading related to stress distribution in the longitudinal direction. **a** specimen; **b** load strain relationship; **c** thermographic images corresponding to the points *a, b, c, d, e, f* from **b**. (Okumura et al. 1996; courtesy of Wood Research Institute, Kyoto University)

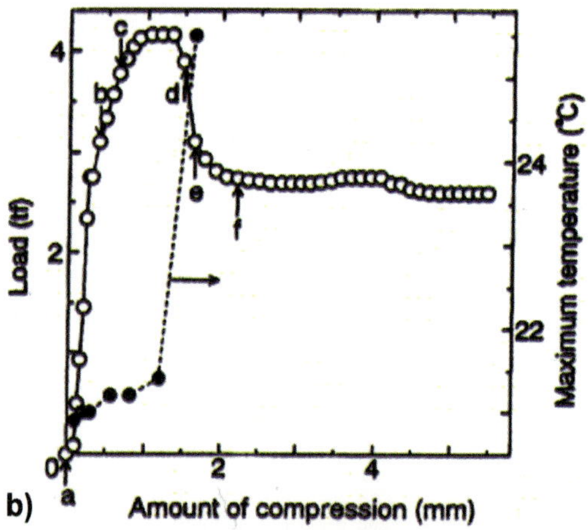

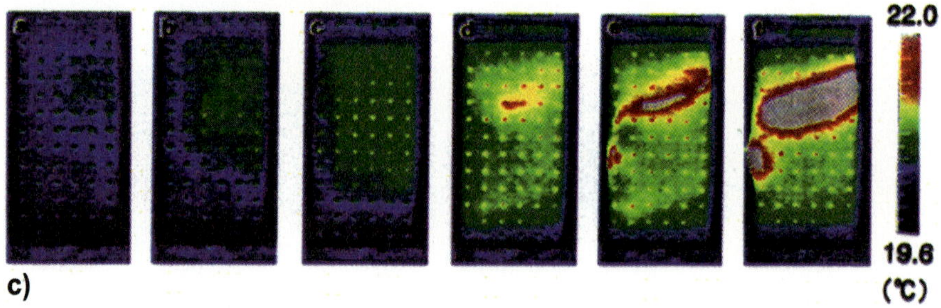

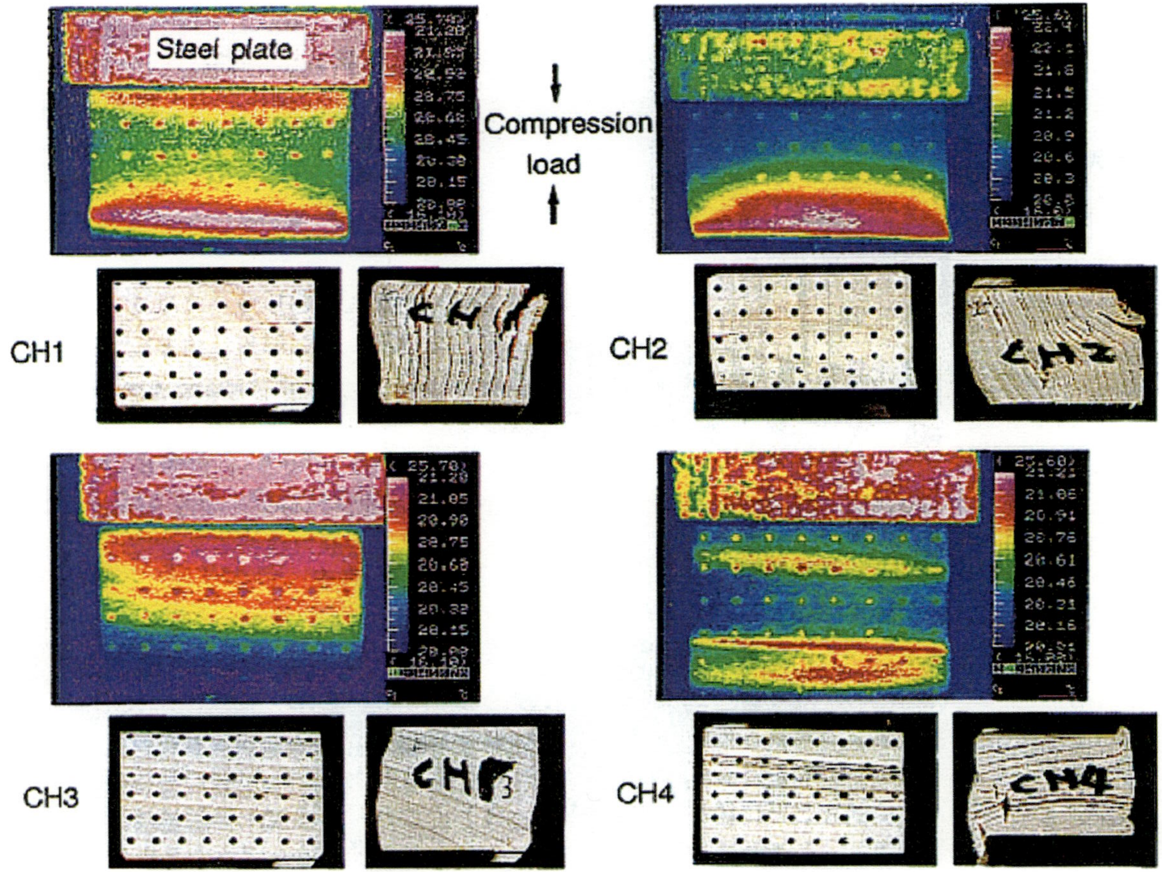

Fig. 3.31. Infrared imaging of specimens under compression loading related to the stress distribution in the transverse plane. Images of specimens at rupture under different compression loads applied in radial (*CH4*) tangential (*CH1*) and inclined directions vs. the annual rings (*CH3* inclined vs. T and *CH2* inclined vs. R). (Okumura et al. 1996; courtesy of Wood Research Institute, Kyoto University)

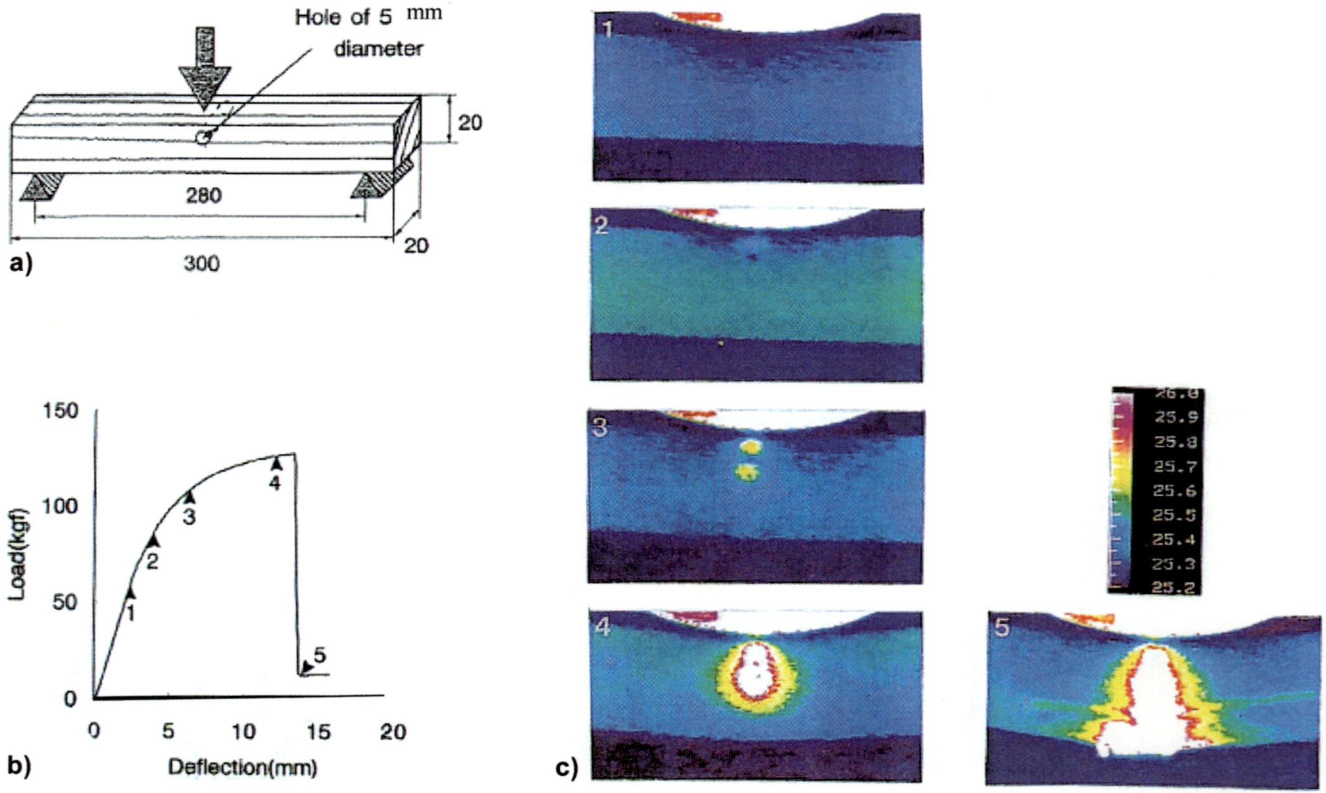

Fig. 3.32a–c. Infrared imaging of specimen under static bending loading. **a** Specimen with a central hole under static bending. **b** Stress strain curve. **c** Temperature distribution as a function of loading and strain, corresponding to points *1*, *2*, *3*, and *4*. (Naito et al.1998; courtesy of Wood Research Institute Kyoto University)

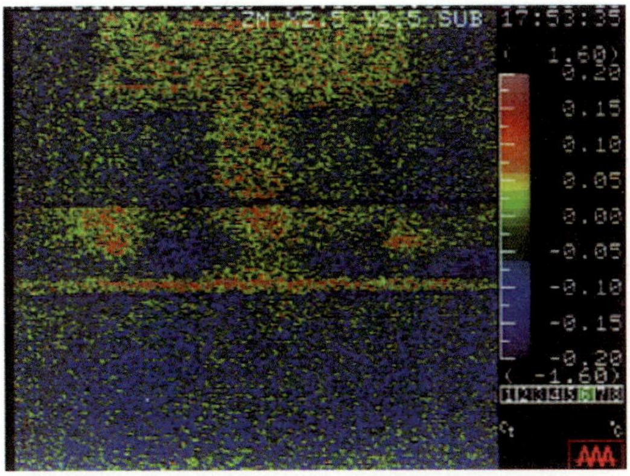

Fig. 3.34. The thermographic image for a specimen with two 16-cm delamination zones. (Masuda and Takahashi 2000, with permission)

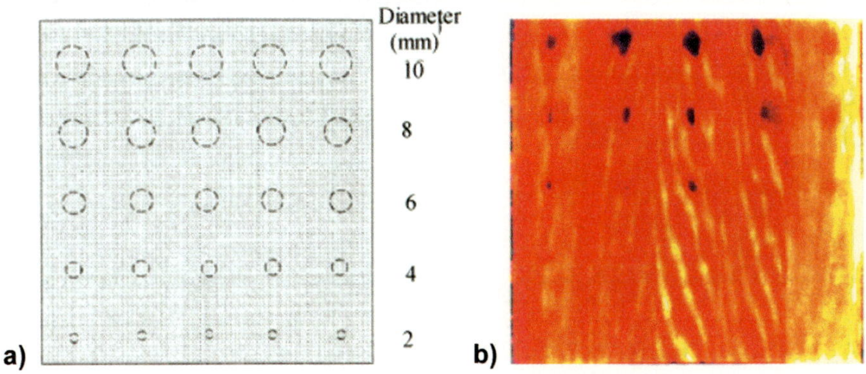

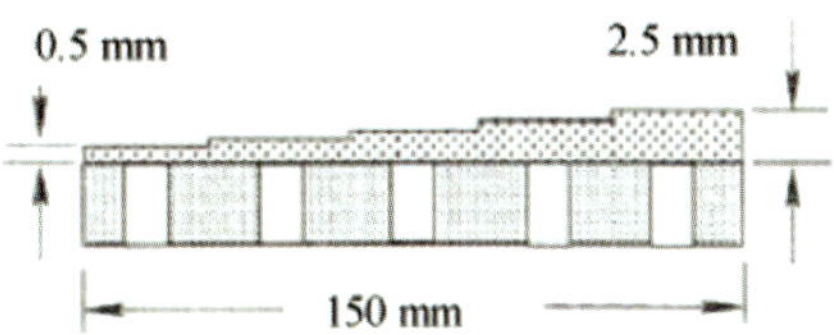

Fig. 3.38a, b. Detection of artificially induced defects in chipboard through the surface veneer sheets of various thicknesses. **a** Geometry of the specimen. **b** Thermal image. (Wu and Busse 1996, with permission)

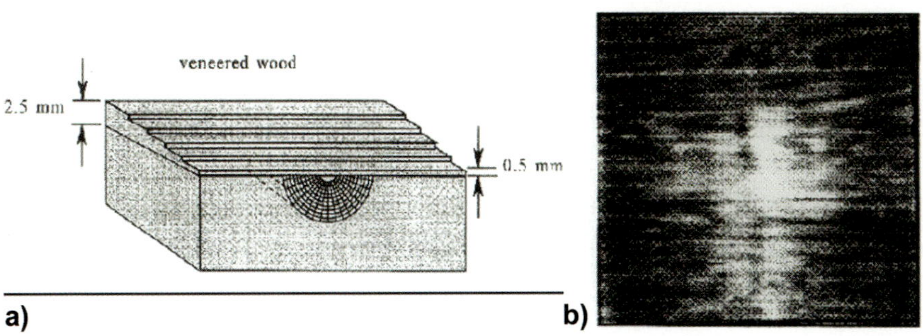

Fig. 3.39a, b. Detection of knots present in solid wood covered with veneer sheets. **a** Geometry of the specimen. **b** Thermal image. (Wu and Busse 1995, with permission)

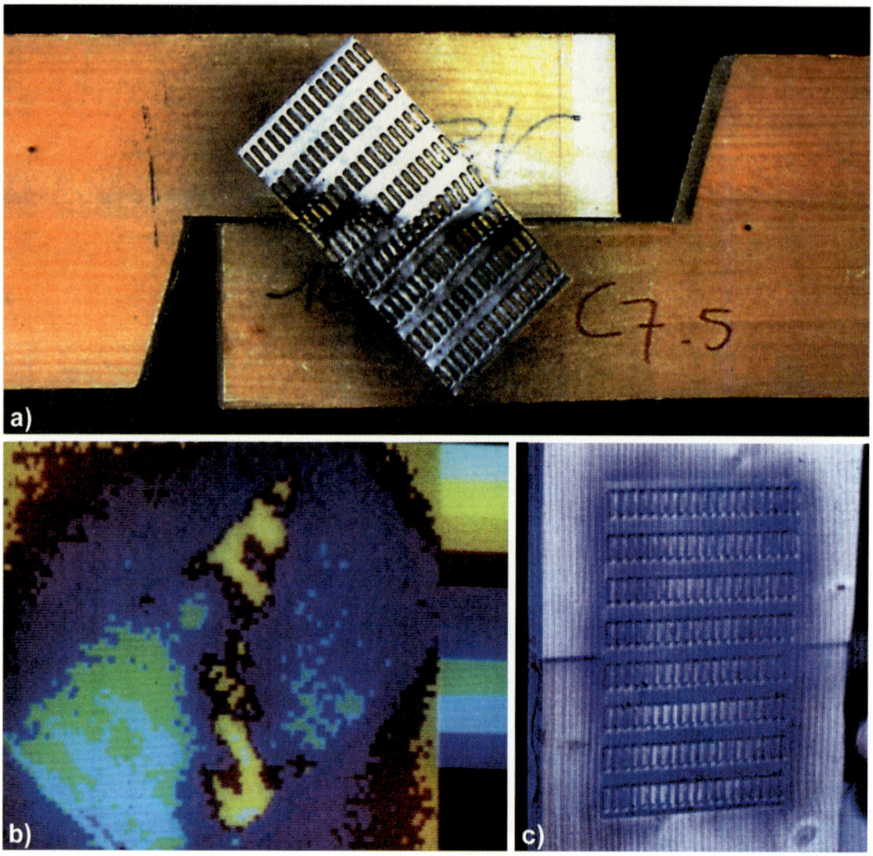

Fig. 3.41a–c. Infrared images of splice joints under shear loading. **a** Before loading. **b** Shear stress before failure. Color hue corresponds to 0.2 °C. (Luong 1996, with permission)

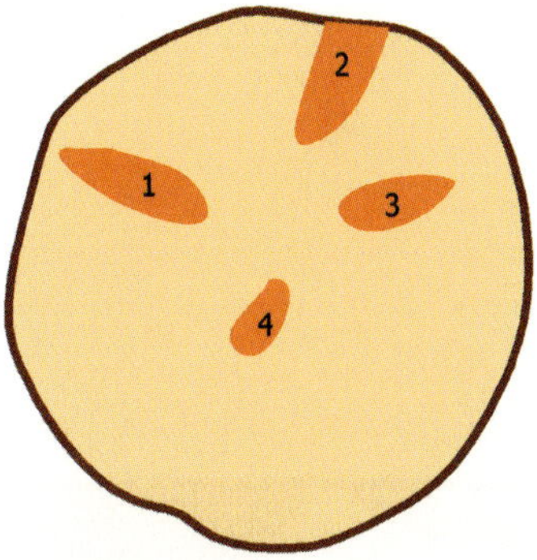

Fig. 4.30. Distribution of knots on the transverse section of the bolt. (Kaestner and Baath 2000, with permission)

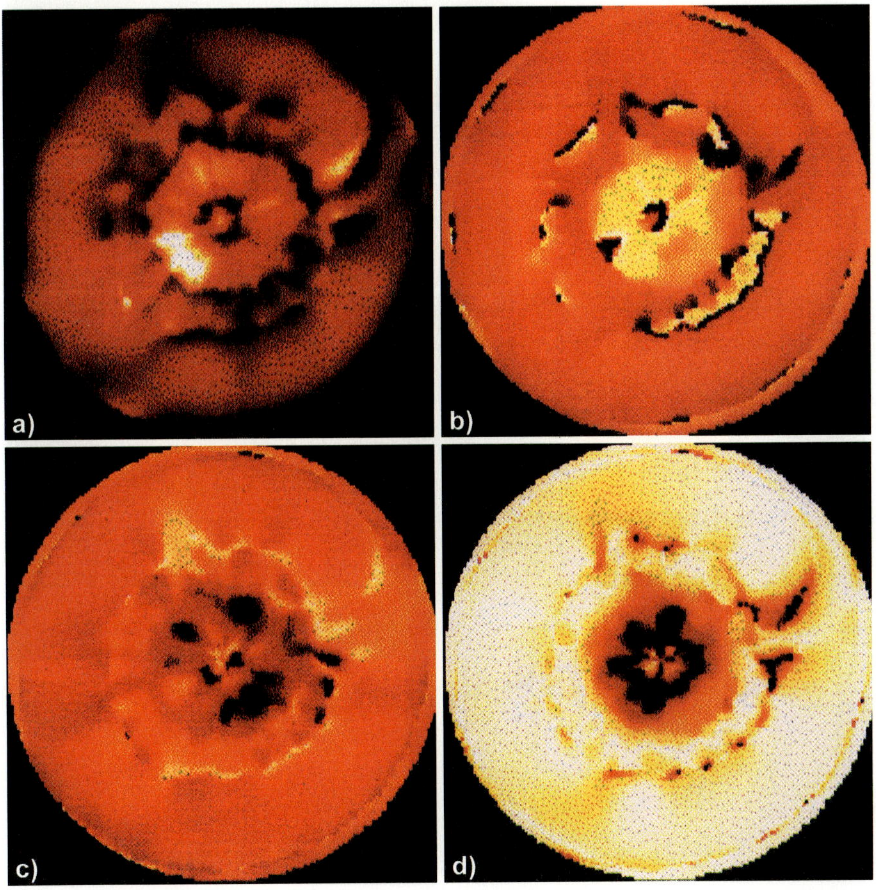

Fig. 4.31a–d. Iso-surface on a topographic slice. **a** $|\rho|$ The amplitude; **b** γ the argument (γ, arg ρ). **c** ε The ellipticity angle which gives the phase difference between two wave components and describes the degree of elliptical polarization. **d** τ The angle between the two components of the wave, corresponding to the tilt of the linear polarization. (Kaestner and Baath 2000, with permission)

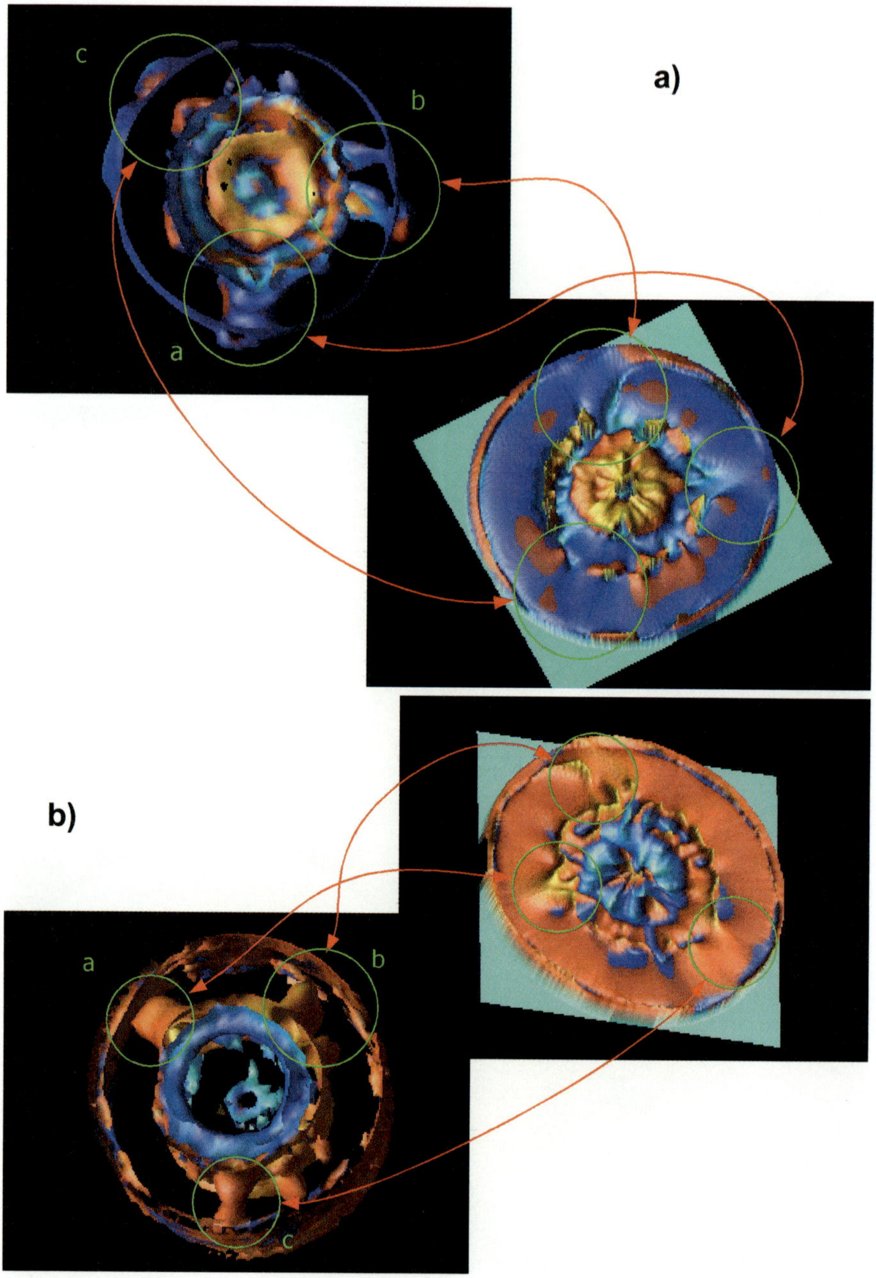

Fig. 4.33a,b. Images of knots reconstructed with **a** the polarization ratio; **b** the phase difference between the two components of the electric field, parallel and perpendicular to the fibers. (Kaestner and Baath 2000, with permission)

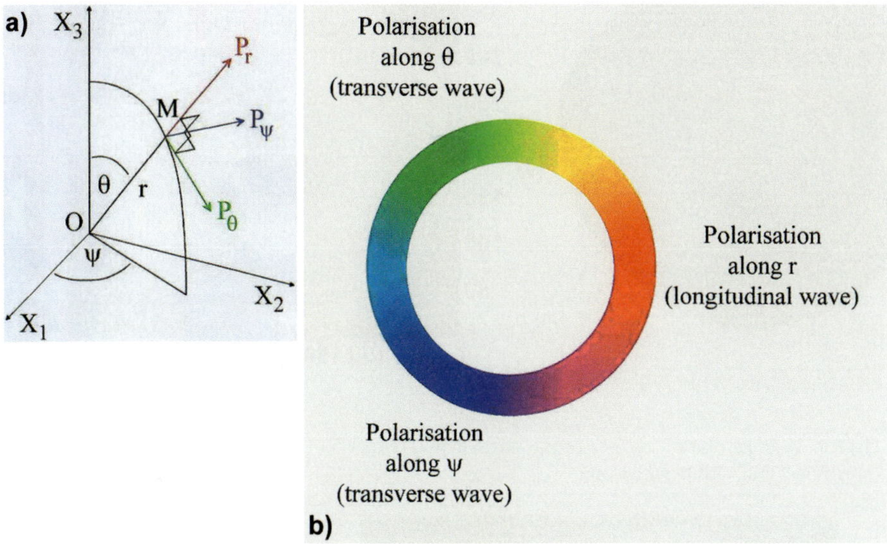

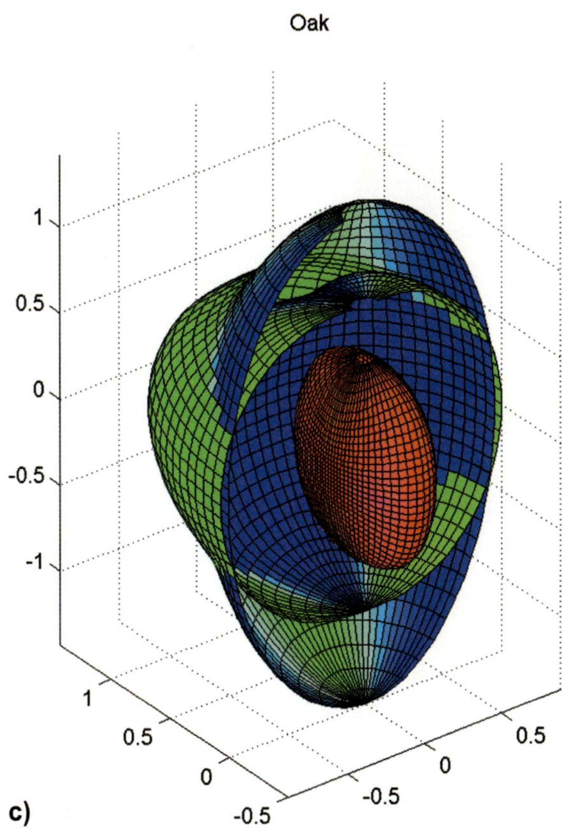

Fig. 5.2a–c. Three-dimensional representation of acoustic properties of oak (**c**). **a** Local basis and color code. **b** Variation of polarization angle on slowness surface for oak. (Bucur et al. 2001, with permission)

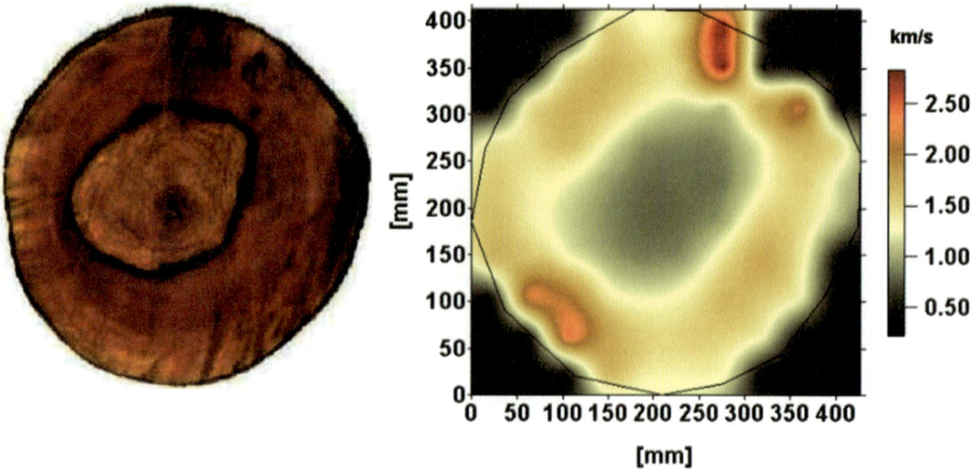

Fig. 5.5. High resolution images of the transverse section of a tree (*Platanus acerifolia*) (Martinis 2002, with permission)

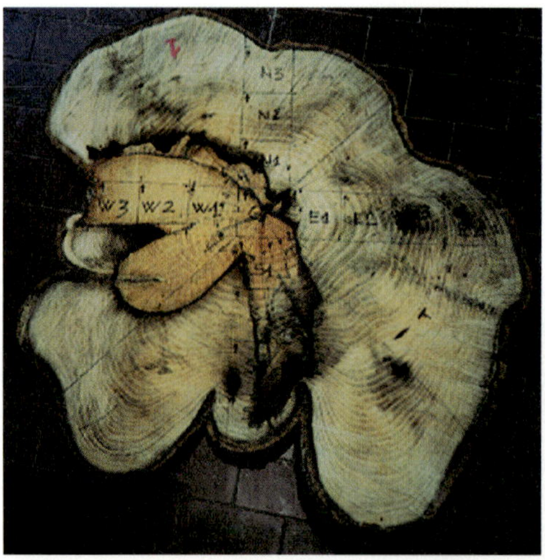

Fig. 5.16. The complex shape of the transverse section of the trunk of hackberry (*Celtis australis*) (Comino et al. 2000, with permission)

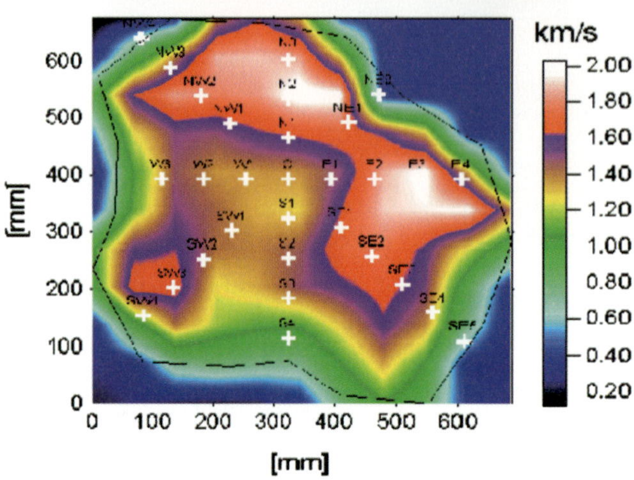

Fig. 5.17. Tomographic images of the transverse section of the trunk in Fig. 5.16 obtained with 120 independent velocity measurements. (Comino et al. 2000, with permission)

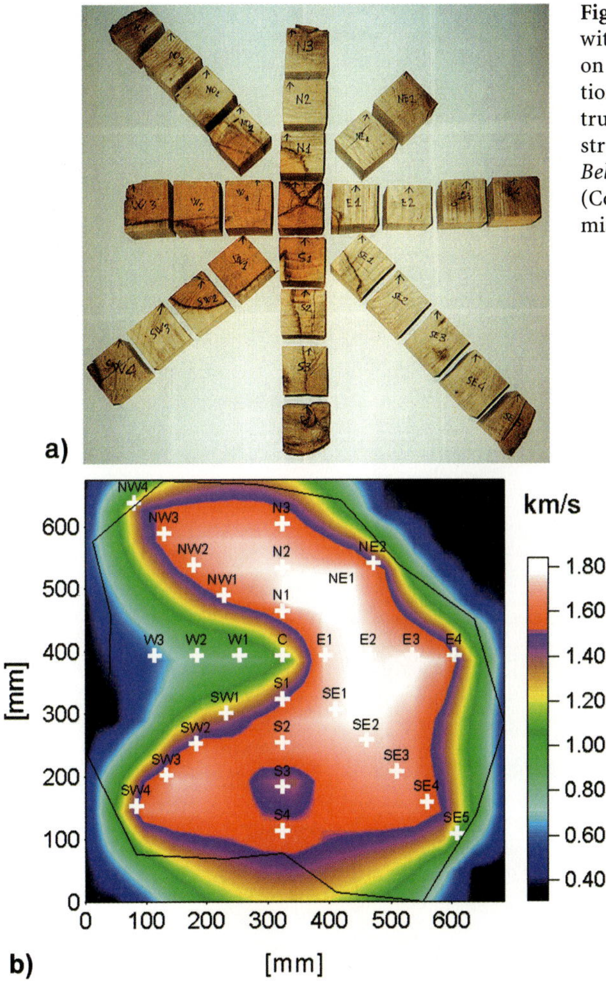

Fig. 5.18. Reconstructed image with velocity values measured on cubic specimens and positions of the specimens in the trunk section. *Above* Reconstructed tomographic image. *Below* Cubic specimens. (Comino et al. 2000, with permission)

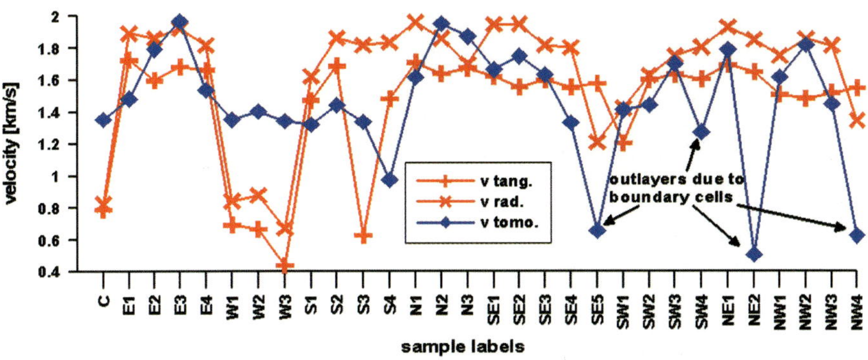

Fig. 5.19. Measured ultrasonic velocities and calculated velocities for three orthotropic directions, L, R and T on selected specimens shown in Fig. 5.18. (Socco et al. 2000, with permission)

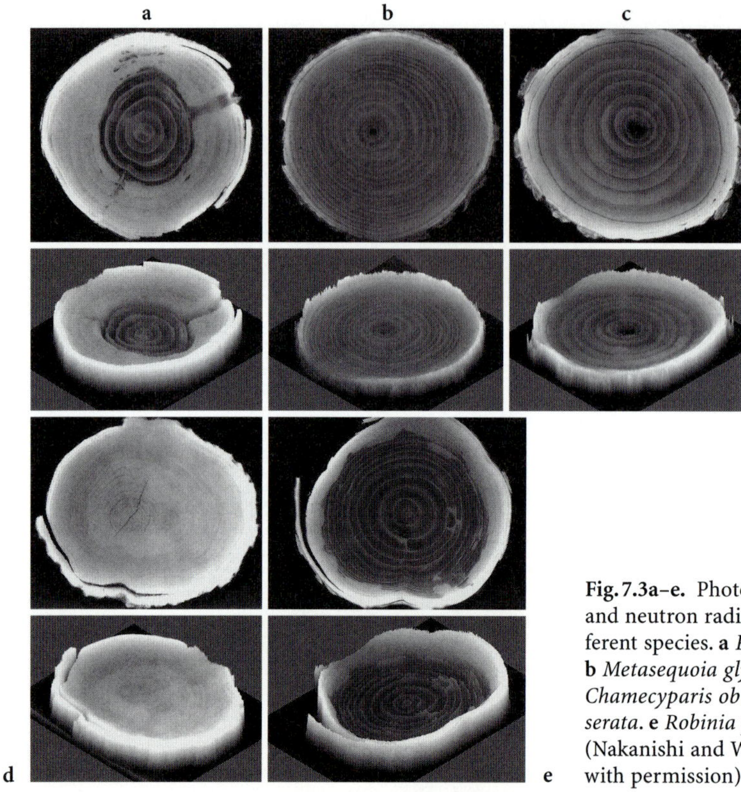

Fig. 7.3a–e. Photographic images and neutron radiography of different species. **a** *Pinus thunbergii*. **b** *Metasequoia glyptostroboides*. **c** *Chamecyparis obtusa*. **d** *Quercus serata*. **e** *Robinia pseudoacacia*. (Nakanishi and Watanabe 1995, with permission)

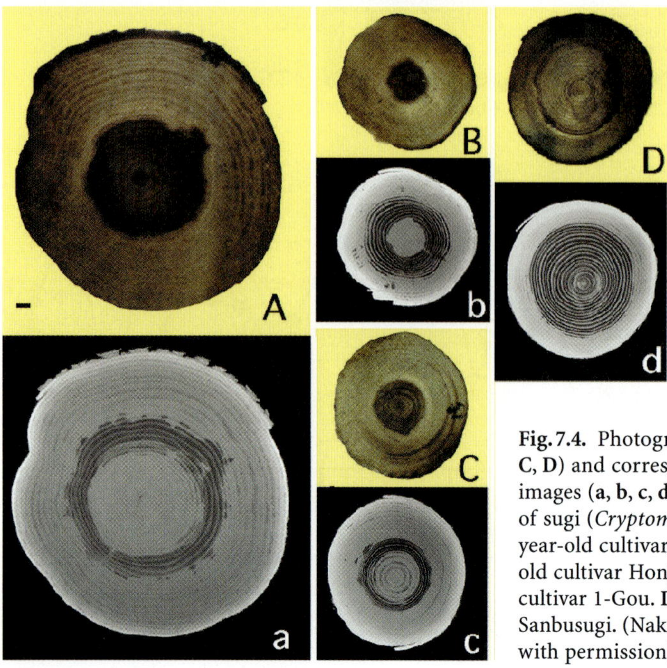

Fig. 7.4. Photographic images (**A**, **B**, **C**, **D**) and corresponding neutron images (**a**, **b**, **c**, **d**) of different cultivars of sugi (*Cryptomeria japonica)*. **A** 24-year-old cultivar, 25 Gou. **B** 25-year-old cultivar Honjiro. **C** 29-year-old cultivar 1-Gou. **D** 30-year-old cultivar, Sanbusugi. (Nakanishi et al. 1998b, with permission)